Molecular Mechanism Discovery of the Bioactive Phytochemicals Against Different Diseases Based on Network Pharmacology and Molecular Docking

Molecular Mechanism Discovery of the Bioactive Phytochemicals Against Different Diseases Based on Network Pharmacology and Molecular Docking

Guest Editor

Baojun Xu

MDPI

Basel • Beijing • Wuhan • Barcelona • Belgrade • Novi Sad • Cluj • Manchester

Guest Editor
Baojun Xu
Department of Life Sciences
Beijing Normal-Hong Kong
Baptist University
Zhuhai
China

Editorial Office
MDPI AG
Grosspeteranlage 5
4052 Basel, Switzerland

This is a reprint of the Special Issue, published open access by the journal *International Journal of Molecular Sciences* (ISSN 1422-0067), freely accessible at: https://www.mdpi.com/journal/ijms/special_issues/6TX9X25N00.

For citation purposes, cite each article independently as indicated on the article page online and as indicated below:

Lastname, A.A.; Lastname, B.B. Article Title. *Journal Name* **Year**, *Volume Number*, Page Range.

ISBN 978-3-7258-4507-1 (Hbk)
ISBN 978-3-7258-4508-8 (PDF)
https://doi.org/10.3390/books978-3-7258-4508-8

Contents

About the Editor

Baojun Xu

Dr. Xu is a Chair Professor in Beijing Normal-Hong Kong Baptist University (BNBU, a full English teaching university in China), Fellow of the Royal Society of Chemistry, Zhuhai Scholar Distinguished Professor, Department Head of Department of Life Sciences, Program Director of Food Science and Technology Program, and author of over 400 peer-reviewed papers. Dr. Xu received a Ph.D. in *Food Science* from Chungnam National University, South Korea. He conducted postdoctoral research work in North Dakota State University (NDSU), Purdue University, and Gerald P. Murphy Cancer Foundation in USA during 2005–2009. He conducted short-term visiting research in NDSU in 2012, and University Georgia in 2014, followed by visiting research during his sabbatical leave (7 months) in Pennsylvania State University, USA, in 2016. Dr. Xu is currently serving as the Associate Editor-in-Chief of *Food Science* and *Human Wellness*, Associate Editor of *Food Research International*, and Associate Editor of *Food Frontiers*, as well as being an Editorial Board Member of around 10 international journals. He received the inaugural President's Award for Outstanding Research of UIC in 2016, and the President's Award for Outstanding Service of UIC in 2020. Dr. Xu has been listed among the world's top 2% scientists by Stanford University for the past five consecutive years and was listed as the Best Scientist in the world in the field of Biology and Biochemistry at Research.com in 2023 and 2024. Prof. Xu was named a Highly Ranked Scholar (top 0.05%) by ScholarGPSTM, ranked at #8 in *Food Science* and Technology in the world, and ranked at #14 in *Agricultural and Natural Sciences*.

International Journal of
Molecular Sciences

MDPI

Editorial

Special Issue: Discovery of Bioactive Phytochemicals' Molecular Mechanisms Against Different Diseases Based on Network Pharmacology and Molecular Docking

Baojun Xu

Food Science and Technology Program, Department of Life Sciences, Beijing Normal-Hong Kong Baptist University, Zhuhai 519087, China; baojunxu@uic.edu.cn

1. Introduction

Natural bioactive phytochemicals have shown great potential in disease management [1,2]. With the development of bioinformatics and the deepening of research on active phytochemicals, network pharmacology and molecular docking techniques have been proposed to study the molecular mechanisms of phytochemicals in treating different diseases [3–5]. Furthermore, in vivo and in vitro studies can further validate these findings. These studies provide new strategies for disease treatment and management and lay the foundation for developing new natural medicines [6]. With the continuous development and improvement of technology, network pharmacology and molecular docking are playing a greater role in discovering natural bioactive substances, helping to discover more drug therapeutic targets, accelerating the drug development process, and promoting the development of the biopharmaceutical industry [7]. This Special Issue delves into the latest research findings from combining network pharmacology with molecular docking to reveal the molecular mechanisms of active phytochemicals in combating various diseases, revealing recent studies that have attracted widespread attention and made significant discoveries.

2. Overview of Published Articles

Bioinformatics analysis is a technical approach that can be combined with network pharmacology and molecular docking, emphasizing the clinical significance of targets in treating related diseases in network pharmacology. Wu's group revealed the molecular mechanism by which Geniposide inhibits colorectal cancer by regulating oxidative stress using network pharmacology, molecular docking, and bioinformatics [8]. More specifically, the anticancer effect of Geniposide may be achieved through the PI3K–Akt signaling pathway, IL-17 signaling pathway, p53 signaling pathway, NF-κB signaling pathway, and NOD-like receptor signaling pathway. The results of molecular docking, bioinformatics, and network pharmacology screening in this study were consistent, demonstrating their effectiveness.

Although this Special Issue emphasizes the combined use of network pharmacology and molecular docking, both techniques can be used separately for scientific research. Martínez-Alarcón et al. discovered a recombinant Tepary bean (*Phaseolus acutifolius*) lectin (rTBL-1) which can induce apoptosis in colon cancer cell lines [9]. This bioactive effect was related to the differential recognition of β1-6 branched N-glycans and studied using molecular docking. This research has illustrated how the Tepary bean lectin affects the microenvironment within which cancerous cells develop, creating new opportunities to uncover the mechanisms by which this lectin exerts its anticancer effects.

Combining molecular docking and molecular dynamics is also a feasible and efficient research strategy. Alotaiq et al. utilized 3D structure modeling, protein–protein docking, molecular dynamics simulations, and MM/PBSA calculations to assess the binding affinities and functional implications of Lumbricus-derived proteins in relation to SOCS2 activity [10]. This study suggests that targeting the SOCS2 protein may stimulate the biological activity of Lumbricus-derived proteins in the treatment of cardiovascular diseases. Further in vitro and in vivo studies are needed to explore the biological activity of Lumbricus-derived proteins and their potential commercial value.

Yu et al. successfully constructed an engineered *S. cerevisiae* strain (BY-Ro) for ginsenoside Ro biosynthesis using S. cerevisiae BY4741 as the host cell by sequentially transferring these DNA fragments into chemically competent cells of engineering strains and conducting screening and target product detection [11]. Moreover, the ginsenoside Ro is one of the few oleanane-type ginsenosides, which is well known for its unique molecular structure and biological activities.

Ruiz-Alcaraz et al. studied the in vitro anti-inflammatory effect of serious phytochemicals, including phenolic di-caffeoylquinic acid (Di-CQA), flavonoid cyanidin-3,5-diglucoside (Cy3,5DiG), aromatic isothiocyanate sinalbin (SNB), and aliphatic isothiocyanate sulforaphane (SFN), which are sourced from vegetables and fruits [12]. These results indicate that plant compounds from dietary sources can significantly reduce the regulation of the LPS-induced inflammatory factors TNF-α and IL6. This study highlights the potential of plant-derived bioactive compounds as valuable contributors to the prevention or treatment of chronic inflammatory diseases.

Verifying the findings of network pharmacology and molecular docking through in vivo animal experiments is a condensed research strategy. Shu et al. demonstrated the potential of hesperidin to treat CPT-11-induced diarrhea through a combination of network pharmacology, molecular docking, and experimental validation [13]. This study demonstrated that hesperidin significantly alleviated chronic diarrhea in mice and further validated the relevant signaling pathways predicted by network pharmacology. In addition, the interaction between hesperidin and seven key targets regulating cell apoptosis (AKT1, ANXA5, CASP3, HSP90AA1, IGF1, MMP9, and PPARG), as demonstrated by molecular docking, has also been validated through in vivo experiments.

One of the reviews presented in this Special Issue provides a detailed analysis of COVID-19 in children and its relationship with vitamin D, highlighting the impact of pandemic-related confinement on children's vitamin D levels [14]. It explores how vitamin D acts as a critical regulator of immune responses and its potential role in mitigating the cytokine storm associated with severe COVID-19. The review also evaluates the limited evidence supporting vitamin D supplementation as an adjuvant treatment for COVID-19, emphasizing the need for further large-scale randomized controlled trials to confirm its benefits.

Another review comprehensively examines the therapeutic potential of aminoflavonoids, focusing on their synthesis, biological activities, and the influence of structural modifications on their efficacy [15]. It highlights the anticancer, antimicrobial, antiviral, and anti-inflammatory properties of aminoflavonoids and their metal complexes, emphasizing the importance of the nitrogen atom in enhancing their biological activities. The review also identifies promising candidates like AFP464, which has shown potential in clinical trials, and underscores the need for further research into the mechanisms of action and safety profiles of these compounds to facilitate the development of novel therapeutic agents.

3. Conclusions

This Special Issue focuses on the discovery of molecular mechanisms for bioactive phytochemicals against different diseases based on network pharmacology and molecular docking, selecting eight academic papers that delve into the complex molecular mechanisms behind these topics. These papers also highlight the enormous potential for further exploration and discovery in the field of phytomedicine. Looking ahead to the future, the advancement of network pharmacology combined with molecular simulation, bioinformatics, transcriptomics, proteomics, and other related technologies is expected to promote breakthrough research and innovation in the field of phytomedicine.

Conflicts of Interest: The author declares no conflicts of interest.

References

1. Bernardini, S.; Tiezzi, A.; Laghezza Masci, V.; Ovidi, E. Natural products for human health: An historical overview of the drug discovery approaches. *Nat. Prod. Res.* **2018**, *32*, 1926–1950. [CrossRef] [PubMed]
2. Krzyzanowska, J.; Czubacka, A.; Oleszek, W. Dietary phytochemicals and human health. *Adv. Exp. Med. Biol.* **2010**, *698*, 74–98. [CrossRef] [PubMed]
3. Kibble, M.; Saarinen, N.; Tang, J.; Wennerberg, K.; Makela, S.; Aittokallio, T. Network pharmacology applications to map the unexplored target space and therapeutic potential of natural products. *Nat. Prod. Rep.* **2015**, *32*, 1249–1266. [CrossRef] [PubMed]
4. Hopkins, A.L. Network pharmacology: The next paradigm in drug discovery. *Nat. Chem. Biol.* **2008**, *4*, 682–690. [CrossRef] [PubMed]
5. Zhang, P.; Zhang, D.; Zhou, W.; Wang, L.; Wang, B.; Zhang, T.; Li, S. Network pharmacology: Towards the artificial intelligence-based precision traditional Chinese medicine. *Brief. Bioinform.* **2023**, *25*, bbad518. [CrossRef] [PubMed]
6. Wang, Z.Y.; Wang, X.; Zhang, D.Y.; Hu, Y.J.; Li, S. Traditional Chinese medicine network pharmacology: Development in new era under guidance of network pharmacology evaluation method guidance. *Zhongguo Zhong Yao Za Zhi* **2022**, *47*, 7–17. [CrossRef] [PubMed]
7. Jiao, X.; Jin, X.; Ma, Y.; Yang, Y.; Li, J.; Liang, L.; Liu, R.; Li, Z. A comprehensive application: Molecular docking and network pharmacology for the prediction of bioactive constituents and elucidation of mechanisms of action in component-based Chinese medicine. *Comput. Biol. Chem.* **2021**, *90*, 107402. [CrossRef] [PubMed]
8. Wu, Y.; Luo, J.; Xu, B. Network pharmacology and bioinformatics study of Geniposide regulating oxidative stress in colorectal cancer. *Int. J. Mol. Sci.* **2023**, *24*, 15222. [CrossRef] [PubMed]
9. Martinez-Alarcon, D.; Castro-Guillen, J.L.; Fitches, E.; Gatehouse, J.A.; Przyborski, S.; Moreno-Celis, U.; Blanco-Labra, A.; Garcia-Gasca, T. Directed mutagenesis for arginine substitution of a *Phaseolus acutifolius* recombinant lectin disrupts its cytotoxic activity. *Int. J. Mol. Sci.* **2024**, *25*, 13258. [CrossRef] [PubMed]
10. Alotaiq, N.; Dermawan, D.; Elwali, N.E. Leveraging therapeutic proteins and peptides from Lumbricus earthworms: Targeting SOCS2 E3 ligase for cardiovascular therapy through molecular dynamics simulations. *Int. J. Mol. Sci.* **2024**, *25*, 10818. [CrossRef] [PubMed]
11. Yu, X.; Yu, J.; Wang, D.; Liu, S.; Wang, K.; Zhao, M.; Chen, P.; Wang, Y.; Wang, Y.; Zhang, M. A novel biosynthetic strategy for ginsenoside Ro: Construction of a metabolically engineered *Saccharomyces cerevisiae* strain using a newly identified UGAT gene from *Panax ginseng* as the key enzyme gene and optimization of fermentation conditions. *Int. J. Mol. Sci.* **2024**, *25*, 11331. [CrossRef] [PubMed]
12. Ruiz-Alcaraz, A.J.; Baquero, L.; Perez-Munar, P.M.; Oliva-Bolarin, A.; Sanchez-Martinez, M.A.; Ramos-Molina, B.; Nunez-Sanchez, M.A.; Moreno, D.A. In vitro study of the differential anti-inflammatory activity of dietary phytochemicals upon human macrophage-like cells as a previous step for dietary intervention. *Int. J. Mol. Sci.* **2024**, *25*, 10728. [CrossRef] [PubMed]
13. Shu, X.; Xu, R.; Xiong, P.; Liu, J.; Zhou, Z.; Shen, T.; Zhang, X. Exploring the effects and potential mechanisms of hesperidin for the treatment of CPT-11-induced diarrhea: Network pharmacology, molecular docking, and experimental validation. *Int. J. Mol. Sci.* **2024**, *25*, 9309. [CrossRef] [PubMed]
14. Dura-Trave, T.; Gallinas-Victoriano, F. COVID-19 in children and vitamin, D. *Int. J. Mol. Sci.* **2024**, *25*, 2205. [CrossRef] [PubMed]
15. Stompor-Goracy, M.; Bajek-Bil, A.; Potocka, N.; Zawlik, I. Therapeutic perspectives of aminoflavonoids—A review. *Int. J. Mol. Sci.* **2025**, *26*, 2014. [CrossRef] [PubMed]

International Journal of *Molecular Sciences*

MDPI

Article

Directed Mutagenesis for Arginine Substitution of a *Phaseolus acutifolius* Recombinant Lectin Disrupts Its Cytotoxic Activity

Dania Martínez-Alarcón [1,†], José Luis Castro-Guillén [2,†], Elaine Fitches [3], John A. Gatehouse [3,‡], Stefan Przyborski [3], Ulisses Moreno-Celis [4], Alejandro Blanco-Labra [1] and Teresa García-Gasca [4,*]

1 Centro de Investigación y de Estudios Avanzados Unidad Irapuato, Departamento de Biotecnología y Bioquímica, Irapuato 36821, Guanajuato, Mexico; dania.martinez.alarcon@gmail.com (D.M.-A.); alejandroblancolabra@gmail.com (A.B.-L.)

2 Tecnológico Nacional de México/Instituto Tecnológico Superior de Irapuato (ITESI), Km. 12.5, Carretera Irapuato-Silao, El Copal, Irapuato 36821, Guanajuato, Mexico; jose.cg1@irapuato.tecnm.mx

3 Department of Biosciences, Durham University, Durham DH1 3LE, UK; e.c.fitches@durham.ac.uk (E.F.); stefan.przyborski@durham.ac.uk (S.P.)

4 Facultad de Ciencias Naturales, Universidad Autónoma de Querétaro, Querétaro 76230, Querétaro, Mexico; ulisses.moreno@uaq.mx

* Correspondence: tggasca@uaq.edu.mx,

† These authors contributed equally to this work.

‡ This paper is dedicated to the memory of Professor John Gatehouse (1951–2022).

Abstract: Recently, we reported that a recombinant Tepary bean (*Phaseolus acutifolius*) lectin (rTBL-1) induces apoptosis in colon cancer cell lines and that cytotoxicity was related to differential recognition of β1-6 branched *N*-glycans. Sequencing analysis and resolution of the rTBL-1 3D structure suggest that glycan specificity could be strongly influenced by two arginine residues, R103 and R130, located in the carbohydrate binding pocket. The aim of this work was to determine the contribution of these residues towards cytotoxic activity. Two rTBL-1 mutants were produced in *Pichia pastoris*, biochemically characterized, and cytotoxic effects were evaluated on human colorectal cancer cells (HT-29). Substitution of either of the arginine residues with glutamines resulted in significant reductions in cytotoxic activity, with losses of 1.5 and 3 times for R103 and R130, respectively. Docking analysis showed that the mutations decreased lectin affinity binding to some Epidermal Growth Factor Receptor (EGFR)-related *N*-glycans. Together, these findings confirm that both of the selected arginine residues (R103 and R130) play a key role in the recognition of tumor cell glycoconjugates by rTBL-1.

Keywords: recombinant lectins; tepary bean; branched *N*-glycans; cancer

1. Introduction

Anticancer lectins are a specialized group of proteins that are able to recognize and interact with specific carbohydrate structures of cancer cells [1]. By specifically targeting saccharide structures, these lectins can disrupt crucial cellular processes, interfere with signaling pathways, and ultimately exert cytotoxic and antiproliferative effects on cancer cells. As a consequence, these proteins have gained considerable attention in cancer research due to their potential applications in cancer therapies [2]. One notable example of an anticancer lectin is the Phytohemagglutinin (PHA) derived from kidney bean (*Phaseolus vulgaris*) seeds. PHA has demonstrated cytotoxic effects on cancer cells, leading to cell death and reduced tumor growth [3]. PHA's anticancer effect stems from its ability to recognize and bind to specific carbohydrate molecules displayed on the surface of cancer cells. By binding to these structures, PHA can interfere with vital cellular processes, disrupt signaling pathways, and induce various anti-tumor effects, including the induction of apoptosis, inhibition of cell proliferation, modulation of immune responses, and inhibition of angiogenesis [4–7].

In 2012, we reported that a group of PHA-homologous native lectins derived from Tepary bean (*Phaseolus acutifolius*) exhibited cytotoxic effects on various types of cancer cell lines [8] by inducing apoptosis and cell cycle arrest [9]. It was found that the Tepary Bean Lectin Fraction (TBLF) contained at least two cytotoxic lectins that were characterized at molecular and bioactive levels [10,11]; furthermore, when administered orally, TBLF exhibited the ability to inhibit early colon tumorigenesis in rats [12]. As Tepary Bean Lectin-1 was identified as the main lectin responsible for this cytotoxic effect, we subsequently produced it as a recombinant protein by bench-top fermentation of transformed *Pichia pastoris* and determined its three-dimensional structure [13,14]. The recombinant lectin-1 (rTBL-1) was found to recognize β1-6 branched *N*-glycans, which are overrepresented in several cancer types, including colon cancer. Recognition of β1-6 branched *N*-glycans does not appear to be affected by elongation with more units as sialic acid [13]. We also found that rTBL-1 retains its cytotoxic apoptosis-inducing effect on colon cancer cells via influencing epidermal growth factor cell receptor (EGFR) signaling pathways and principally by causing partial receptor degradation and activation of p38 MAPK signaling [15].

The structural characterization of rTBL-1 enabled us to compare its putative binding pocket and residues responsible for binding to the trisaccharide unit Gal(β1-4)GlcNAc(β1-2)Man with that previously reported for PHA [16]. We identified that, unlike other closely related legume agglutinins with anticancer activity, rTBL-1 has an arginine residue instead of lysine at position 103, which could be involved in the formation of hydrogen bonding to carbohydrates. Additionally, we identified the presence of an additional arginine within the binding pocket and adjacent to other residues involved in the recognition of carbohydrates reported for PHA. Therefore, as a highly conserved arginine residue is crucial for carbohydrate recognition in other families of lectins, such as the siglectins (sialic acid immunoglobilin-superfamily lectins) [17,18], we hypothesize that the sequence discrepancy in rTBL-1 could explain its cytotoxic effect as compared to the legume leucoagglutinins. To investigate this further, we have carried out a conservative substitution of two arginine residues with the uncharged amino acid glutamine in order to evaluate the impact of these modifications on the cytotoxic effect of rTBL-1 lectin.

2. Results and Discussions

2.1. Production of rTBL-1 Mutants

Multiple alignments of leucoagglutinin sequences, along with the findings of Nagae et al. [16], revealed that all sequences have identical carbohydrate binding pockets (CBP) (Figure 1A, blue triangles), except *P. acutifolius* and rTBL-1, where an arginine (R) instead of lysine (K) is located at position 103. This change is particularly interesting because, despite the similarity between these amino acids, conservative substitutions such as replacing R with K in other lectins, such as the animal I-type lectins (siglectins), have shown a conserved arginine residue to be essential for sialic-binding [17,19]. Sialic acid is particularly abundant in tumor cells where abnormal sialylation is considered as a hallmark of cancer [20], so it has been studied for cancer-involved membrane receptors such as EGFR [21].

An additional arginine residue located at position 130 (R130) in TBL-1 differs from the presence of a lysine in the *P. vulgaris* lectin or tryptophan in *P. costaricensis* or *P coccineus* lectins. R130 does not appear to have direct interaction with the ligand but is located adjacent to two of the residues involved in carbohydrate recognition and also for metal-bonding cations. These bonds are essential for interactions between monomers and, therefore, for the recognition of complex carbohydrates. It is thus possible that R130 may contribute to enhancing the cytotoxic effects of TBL-1. It was also found that the sequon, unique consensus sequence for *N*-glycosylation (N-X-S/T, where X represents any amino acid except P) [22] was located between residues 13 and 15 (N-E-T).

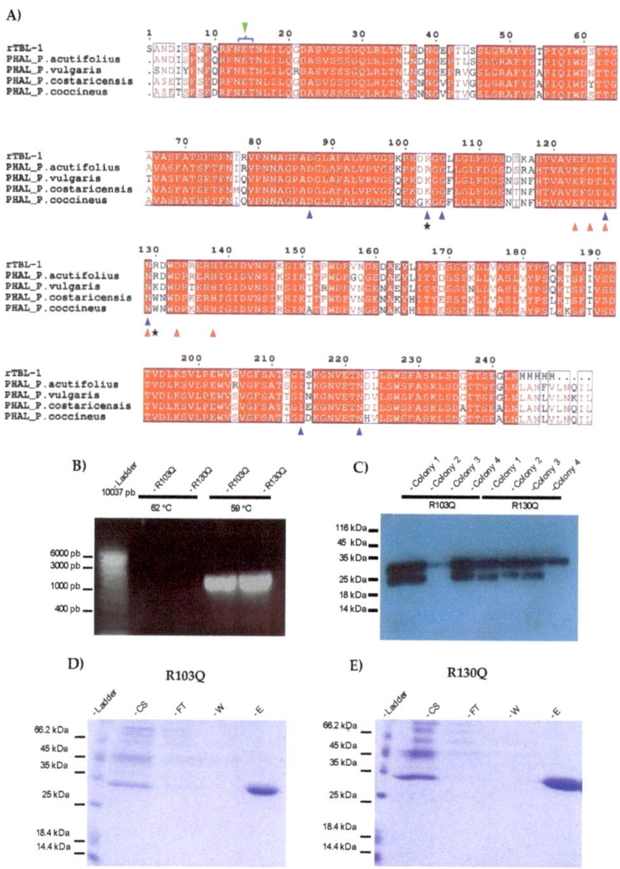

Figure 1. Mutants' design and production. (**A**) Sequence alignment of rTBL-1 homologous lectins. PHAL_*Phaseolus costaricensis* (Q5ZF34_9FABA), PHAL_*Phaseolus coccineus* (Q84RP8_PHACN), PHAL_*Phaseolus vulgaris* (PHAL_PHAVU), PHAL_*Phaseolus acutifolius* (Q40750_PHAAT), rTBL-1 sequence [14]. White letters represent identical residues, semiconserved residues are displayed in red enclosed into empty boxes and non-conserved residues are depicted in black. Blue triangles show residues that correspond to the CBP similarly to PHA-L from *P. Vulgaris*; red triangles show residues interacting with metal cations; green triangle depicts sequence for *N*-glycosylation marked with a blue bracket; black stars show residues presumably responsible for the cytotoxic effect of rTBL-1 (R103 and R130). (**B**) Agarose gel electrophoresis showing pGAPαZB-rTBL-1 vector amplification. Lane 1, ladder; lanes 2–5, amplification products. (**C**) Western blot screening using anti-His antibodies for transformed colonies. (**D**,**E**) R103Q and R130Q SDS-PAGE electrophoretic profiles stained for total protein. Lane 1, ladder; lane 2, 25 μL of culture supernatant (CS); lane 3, 25 μL of flow through (FT); lane 4, 25 μL wash with 10 mM imidazole buffer (W); and lane 5, 25 μL elution with 200 mM imidazole buffer (E). PHAL, Leukocyte phytohemagglutinin.

Two rTBL-1 mutants were generated by substituting either R-103 or R-130 within the CBP with a non-charged glutamine (Q) residue. The resulting proteins were designated as R103Q and R130Q mutant lectins, respectively. Site-directed mutagenesis was performed on the original coding sequence of rTBL-1 by inverse amplification of the pGAPZαB-rTBL-1 plasmid expression vector [13] to obtain amplicons of approximately 3000 bp (Figure 1B). Whilst amplification parameters were tested at two different temperatures (62 and 59 °C), viable products were only obtained at 59 °C (Figure 1B).

The vectors were gel purified, re-ligated, and subsequently cloned into Top10 *Escherichia coli* cells. Sequence verified plasmid DNAs were linearized and integrated into the GAPDH locus of *P. pastoris* through homologous recombination. Four colonies of each mutant were selected for expression screening via western analysis of supernatants derived from shake flask cultures. As shown in Figure 1C, all selected colonies secreted a recombinant protein of approximately 32 kDa consistent with the predicted mass for the mutants. The presence of an additional immunoreactive protein (ca. 25 kDa) observed in 6 of the 8 culture supernatants suggests degradation at the *N*-terminus of the mutants, since the His tag is at the C-terminus of the protein. Based on the performance of the small-scale cultures, colony number #1 for each mutant was selected to scale up production in a 7 L bench top fermenter.

The electrophoretic profile of the culture media obtained for both fermentations was very similar to the one previously reported for the rTBL-1 fermentation, where the recombinant product is presented as the most predominant band. The final yields per liter of culture supernatant following bench-top fermentation were 205 mg and 184 mg, respectively, for R103Q and R130Q. Proteins were purified by nickel affinity chromatography, and fractions were collected after elution with 200 mM imidazole. Analysis of purified proteins by SDS-PAGE electrophoresis showed the presence of a single band of ~30 kDa in Coomassie-stained gels (Figure 1D,E).

2.2. R103Q and R130Q Biochemical Characterization

rTBL-1 and the purified mutants were all separated as proteins of approx. 30 kDa under denaturing SDS-PAGE gel electrophoresis in line with predicted masses for the recombinant lectins (Figure 2A). However, it is well-known that rTBL-1 folds into a homo-tetramer [13]. To assess the oligomeric forms of R103Q and R130Q mutant lectins, their native molecular weights were determined using gel filtration chromatography using a high-resolution size exclusion column (ENrich™ SEC650, BioRad. Hercules, CA, USA). The elution profiles exhibited a single peak at approximately 120 kDa, indicating that both mutants are able to conform homo-tetramers in solution and that the K to Q substitutions did not disrupt their ability to oligomerize.

A single *N*-glycosylation site near their *N*-terminal region was found for rTBL-1; this sequence leads to the addition of an *N*-glycosidic antenna of approximately ~2.48 kDa on asparagine-13. All recombinant proteins were treated with various combinations of glycosidases in order to compare the type and estimated size of attached glycosidic antennas. Subsequently, the samples were subjected to SDS-PAGE, and the gels were stained using the Schiff-PASS technique, enabling the identification of protein-bound carbohydrates (Figure 2B).

Both R103Q and R130Q mutants showed similar molecular weights (Figure 2A). As rTBL-1 and the two mutants showed no changes in the protein sequon, an analysis of the glyosidic antennas was performed using several glycosidases. Upon treatment with *N*-glycosidase-F, both mutants and rTBL-1 exhibited a reduction in size of between 1.84 and 2.48 kDa, accompanied by the complete carbohydrate removal. No changes in intensity or molecular size were observed after treatment with *O*-glycosidases indicating that rTBL-1 and the mutants contain only *N*-linked glycans. Nano-LC-MS/MS analysis depending on standard data of tryptic peptides, was used to confirm the position of the carbohydrate antennas from samples treated with deglycosylating enzymes. This analysis confirmed the presence of an *N*-deamidated asparagine in the NETN peptide, which corresponds to positions 13–16 of the proteins. Deamination is known to be a byproduct of PNGaseF treatments by removing *N*-linked sugars.

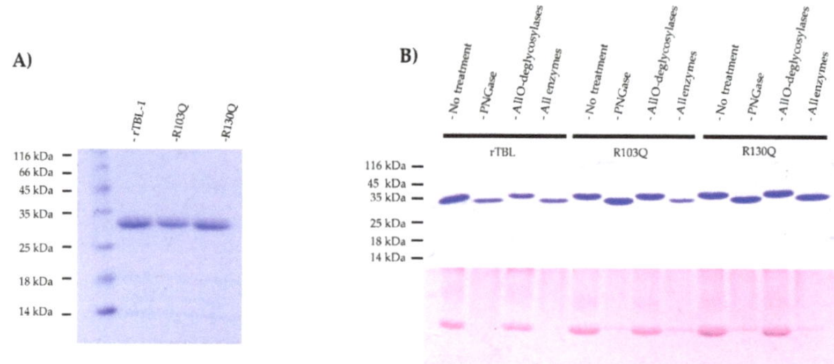

Figure 2. Characterization of the mutants. (**A**) Molecular size comparison between rTBL-1 and the mutants. SDS-PAGE of the electrophoretic profiles of rTBL-1, R103Q, and R130Q. Lane 1, ladder; lane 2, 5 μg of rTBL; lane 3, 5 μg of R103Q; lane 4, 5 μg of R130Q. (**B**) Analysis of glycosylation. Top panel shows a coomassie R-250-stained SDS-PAGE. *N*-Glycosidase F (PGNaseF), all *O*-glycosidases (β1,4-Galactosidase, endo-α-*N*-acetylgalactosaminidase, α2-3,6,8,9-Neuraminidase, and β-*N*-Acetylglucosaminidase), all *N*- and *O*-glycosidases (*N*-Glycosidase F, β1,4-Galactosidase, endo-α-*N*-acetylgalactosaminidase, α2-3,6,8,9-Neuraminidase, and β-*N*-Acetylglucosaminidase).

2.3. Cytotoxic Effect of R103Q and R130Q

The cytotoxicity of rTBL-1 and its mutants was assessed on HT-29 colorectal cancer cells through dose–response curves using concentrations ranging from 0 to 50 μg/mL (Figure 3). The percentage of cell survival was normalized against the negative PBS-treated control treatment. The results show concentration-dependent responses for all proteins. However, rTBL-1 showed significantly greater cytotoxic effects as compared to either of the mutants, as evidenced by significant differences ($p < 0.05$) in lethal concentrations (LC_{50} values) derived from dose–response assays. LC_{50} values were 2.37 μg/mL ($y = -16 - 317x + 88.654$) for rTBL-1 as compared to 4.94 μg/mL ($y = -12.702x + 112.59$) and 7.22 μg/mL ($y = -8.6725x + 112.7$) for R103Q and R130Q, respectively.

These results show that each arginine substitution plays a role in exerting the cytotoxic effects of rTBL-1. R130Q exhibited the greatest inhibition of cytotoxicity, being 3 times less cytotoxic as compared to rTBL-1 and 1.5 times lower than R103Q. No statistical differences ($p > 0.05$) were observed for all tested concentrations of R130 as compared to control cells. For R103, significant differences ($p < 0.05$) were observed for the 5 to 50 μg/mL treatments as compared to the control cells. These results highlight the differential role of arginine residues 103 or 130 in the lectin's carbohydrate binding activity.

2.4. Docking Analysis

In order to know more about the participation of R103 and R130 residues, a docking analysis was performed using the rTBL-1 sequence (PDB ID: 6tt9); this sequence has one extra alanine beside the first serine of the rTBL-1 sequence [14]. This change did not affect the tridimensional structure, but it was necessary to go through the numbering one position, changing R103Q to R104Q and R130Q to R131Q only for the docking analysis. On the other hand, all docking procedures were performed using the LZERD web server according to its specifications (https://lzerd.kiharalab.org accessed on 28 January 2023) [23].

2.4.1. Docking Analysis with Experimentally Recognized *N*-glycans

Previously we have shown that rTBL-1 recognized experimental complex *N*-glycans from the Mammalian Glycan Array (GA) 5.4; therefore, a docking analysis was performed considering only those whose relative signal was above 40% [13]. Figure 4 shows the interaction between rTBL-1 and experimental *N*-glycans, where GA-01, GA-02, GA-03, GA-04, GA-06, GA-07, and GA-09 exhibited a docking interaction in the upper zone between monomers 1 and 2 of tetrameric rTBL-1 located near the CBP, confirming recognition and suitable interaction positions of the experimentally recognized *N*-glycans. Only those complex glycans having four sialylated branches (GA-08) or highly elongated branches (GA-05) interacted directly in the CBP with a large part of the glycan structure. These results provide the first two selection criteria for those *N*-glycans that had direct and considerable interactions with the lectin CBP, namely the presence of at least four-ended sialylated branches and/or elongated branches.

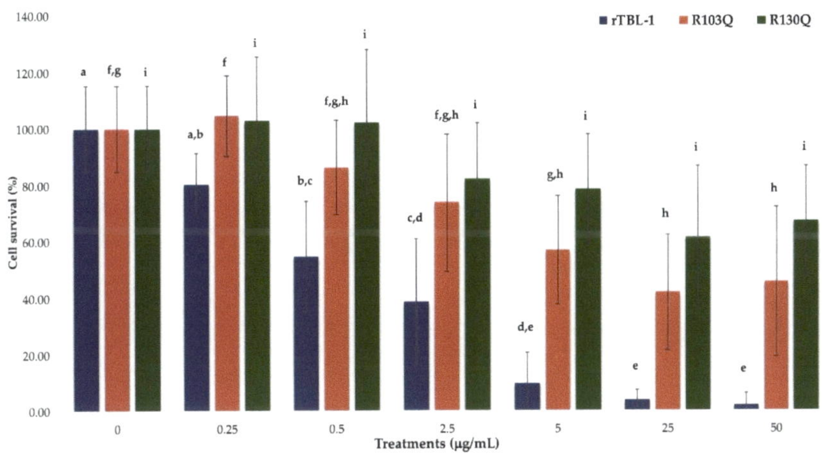

Figure 3. Cytotoxic effect of *r*TBL-1 and its mutants on HT-29 colon cancer cells. Cells were treated with rTBL-1, R103Q, or R130Q (0, 0.25, 0.5, 2.5, 5, 25, and 50 μg/mL) for 24 h and cell survival percentages were calculated from hemocytometer counts. Small letters show significant differences for each separate treatment at different concentrations (Tukey $p < 0.05$), where differences between treatments for rTBL-1 were pointed out with letters a–e, for R103Q with letters f–h, and for R130Q with letter i.

2.4.2. Docking Analysis with EGFR-Related *N*-glycans

EGFR is an important glycoprotein for cancer biology, being highly expressed and considered as a proteomic marker in different types of cancers [24–26]. EGFR-specific *N*-glycosylations are correlated with disease state/progression of colorectal cancer [27], and, for this reason, EGFR-related *N*-glycans were considered in this work for docking analysis using the structures of rTBL-1 and the in silico mutated versions R104Q and R131Q.

Complexity in the structure of *N*-glycans was also considered to select the EGFR-related glycans for the docking analysis, since rTBL-1 previously showed high affinity for complex glycans [13]. Furthermore, the protein–protein interaction of rTBL-1 with the EGFR has been observed [15]. On the other hand, due to the possible positions of its flexible structure, it was necessary to use several EGFR-related *N*-glycan conformers generated on the GLYCAM web server [29] in each docking analysis. This allowed us to screen a wide variety of docking results and select those within the CBP, and specifically those whose interactions were related to the mutated residues. Thirty-two of the 52 EGFR-related *N*-glycans (GlyConnect-https://glyconnect.expasy.org accessed on 8 February 2023) [30] were used for docking analysis with rTBL-1.

Only eight *N*-glycans were not recognized by rTBL-1 (considering that the resulting interactions were located between monomers [11]), and 24 EGFR-related *N*-glycans had reliable docking positions, including six interacting directly with CBP (Figure 5). Interestingly, when the reported molecular mass of each selected *N*-glycan (GlyConnect platform, https://glyconnect.expasy.org/ accessed on 8 February 2023) [30] was compared, a molecular mass-lectin recognition correlation was found (Supplementary Material, Table S1). The unrecognized *N*-glycans (those *N*-glycans with unsuitable docking interactions) had an average mass of 1.77 ± 0.33 kDa and included mainly complex two-branched glycans and high-mannose *N*-glycans. The *N*-glycans whose docking interaction was appropriate but without considerable interactions with CBP had 2.26 ± 0.25 kDa as average mass. Some of these *N*-glycans have three branches, two of which are located directly on the mannose (α1-6) of the *N*-acetyl-glucosamine-trimannoside core. Most of these *N*-glycans were fucosylated. Finally, the EGFR-related *N*-glycans that had multiple interaction points with the CBP of rTBL-1 showed an average mass of 4.05 ± 0.25 kDa, having four branches containing sialic acid (*N*-acetylneuraminic acid [NeuAc(α2-3)]) at the end of each branch and presenting a *N*-glycan core fucosylation at the innermost *N*-acetylglucosamine residue of the chitobiose ($GlcNAc_2$) core. An interesting feature of these *N*-glycans is that their structure was formed by four sialylated antennae and a core-fucosylated structure.

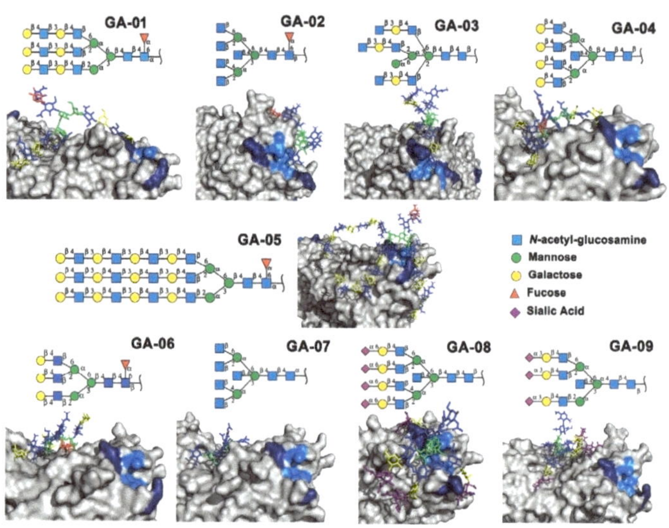

Figure 4. Molecular docking of the experimental *N*-glycans from the Mammalian Glycan Array 5.4 with rTBL-1 structures. The graphical 2D representation of *N*-glycans is shown according to the Symbol Nomenclature for Glycans (SNFG) [28]. The *N*-glycans from the mammalian glycan array "Version 5.4" are labeled as GA-##, where ## is the number described by Martínez-Alarcón et al. (2020) [13] and is related to the signal's relative intensity of lectin recognition. Docking results are shown as topological representations of interactions at the same surface location between the *N*-glycans with the recombinant lectins: rTBL-1, R104Q, and with the R131Q lectin. The *N*-glycan in topological representations is presented as sticks colored according to the SNFG (as indicated by geometric codes), the protein surface in gray, highlighting the CBP in light blue and the arginine residues in dark blue. GA, glycan array.

It is worth mentioning that all those EGFR-related *N*-glycans whose interaction with Tepary bean lectin-related structures (rTBL-1, R104Q, and R131Q) have multiple contact points on the CBP region, especially for core-fucosylated and highly sialylated *N*-glycans. This fact is important since core fucosylation has been associated with a tumorigenic phenotype, and direct inhibition is able to inhibit cellular signaling for cell proliferation, migration, and tumor growth [31,32]. On the other hand, aberrant sialylation in cancer cells plays an important role in tumor progression, favoring the evasion of cell death and promoting metastasis, immune evasion, and drug resistance. Aberrant sialylation, specifically α2-3- and α2-6-linked sialic acids, has been reported in highly metastatic and drug-resistant colorectal cancer and other tumor cells [33–38]. Highly sialylated *N*-glycans of EGFR have also been shown to disrupt ligand interactions by avoiding dimerization and subsequent phosphorylation [39], thereby protecting tumors against drug-induced cell death [40]; therefore, sialylation is an important target against cancer progression [41]. In consequence, the high affinity of rTBL-1 for highly sialylated EGFR-related *N*-glycans, especially those having α2-3-linked sialic acids, suggest that its interaction directly promotes apoptosis of colon cancer cells, and this preference could explain the anticancer effect of this lectin and potential uses for the treatment of other types of cancer.

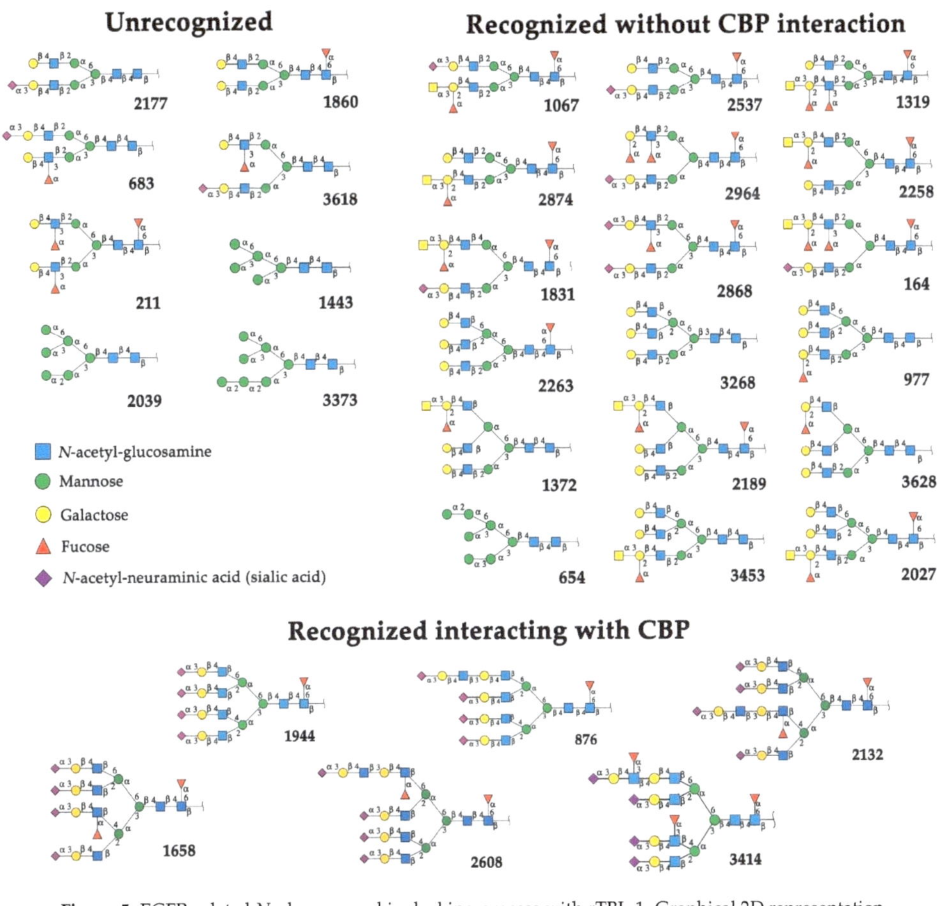

Figure 5. EGFR-related *N*-glycans used in docking process with rTBL-1. Graphical 2D representation of the *N*-glycans was created according to SNFG [28]. All EGFR-related *N*-glycans were named according to the GlyConnect platform [30]. CBP, carbohydrate-binding pocket.

2.4.3. The Role of R104 and R131 Residues in the Affinity rTBL-1 Toward EGFR-Related N-glycans by Docking Analysis

To determine the effect of punctual mutations on the lectin's recognition of EGFR-related *N*-glycans, the high-molecular-mass *N*-glycans that interacted directly with a large part of their structure on the lectin CBP (forming a preliminary docking process using the structures shown in Figure 5) were chosen for docking analysis with in silico-mutated lectin structures (R104Q and R131Q). Considering that all *N*-glycans have a $Man_3GlcNAc_2$ core, in this work the branches or "antennae" of complex *N*-glycans were numbered according to their relative position with respect to the mannose residue to which they are attached, either to Man(α1-6) or Man(α1-3) to simplify the nomenclature of the EGFR-related *N*-glycans (Supplementary Material, Figure S1). For instance, branch 1 corresponded to the antennae attached to carbon 6 of Man(α1-6) of the $Man_3GlcNAc_2$ core, and branch 2 corresponded to the antennae attached to carbon 2 of the same mannose residue. In the same way, branch 3 corresponded to carbon 4 of Man(α1-3) of the $Man_3GlcNAc_2$ core, and branch 4 corresponded to the antennae attached to carbon 2 of the same mannose residue.

Some of the EGFR-related *N*-glycans have elongated antennae. Generally, branches of complex *N*-glycans are composed of NeuAc(α2-3)Gal(β1-4)GlcNAc(β1). However, in elongated antennae, two additional monosaccharides are present, which are *N*-acetylglucosamine with a glycosidic bond β1-3 [GlcNAc(β1-3)] and galactose attached to the *N*-acetylneuraminic acid (sialic acid) [NeuAc(α2-3)] found at the end of the branch, as follows: NeuAc(α2-3)Gal(β1-4)GlcNAc(β1-3)Gal(β1-4)GlcNAc(β1).

Figure 6 shows the topological representation of the docking interactions of the selected EGFR-related *N*-glycans and the three lectin structures: rTBL-1, R104Q, or R131Q. The lectin structure was colored as a gray surface, and the CBP zone was highlighted as a light-blue zone, with residues R104 and R131 as dark-blue zones and mutated residues (Q104 and Q131) as dark-red surfaces, in order to compare the different ways in which the glycans interacted with the CBP of the lectins. When *N*-glycans interacted with rTBL-1, all except *N*-glycan 876 had a considerable interaction area on the CBP surface. Then, when comparing the docking results of the R104Q lectin with the EGFR-related *N*-glycans, *N*-glycans such as 876 and 2132 had no interaction surface in the CBP. In the case of R131Q lectin, the interaction positions were slightly changed compared to those of the rTBL-1 lectin, except for glycan 876, which had a very different location.

When comparing the 3D-interaction maps of all docking interactions (Figure 7), where protein-interacting residues are highlighted as orange sticks with their respective surface and chain position number, it is noteworthy that in the interactions of the *N*-glycans with the lectin rTBL-1, the R131 residue participates in most of them (1944, 1658, 2132, and 3414), and the R104 residue only in the interactions between rTBL-1 and 2132, 2608, and 3414 glycans. Both residues interact by a hydrogen bond (Supplementary Material, Tables S2–S19) and this could explain how the change of arginine to glutamine allowed hydrogen bond interactions by the amino group of the glutamine residue. Additionally, the Q104 residue participated in most of the interactions with *N*-glycans, such as 876, 1944, 1658, 2608, and 3414; in the R104Q lectin—2608 *N*-glycan interaction, it was possible to observe the participation of the R131 residue. Finally, the mutated Q131 residue participated only in the interactions of 1944, 1658, and 2608 *N*-glycans, having the additional participation of the R104 residue in the case of the interactions with 1658 and 2608 *N*-glycans.

At this point, it is remarkable that the arginine to glutamine mutation (R → Q) allows, in most cases, the interaction of the lectin with *N*-glycans. Nevertheless, the R104Q mutation completely changed the interaction profile with the EGFR-related 2132 *N*-glycan, which has an elongated antenna with a second fucosylation at branch 3. This change is evident in the respective docking analysis, as there were no interactions in the CBP zone.

This may be an indication that R104 plays a crucial role in the specific recognition of the lectin to certain *N*-glycan structures, and although the mutated residues may interact with *N*-glycans in most cases, they affect the three-dimensional configuration of the CBP and its surrounding regions and ultimately influence the specific recognition of the lectin and its target *N*-glycans.

In order to characterize the interaction interfaces between the EGFR-related *N*-glycans and the CBP of the recombinant and mutant Tepary bean lectins, the solvation free energy gain upon formation of the interface ($\Delta^{i}G$) was determined for all docking interactions (Table 1). The $\Delta^{i}G$ was calculated as a difference in solvation energies of all residues between interfacing structures, and a negative value corresponds to hydrophobic interfaces or is indicative of positive protein affinity according to the docking methodology of the PDBePISA web server [42].

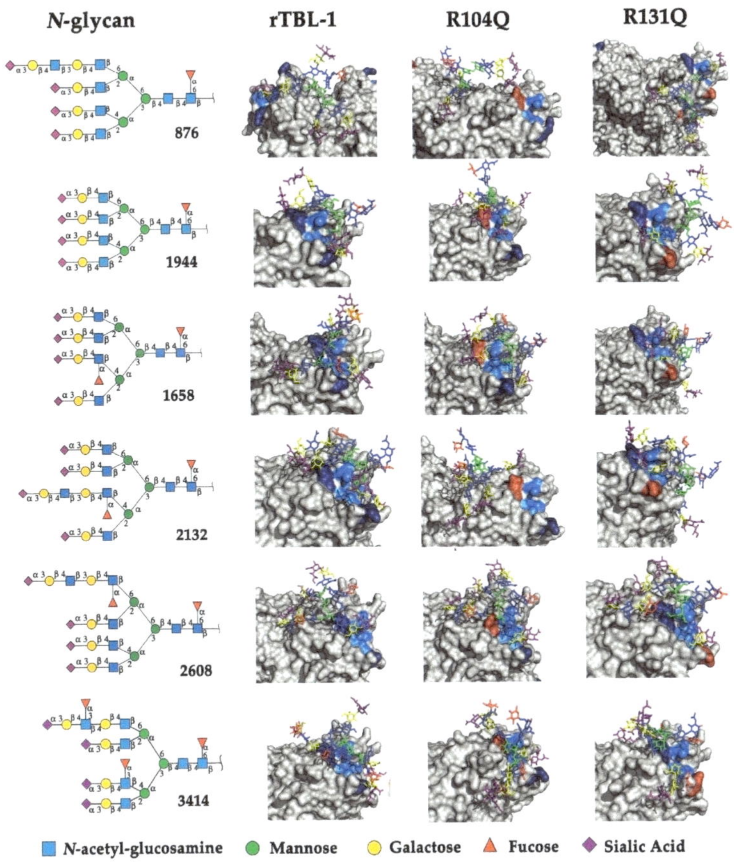

Figure 6. Molecular docking results for EGFR-related *N*-glycans with rTBL-1 and mutants R104Q and R131Q. For each *N*-glycan a graphical 2D representation according to SNFG [28] is shown. Docking results are shown as topological representations of interactions at the same surface location between the *N*-glycans, and the recombinant lectins: rTBL-1, R104Q, and R131Q. The *N*-glycan in topological representations is presented as sticks colored according to the SNFG, the protein surface in gray, highlighting the CBP in light blue, the mutated residues in red, and the original arginine residues in dark blue.

An important finding regarding the role of residues R104 and R131 in the recognition of complex *N*-glycans, considering the $\Delta^{i}G$, is the appreciable loss of affinity in the recognition of EGFR-related *N*-glycan 2132 when R104 is mutated. Although the glycan is recognized and interacts with the lectin, it no longer interacts directly with the CBP and its solvation free energy upon formation of interfaces ($\Delta^{i}G$) decreases considerably, indicating a significant loss of affinity (compare the docking results of glycan 2132 (Figure 5) with the energies of Table 1). This phenomenon also occurs with the interaction between lectins with EGFR-related *N*-glycan 876, which has an elongated antenna at branch 1 without a second fucosylation, but it is not as representative since this glycan interacts with CBP with few residues with the rTBL-1. Additionally, this could also provide insight into the importance of *N*-glycan 2132 as a target molecule for tepary lectin and the effect it has on colon cancer.

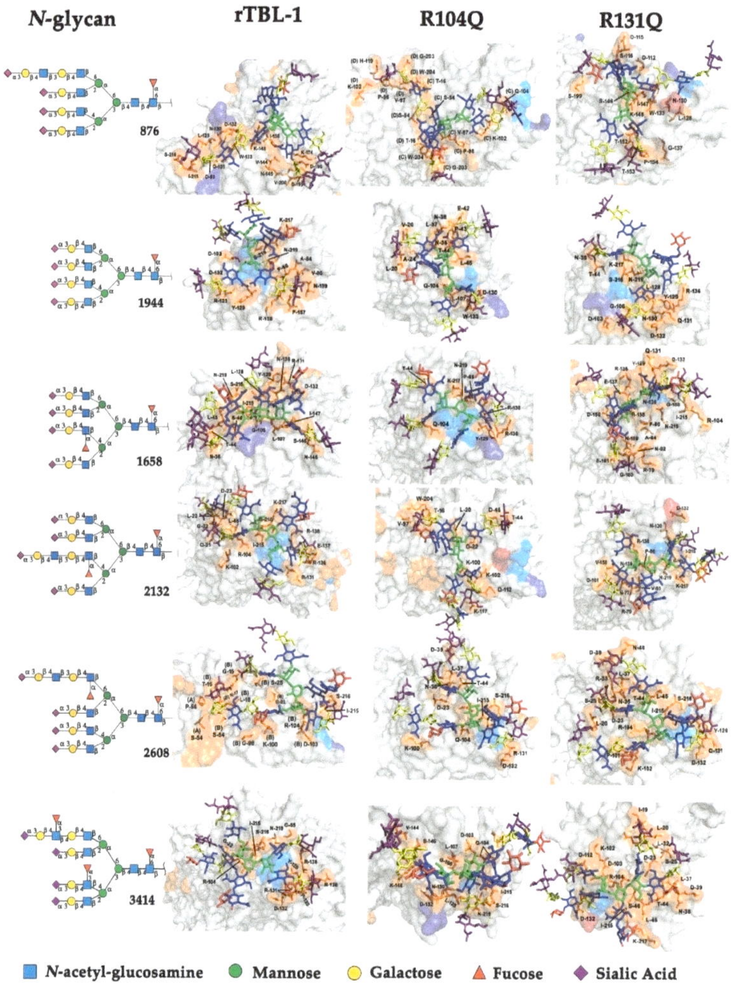

Figure 7. Three-dimensional maps for the interactions between EGFR-related *N*-glycans to rTBL-1, and mutants R104Q and R131Q. For each *N*-glycan, a graphical 2D representation according to SNFG [28] is shown. The interaction maps are shown as topological representations of the interactions of the *N*-glycans and the recombinant lectins: rTBL-1, R104Q, and R131Q. The target *N*-glycans are shown as sticks, and the structure of *N*-glycan residues is colored according to SNFG. The docking-interacting residues of lectins are shown as orange sticks with their respective surface and relative sequence number as labels.

Table 1. Comparison of the solvation free energy gain upon formation of the interface ($\Delta^{i}G$) of the complex EGFR-related *N*-glycans with the rTBL-1 and mutated Tepary bean lectins (R104Q, and R131Q).

2D-Graphical *N*-glycan Structure According to (SNFG) [28]	*N*-glycan ID [30]	Reported Molecular Mass (kDa) [30]	$\Delta^{i}G$ (Kcal/mol)		
			rTBL-1	R104Q	R131Q
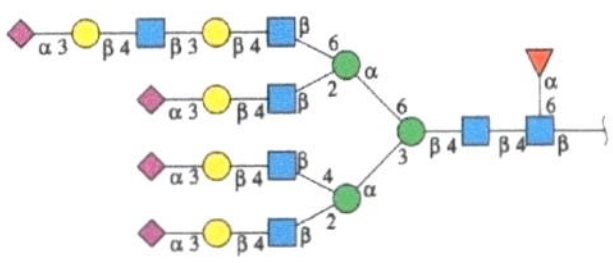	876	4.05	−12.79	−6.57	−8.03
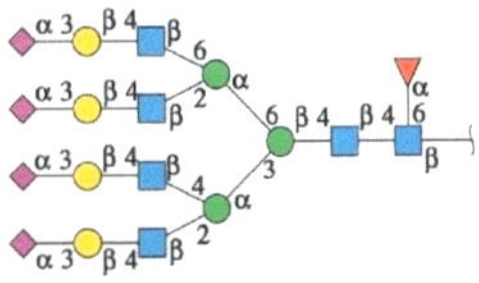	1944	3.68	−7.99	−6.44	−8.20
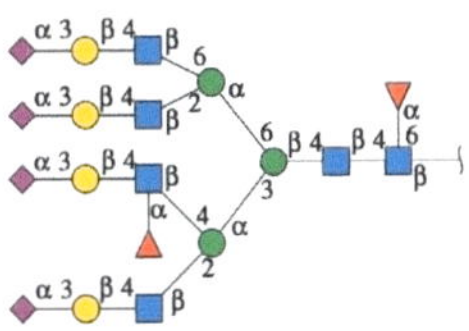	1658	3.83	−10.39	−13.80	−15.09
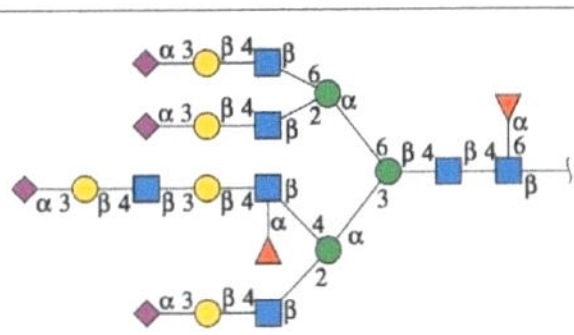	2132	4.19	−10.99	−3.18	−11.75
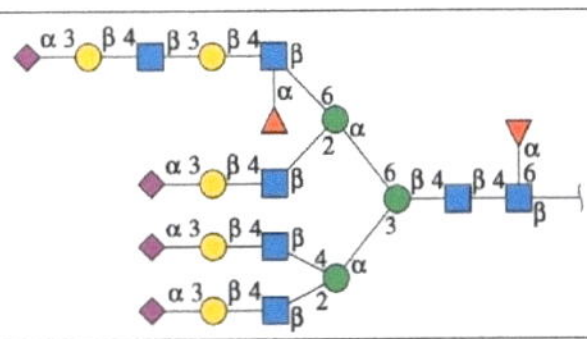	2608	4.19	−7.86	−7.50	−10.99
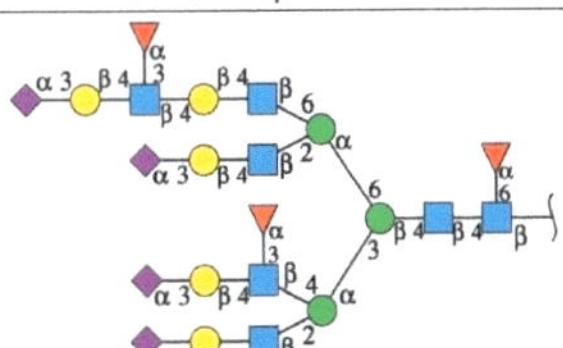	3414	4.34	−12.59	−8.91	−5.83

■ *N*-acetyl-glucosamine ● Mannose ○ Galactose ▲ Fucose ◆ Sialic Acid

rTBL-1, Recombinant lectin; R104Q, Recombinant R104Q mutant lectin; R131Q, Recombinant R131Q mutant lectin.

Another interesting finding is the probable effect of fucosylations on the recognition of *N*-glycans by tepary bean lectins. While the *N*-glycan 876 with elongated branch 1 did not have much interaction with CBP and even lost it when the residues were mutated, the *N*-glycan 2608, which had a similar structure but with a fucosylation on elongated branch 1, did have considerable interaction with CBP, which was maintained (and even apparently enhanced in the R131Q mutant), suggesting that this fucosylation could be important to recognize this structure with an elongated branch 1 (compare Figure 5 and Table 1). Fucosylations in non-elongated branches apparently have no effect (compare interactions with 1944 and 1658 EGFR-related glycans in Figure 5); however, when a fucosylation in branch 3 is present, the values of the Δ^iG increased apparently even more in the interactions with mutated residues, indicating an apparent improvement in affinity between the interfaces of the interaction (compare Δ^iG values of 1944 and 1658 *N*-glycans in Table 1). In addition, the presence of a third fucosylation in N-glycans (i.e., 3414 EGFR-related *N*-glycan) also showed enhanced affinity in the case of the interaction with rTBL-1, but a considerable decrease in the interactions with R104Q and R131Q mutated lectins.

The docking results suggested that the observed loss of cytotoxicity of the mutants was related to the loss of affinity of the studied lectin-*N*-glycan interactions. In fact, each interaction depends on the glycan type suggesting a specific effect. Collectively, the results show that the interactions with *N*-glycans, such as 876, 1658, 2132, and 3414, could promote cytotoxicity since important changes in solvation free energy gain were determined for one or both mutants for each *N*-glycan. In particular, the substitution of R104 with Q104 decreased the affinity and (in some cases) the location of the interaction for three of the six tested *N*-glycans, and an important increase was only observed for one of them. On the other hand, for the substitution of R131 for Q131, an increase in affinity was observed for three *N*-glycans and a decrease for two. Both the decrease and increase in interactions compared to rTBL-1 could influence the properties of the lectin, where a decrease is primarily associated with a loss of affinity, while an increase could be linked to binding dissociation or a regulatory function.

3. Materials and Methods

3.1. Site-Directed Mutagenesis

To generate the mutants, two sets of primers containing the specific mutations were employed to amplify the recombinant plasmid pGAPZαB-rTBL-1, previously generated by Martinez-Alarcon et al. [13]. The modifications were introduced near the 5′ end of the reverse primers (underlined content). R103Q was generated using primers 126_F (5′-GGTCTTCTAGGTCTGTTCGACGGC-3′) and 126_R (5′-GCC<u>TTG</u>GTCTTTGGGCTTAGAGC-3′), while R130Q was generated using primers 152_F (5′-CAGAGAGCGTCATATTGGCATCG-3′) and 152_R (5′-GGGTCCCA<u>GTC</u>CTGGTTGTACAAG-3′). The amplified products were gel purified using a commercial BioLabs kit (Monarch DNA Gel Extraction. Hitchin, UK). Purified plasmids were subsequently ligated at 25 °C for 1 h using T4 ligase. The ligation products were transformed into *Escherichia coli* TOP10 cells (Invitrogen Life Technology, Cat. C4040-06. Waltham, MA, USA) via electroporation (resistance of 100 Ohms, capacitance of 125 μFa, and 1.8 Volts), and cells were plated on Luria–Bertani (LB) medium containing 25 μg/mL zeocin as a selection agent, followed by incubation at 37 °C. After 24 h, transformants were screened by colony PCR (using gene-specific primers or pGAPαB primers), and positive clones were verified by digestion and sequencing.

3.2. Transformation of *Pichia pastoris*

The procedure was followed based on the patent MX/a/2018/009442. Sequence-verified plasmids were digested with *Avr*II at 37 °C for 16 h, and complete linearization of the DNA was confirmed through nucleic acid electrophoresis. Subsequently, the DNA was ethanol precipitated, and the pellet was resuspended in 20 μL of nuclease-free sterile water. The concentration, determined by NanoDrop spectrophotometry (NanoDrop 2000c from Thermo Scientific; Hercules, CA, USA) and electrophoresis, was confirmed to be >5 μg/μL. This DNA material was used to transform *Pichia pastoris* protease-deficient strain SMD1168H (Invitrogen Life Technology, Cat. C18400. Waltham, MA, USA) using the Pichia EasyCompTM Transformation Kit (Thermo Fisher Scientific; Pittsburgh, PA, USA) following the manufacturer's protocol. Transformed cells were recovered and plated on Yeast Extract-Peptone-Glycerol (YPG) medium supplemented with 50 μg/mL zeocin and incubated at 30 °C for 48 h.

Ten colonies of *P. pastoris* were selected, and each one was inoculated into 10 mL of liquid YPG medium supplemented with 25 μg/mL zeocin. The cultures were incubated at 30 °C with shaking for 48 h, after which the supernatants were collected by centrifugation at 11,800× *g* for 10 min. For SDS-PAGE electrophoresis, 25 μL aliquots of each supernatant were loaded and transferred onto nitrocellulose membranes using a semi-dry transfer system (ATTO blotter. Tokyo, Japan) at a constant voltage of 10 V for 1 h in 1X TBS (Tris-Buffered saline, Tris-HCl 15.2 mM, Tris base 46.2 mM, and NaCl 1500 mM). Ponceau red staining was performed to visualize the positions of molecular size markers, followed by blocking the membranes with 5% (*w*/*v*) defatted milk powder in 0.1% (*v*/*v*) Tween TBS for 1 h. The membranes were incubated overnight with anti-histidine monoclonal antibodies at a 1:1000 dilution in blocking solution. After 16 h, the primary antibodies were removed by several washes with blocking solution, and the membranes were further incubated for 2 h in blocking solution containing secondary anti-goat antibodies coupled to alkaline phosphatase. Finally, chemiluminescence was used to visualize the immunoreactivity.

3.3. Heterologous Expression and Purification of R103Q and R130q Mutants

P. pastoris cells expressing recombinant rTBL-1 were cultivated in a 7.5 L vessel Applikon ez-control laboratory fermenter, as described previously [13], with the exception of maintaining the pH at 5.0. After cultivation, the culture was centrifuged at 7500× *g* for 30 min at 4 °C to separate the secreted proteins from the cells. The resulting supernatant was clarified by sequential filtration using 2.7 μM and 0.7 μM glass fiber filters (Whatman. Kent, UK). Purification of the recombinant proteins from the supernatant was performed using nickel-affinity chromatography, followed by dialysis and freeze-drying, following the established protocol [14]. The protein content in the lyophilized samples was determined by SDS-PAGE gels, where the bands corresponding to intact proteins were compared to GNA (Sigma Aldrich. Burlington, MA, USA) standards through visual inspection and by capturing an image of the destained gel using a commercial flat-bed scanner. Image analysis was conducted using a custom-written software "ProQuantify", version 1.0, supplied by Rodrigo Guerrero from Universidad Autónoma de Queretaro, Mexico.

3.4. Size Exclusion Chromatography

Size exclusion chromatography (SEC) was carried out using a High-Resolution ENrich™ SEC 650 column (Bio-Rad. Hercules, CA, USA) on an NGC™ chromatography system (Bio-Rad). Prior to the assay, the column was calibrated with the gel filtration standard #15119001 (Bio-Rad), following the supplier's instructions. Protein samples at a concentration of 10 mg mL-1 were centrifuged at 8000× *g* for 30 min. The column was equilibrated with 50 mL of buffer D (20 mM MES, 100 mM NaCl, pH 6.5), and then 200 μL of the sample were injected into the system, followed by a 40 mL isocratic elution using buffer D. Fractions were monitored at 280 nm for absorbance and collected.

3.5. Identification of the Glycosylation Present in the Recombinant Lectin

To determine the size and composition of the glycosidic antennas on Mut126 and Mut153, the proteins underwent glycan digestion using a combination of five distinct glycosidases that target O and *N*-glycosylations. This process was performed utilizing the Glycoprotein Deglycosylation kit (EMD Millipore. Darmstadt, Germany) according to the manufacturer's instructions.

The glycosidases used were: (i) *N*-Glycosidase F: this enzyme cleaves all *N*-linked oligosaccharides, unless their core is α1-3 fucosylated; (ii) Endo-α-*N*-acetylgalactosaminidase that separates common nucleus (Galβ1,3GalNAca) from all *O*-glycosylations (it only works when no additional sugars are adhered to the main nucleus); (iii) α2-3,6,8,9-Neuraminidase that cleaves all sialic acids that may be attached to the common core of the *O*-glycosylations; (iv) β1,4-Galactosidase that releases only non-reducing terminal galactose, bound in β1-4 to *O*-glycosylations; (v) β-*N*-Acetylglucosaminidase that cleaves all GlcNac residues bound to β-non-reducing terminal *O*-glycosylations. Following digestion, samples were subjected to SDS-PAGE and subsequently stained with Coomassie blue. Additionally, the periodic acid-Schiff (PAS) staining technique (Sigma Chemical, St. Louis, MO, USA) was employed in accordance with the prescribed procedure based on the method originally proposed by Hotchkiss [43].

3.6. Cytotoxic Activity

To assess the cytotoxic activity of the mutants, HT-29 human colon cancer cells (ATCC, HTB-38. Manassas, VA, USA) were employed. Initially, 10,000 cells were seeded per well in 48-well plates containing 0.5 mL of medium A, which consisted of DMEM (Sigma-Aldrich, D-5796. Burlington, MA, USA) supplemented with 10% *v*/*v* fetal bovine serum (FBS) (Sigma-Aldrich, 12003C) and 1% *v*/*v* L-glutamine-antibiotics (Sigma-Aldrich, G6784). The plates were incubated at 37 °C with 5% CO_2 and 95% relative humidity until cells reached 80% confluence. After 48 h, the culture medium was replaced with medium B, composed of DMEM supplemented with 2% FBS *v*/*v* and 1% L-glutamine *v*/*v*, to synchronize the cells. Following an additional 24 h incubation, the medium was removed, and the cells were treated with DMEM supplemented with 0.5% BSA *w*/*v* (AlbuMax. Life Technologies, 11020021. Carlsbad, CA, USA) 1% L-glutamine *v*/*v*, and various concentrations of the respective treatments. PBS was used as the negative control, while rTBL served as the positive control. After 24 h incubation, the cells were harvested using 0.15% trypsin *w*/*v* and counted using a hemocytometer; triplicate wells were counted per treatment to determine cell numbers. Cell survival was calculated considering the negative control to be 100%. Lethal concentration 50 (LC_{50}) was calculated by plotting the $\log_{10}$ of the treatment's concentration vs. cell survival % and taking the value of the ordered pair (y) as 50%.

Statistical analysis was determined by comparing the mean values for each treatment using a one-way ANOVA with a Tukey post hoc analysis ($p < 0.05$) to determine which means were significantly different.

3.7. Docking Analysis Procedure

EGFR-related *N*-glycans were identified using the GlyConnect platform (https://glyconnect.expasy.org/ accessed on 8 February 2023) [30] from the Expasy portal of the SIB Swiss Institute of Bioinformatics. The representative structures of the 52 EGFR-related glycans were chosen. Structures of *N*-glycan models were created using a "Build via Text" tool from the GLYCAM web server (https://glycam.org accessed on 8 February 2023) [29], based on the monosaccharide sequence on the GlyConnect platform. Structures from the main conformers were selected according to differences in the position of their glycan branches and were considered for the docking process.

To test the effect of mutations on lectin-glycan docking, two rTBL mutant structures were generated using the PyMOL software version 2.5.1 [44], considering the R103Q and R130Q experimentally obtained mutants. As the available rTBL-1 sequence (PDB ID: 6tt9) has one extra alanine beside the first serine of the rTBL-1 sequence [14], it was necessary to go through the numbering one position. Therefore, only for the docking analysis, R103 changes to R104, and R130 changes to R131.

Docking analysis of *N*-glycans with rTBL-1 and its two in silico-mutated structures was performed on the LZERD platform [23] according to its requirements. A list of 500 ranked models with their respective structure files was obtained for each docking, and those with higher Ranksum values were selected. However, docking models whose interactions were located in the tetrameric zone of the quaternary structure of rTBL-1 were discarded because of steric hindrance present between monomers [11]. Also, docking results were compared with the glycosylated recombinant Tepary bean lectin structure (PDB ID: 6tt9) in order to discard those whose interactions were located in the natural glycosylation site. Interactions near the mutation sites (R104Q and R131Q) and near the CBP were preferentially selected.

Interaction residues for each docking result were analyzed using the Protein-Ligand Interaction Profiler server [45,46] (PLIP-server, https://plip-tool.biotec.tu-dresden.de/plip-web/plip/index accessed on 20 February 2023) and the software Discovery Studio Visualizer version 21.1.0.20298 (Dassault Systèmes Biovia Corp. Velizy-Villacoublay, France) [47]. In addition, the types of interactions and the solvation free energy gain of the interfaces formed in these interactions were analyzed. using the PDBePISA web server [42] (http://www.ebi.ac.uk/pdbe/prot_int/pistart.html accessed on 20 February 2023). Lectin-interacting residues of each docking result were identified and corroborated by PLIP-server [45]. Discovery Studio software [47] and the PDBePISA web server [42]. The total solvation energy gain upon formation of the interface of the interactions was obtained as the sum of the solvation energy gain corresponding to each of the protein-glycan interacting residues, expressed in kCal/mol, upon formation of the interface (Supplementary Material, Tables S2–S19).

4. Conclusions

The analysis of lectin sequences revealed that the *P. acutifolius* lectin, and consequently rTBL-1 have arginine residues at positions 103 and 130 (104 and 131 for docking analysis, respectively). The conservative substitution of either of these residues by glutamine in recombinant rTBL-1 mutants led to significantly decreased cytotoxicity towards colon cancer cells, suggesting that both residues are important for the establishment of lectin-*N*-glycan interactions. Moreover, R130Q inhibited cytotoxicity towards HT-29 human colon cancer cells by two-fold more than R103Q, which is indicative of distinct roles for each of the arginine residues.

Docking analysis for previously studied experimental *N*-glycans recognized by rTBL-1, showed correlation for the interaction in the CBP (close to R104 and R131). Just two of the experimentally recognized *N*-glycans interacted with the CBP, and these were highly sialylated and core-fucosylated *N*-glycans. The same pattern was observed for 32 of 52 EGFR-related *N*-glycans, where the low molecular mass (average mass: 1.77 ± 0.33 kDa) bisected *N*-glycans have no interaction at all, and 18 EGFR-related *N*-glycans with an average mass of 2.26 ± 0.25 kDa had similar interaction zones with rTBL-1 without interacting with CBP, such as those shown by the experimentally recognized *N*-glycans. Only the high molecular mass *N*-glycans (average mass: 4.05 ± 0.25 kDa) with sialylated four antennae and core-fucosylated structure had a direct docking interaction in the CBP of rTBL-1.

The interaction between the recombinant lectin and certain EGFR *N*-glycans occurs within a complex biological context, where the final effect depends on multifactorial parameters. However, this mutant study provides a better understanding of the molecular mechanism of action. It is also likely that EGFR is one of the target molecules for the rTBL-1, and many other membrane glycoproteins could be interacting, as aberrant sialylation and fucosylation are common in other glycoproteins involved in the development of many types of cancers, including colon cancer.

This research has expanded our understanding of how the Tepary bean lectin affects the microenvironment within which cancerous cells develop, creating new opportunities to uncover the mechanisms by which this lectin exerts its anticancer effects. Further studies are needed to gain a deeper understanding of the molecular interactions related to the lectin's cytotoxic effects to develop new strategies against EGFR-dependent cancers, particularly colon cancer.

Supplementary Materials: The following supporting information can be downloaded at: https://www.mdpi.com/article/10.3390/ijms252413258/s1.

Author Contributions: Conceptualization, D.M.-A., E.F., A.B.-L. and T.G.-G.; methodology, D.M.-A., J.L.C.-G., E.F., J.A.G. and S.P.; formal analysis, D.M.-A., E.F., S.P., A.B.-L. and T.G.-G.; investigation, D.M.-A., J.L.C.-G., E.F., A.B.-L. and T.G.-G.; data curation, D.M.-A., J.L.C.-G. and T.G.-G.; writing—original draft preparation, D.M.-A., J.L.C.-G. and T.G.-G.; writing—review and editing, J.L.C.-G., U.M.-C., A.B.-L. and T.G.-G.; visualization, J.L.C.-G., U.M.-C. and T.G.-G.; supervision, E.F., J.A.G., S.P., A.B.-L. and T.G.-G.; project administration, T.G.-G.; funding acquisition, T.G.-G. All authors have read and agreed to the published version of the manuscript.

Funding: This research was funded by CONAHCYT Ciencia de Frontera (171873). D.M.-A. received a Master's scholar fellowship from CONAHCYT. Partial publication fee support was provided by the Universidad Autónoma de Querétaro.

Institutional Review Board Statement: The study was approved by the Ethics Committee of Facultad de Ciencias Naturales, Universidad Autónoma de Querétaro (protocol code 29FCN2021 on 29 April 2021).

Informed Consent Statement: Not applicable.

Data Availability Statement: Data of mutants generation can be found in the repository link https://repositorio.cinvestav.mx/handle/cinvestav/1634.

Acknowledgments: We thank the technical support of Josué López-Martínez and the administrative support of Juan Trejo-Vega.

Conflicts of Interest: The authors declare no conflicts of interest.

References

1. Yau, T.; Dan, X.; Ng, C.C.; Ng, T.B. Lectins with potential for anti-cancer therapy. *Molecules* **2015**, *20*, 3791–3810. [CrossRef]
2. Mazalovska, M.; Kouokam, J.C. Plant-Derived Lectins as Potential Cancer Therapeutics and Diagnostic Tools. *Biomed. Res. Int.* **2020**, *2020*, 1631394. [CrossRef]
3. Wang, P.; Leng, X.; Duan, J.; Zhu, Y.; Wang, J.; Yan, Z.; Min, S.; Wei, D.; Wang, X. Functional Component Isolated from *Phaseolus vulgaris* Lectin Exerts In Vitro and In Vivo Anti-Tumor Activity Through Potentiation of Apoptosis and Immunomodulation. *Molecules* **2021**, *26*, 498. [CrossRef]
4. Kiss, R.; Camby, I.; Duckworth, C.; De Decker, R.; Salmon, I.; Pasteels, J.L.; Danguy, A.; Yeaton, P. In vitro influence of *Phaseolus vulgaris*, *Griffonia simplicifolia*, concanavalin A, wheat germ, and peanut agglutinins on HCT-15, LoVo, and SW837 human colorectal cancer cell growth. *Gut* **1997**, *40*, 253–261. [CrossRef]
5. Bardocz, S.; Grant, G.; Duguid, T.J.; Brown, D.S.; Pusztai, A.; Pryme, I.F. Intracellular levels of polyamines in Krebs II lymphosarcoma cells in mice fed phytohaemagglutinin-containing diets are coupled with altered tumour growth. *Cancer Lett.* **1997**, *121*, 25–29. [CrossRef]
6. Pryme, I.F.; Pusztai, A.; Bardocz, S.; Ewen, S.W. A combination of dietary protein depletion and PHA-induced gut growth reduce the mass of a murine non-Hodgkin lymphoma. *Cancer Lett.* **1999**, *139*, 145–152. [CrossRef]
7. Pryme, I.F.; Bardocz, S.; Pusztai, A.; Ewen, S.W. The growth of an established murine non-Hodgkin lymphoma tumour is limited by switching to a phytohaemagglutinin-containing diet. *Cancer Lett.* **1999**, *146*, 87–91. [CrossRef]
8. Valadez-Vega, C.; Morales-González, J.A.; Sumaya-Martínez, M.T.; Delgado-Olivares, L.; Cruz-Castañeda, A.; Bautista, M.; Sánchez-Gutiérrez, M.; Zuñiga-Pérez, C. Cytotoxic and antiproliferative effect of tepary bean lectins on C33-A, MCF-7, SKNSH, and SW480 cell lines. *Molecules* **2014**, *19*, 9610–9627. [CrossRef]
9. Moreno-Celis, U.; López-Martínez, F.J.; Cervantes-Jiménez, R.; Ferríz-Martínez, R.A.; Blanco-Labra, A.; García-Gasca, T. Tepary Bean (*Phaseolus acutifolius*) Lectins Induce Apoptosis and Cell Arrest in G0/G1 by P53(Ser46) Phosphorylation in Colon Cancer Cells. *Molecules* **2020**, *25*, 1021. [CrossRef]

10. Torres-Arteaga, I.; Castro-Guillén, J.L.; Mendiola-Olaya, E.; García-Gasca, T.; Ángeles-Zaragoza, M.V.; García-Santoyo, V.; Torres-Castillo, J.A.; Aguirre, C.; Phinney, B.; Blanco-Labra, A. Characterization of Two Non-fetuin-binding Lectins from Tepary Bean (*Phaseolus acutifolius*) Seeds with Differential Cytotoxicity on Colon Cancer Cells. *J. Glycobiol.* **2016**, *5*, 1000117. [CrossRef]
11. Torres-Arteaga, I.; Blanco-Labra, A.; Mendiola-Olaya, E.; García-Gasca, T.; Aguirre-Mancilla, C.; Ortega-de-Santiago, A.L.; Barboza, M.; Lebrilla, C.B.; Castro-Guillén, J.L. Comparative study, homology modelling and molecular docking with cancer associated glycans of two non-fetuin-binding Tepary bean lectins. *Glycoconj. J.* **2022**, *40*, 69–84. [CrossRef]
12. Moreno-Celis, U.; López-Martínez, J.; Blanco-Labra, A.; Cervantes-Jiménez, R.; Estrada-Martínez, L.A.; García-Pascalin, A.E.; Guerrero-Carrillo, M.J.; Rodríguez-Méndez, A.J.; Mejía, C.; Ferríz-Martínez, R.A.; et al. *Phaseolus acutifolius* Lectin Fractions Exhibit Apoptotic Effects on Colon Cancer: Preclinical Studies Using Dimethilhydrazine or Azoxi-Methane as Cancer Induction Agents. *Molecules* **2017**, *22*, 1670. [CrossRef]
13. Martínez-Alarcón, D.; Varrot, A.; Fitches, E.; Gatehouse, J.A.; Cao, M.; Pyati, P.; Blanco-Labra, A.; Garcia-Gasca, T. Recombinant Lectin from Tepary Bean (*Phaseolus acutifolius*) with Specific Recognition for Cancer-Associated Glycans: Production, Structural Characterization, and Target Identification. *Biomolecules* **2020**, *10*, 654. [CrossRef]
14. Palmerín-Carreño, D.; Martínez-Alarcón, D.; Dena-Beltrán, J.L.; Vega-Rojas, L.J.; Blanco-Labra, A.; Escobedo-Reyes, A.; García-Gasca, T. Optimization of a Recombinant Lectin Production in *Pichia pastoris* Using Crude Glycerol in a Fed-Batch System. *Processes* **2021**, *9*, 876. [CrossRef]
15. Dena-Beltrán, J.L.; Nava-Domínguez, P.; Palmerín-Carreño, D.; Martínez-Alarcón, D.; Moreno-Celis, U.; Valle-Pacheco, M.; Castro-Guillén, J.L.; Blanco-Labra, A.; García-Gasca, T. EGFR and p38MAPK Contribute to the Apoptotic Effect of the Recombinant Lectin from Tepary Bean (*Phaseolus acutifolius*) in Colon Cancer Cells. *Pharmaceuticals* **2023**, *16*, 290. [CrossRef]
16. Nagae, M.; Soga, K.; Morita-Matsumoto, K.; Hanashima, S.; Ikeda, A.; Yamamoto, K.; Yamaguchi, Y. Phytohemagglutinin from *Phaseolus vulgaris* (PHA-E) displays a novel glycan recognition mode using a common legume lectin fold. *Glycobiology* **2014**, *24*, 368–378. [CrossRef]
17. Radhakrishnan, A.; Chellapandian, H.; Ramasamy, P.; Jeyachandran, S. Back2Basics: Animal lectins: An insight into a highly versatile recognition protein. *J. Proteins Proteom.* **2023**, *14*, 43–59. [CrossRef]
18. Bianchet, M.A.; Ahmed, H.; Vasta, G.R.; Amzel, L.M. Structural aspects of lectin-ligand interactions. In *Animal Lectins: A Functional View*; Vasta, G.R., Ahmed, H., Eds.; CRC Press Taylor & Francis Group: Boca Raton, FL, USA, 2008; pp. 25–26. [CrossRef]
19. Yoshimura, A.; Hatanaka, R.; Tanaka, H.; Kitajima, K.; Sato, C. The conserved arginine residue in all siglecs is essential for Siglec-7 binding to sialic acid. *Biochem. Biophys. Res. Commun.* **2021**, *534*, 1069–1075. [CrossRef]
20. Ren, X.; Lin, S.; Guan, F.; Kang, H. Glycosylation Targeting: A Paradigm Shift in Cancer Immunotherapy. *Int. J. Biol. Sci.* **2024**, *20*, 2607–2621. [CrossRef]
21. Stroop, C.J.; Weber, W.; Gerwig, G.J.; Nimtz, M.; Kamerling, J.P.; Vliegenthart, J.F. Characterization of the carbohydrate chains of the secreted form of the human epidermal growth factor receptor. *Glycobiology* **2000**, *10*, 901–917. [CrossRef]
22. Luke, B. *Beyond the Sequon: Sites of N-Glycosylation*; IntechOpen: London, UK, 2012. [CrossRef]
23. Venkatraman, V.; Yang, Y.D.; Sael, L.; Kihara, D. Protein-protein docking using region-based 3D Zernike descriptors. *BMC Bioinform.* **2009**, *10*, 407. [CrossRef]
24. Khorana, A.A.; Ryan, C.K.; Cox, C.; Eberly, S.; Sahasrabudhe, D.M. Vascular endothelial growth factor, CD68, and epidermal growth factor receptor expression and survival in patients with Stage II and Stage III colon carcinoma: A role for the host response in prognosis. *Cancer* **2003**, *97*, 960–968. [CrossRef]
25. Luque-Garcia, J.L.; Martinez-Torrecuadrada, J.L.; Epifano, C.; Canamero, M.; Babel, I.; Casal, J.I. Differential protein expression on the cell surface of colorectal cancer cells associated to tumor metastasis. *Proteomics* **2010**, *10*, 940–952. [CrossRef]
26. Gan, Y.; Chen, D.; Li, X. Proteomic analysis reveals novel proteins associated with progression and differentiation of colorectal carcinoma. *J. Cancer Res. Ther.* **2014**, *10*, 89–96. [CrossRef]
27. Sethi, M.K.; Fanayan, S. Mass Spectrometry-Based *N*-Glycomics of Colorectal Cancer. *Int. J. Mol. Sci.* **2015**, *16*, 29278–29304. [CrossRef]
28. Varki, A.; Cummings, R.D.; Aebi, M.; Packer, N.H.; Seeberger, P.H.; Esko, J.D.; Stanley, P.; Hart, G.; Darvill, A.; Kinoshita, T.; et al. Symbol Nomenclature for Graphical Representations of Glycans. *Glycobiology* **2015**, *25*, 1323–1324. [CrossRef]
29. Woods Group. GLYCAM Web. Complex Carbohydrate Research Center, University of Georgia, Athens, GA. 2005–2024. Available online: http://legacy.glycam.org (accessed on 8 February 2023).
30. Duvaud, S.; Gabella, C.; Lisacek, F.; Stockinger, H.; Ioannidis, V.; Durinx, C. Expasy, the Swiss Bioinformatics Resource Portal, as designed by its users. *Nucleic Acids Res.* **2021**, *49*, W216–W227. [CrossRef]
31. Zhou, Y.; Fukuda, T.; Hang, Q.; Hou, S.; Isaji, T.; Kameyama, A.; Gu, J. Inhibition of fucosylation by 2-fluorofucose suppresses human liver cancer HepG2 cell proliferation and migration as well as tumor formation. *Sci. Rep.* **2017**, *7*, 11563. [CrossRef]
32. De Freitas Junior, J.C.M.; Morgado-Díaz, J.A. The role of *N*-glycans in colorectal cancer progression: Potential biomarkers and therapeutic applications. *Oncotarget* **2016**, *7*, 19395–19413. [CrossRef]

33. Kim, H.J.; Kim, S.C.; Ju, W.; Kim, Y.H.; Yin, S.Y.; Kim, H.J. Aberrant sialylation and fucosylation of intracellular proteins in cervical tissue are critical markers of cervical carcinogenesis. *Oncol. Rep.* **2014**, *31*, 1417–1422. [CrossRef]
34. Yang, S.T.; Liu, C.H.; Chao, W.T.; Liu, H.H.; Lee, W.L.; Wang, P.H. The role of sialylation in gynecologic cancers. *Taiwan J. Obstet. Gynecol.* **2023**, *62*, 651–654. [CrossRef]
35. Park, J.-J.; Lee, M. Increasing the α 2,6 Sialylation of Glycoproteins May Contribute to Metastatic Spread and Therapeutic Resistance in Colorectal Cancer. *Gut Liver* **2013**, *7*, 629–641. [CrossRef]
36. Pearce, O.M.T.; Läubli, H. Sialic acids in cancer biology and immunity. *Glycobiology* **2016**, *26*, 111–128. [CrossRef]
37. Dorsett, K.A.; Marciel, M.P.; Hwang, J.; Ankenbauer, K.E.; Bhalerao, N.; Bellis, S.L. Regulation of ST6GAL1 sialyltransferase expression in cancer cells. *Glycobiology* **2021**, *31*, 530–539. [CrossRef]
38. Munkley, J. Aberrant Sialylation in Cancer: Therapeutic Opportunities. *Cancers* **2022**, *14*, 4248. [CrossRef]
39. Yen, H.Y.; Liu, Y.C.; Chen, N.Y.; Tsai, C.F.; Wang, Y.T.; Chen, Y.J.; Hsu, T.L.; Yang, P.C.; Wong, C.H. Effect of sialylation on EGFR phosphorylation and resistance to tyrosine kinase inhibition. *Proc. Natl. Acad. Sci. USA* **2015**, *112*, 6955–6960. [CrossRef]
40. Britain, C.M.; Holdbrooks, A.T.; Anderson, J.C.; Willey, C.D.; Bellis, S.L. Sialylation of EGFR by the ST6Gal-I sialyltransferase promotes EGFR activation and resistance to gefitinib-mediated cell death. *J. Ovarian Res.* **2018**, *11*, 12. [CrossRef]
41. Dobie, C.; Skropeta, D. Insights into the role of sialylation in cancer progression and metastasis. *Br. J. Cancer* **2021**, *124*, 76–90. [CrossRef]
42. Krissinel, E.; Henrick, K. Inference of macromolecular assemblies from crystalline state. *J. Mol. Biol.* **2007**, *372*, 774–797. [CrossRef]
43. Hotchkiss, R.D. A microchemical reaction resulting in the staining of polysaccharide structures in fixed tissue preparations. *Arch. Biochem.* **1948**, *16*, 131–141. Available online: https://pubmed.ncbi.nlm.nih.gov/18921848/ (accessed on 22 January 2021).
44. DeLano, W.L. *PyMOL*, version 2.3.4; The PyMOL Molecular Graphics System; DeLano Scientific LLC: San Carlos, CA, USA, 2022. Available online: https://pymol.org/2/ (accessed on 20 December 2022).
45. Adasme, M.F.; Linnemann, K.L.; Bolz, S.N.; Kaiser, F.; Salentin, S.; Haupt, V.J.; Schroeder, M. PLIP 2021: Expanding the scope of the protein–ligand interaction profiler to DNA and RNA. *Nucl. Acids Res.* **2021**, *49*, W530–W534. [CrossRef]
46. Salentin, S.; Schreiber, S.; Haupt, V.J.; Adasme, M.F.; Schroeder, M. PLIP: Fully automated protein–ligand interaction profiler. *Nucl. Acids Res.* **2015**, *43*, W443–W447. [CrossRef]
47. Dassault Systèmes. *BIOVIA Discovery Studio Visualizer*, version 21.1.0.20298; Dassault Systèmes: San Diego, CA, USA, 2020.

International Journal of
Molecular Sciences

Article

A Novel Biosynthetic Strategy for Ginsenoside Ro: Construction of a Metabolically Engineered *Saccharomyces cerevisiae* Strain Using a Newly Identified UGAT Gene from *Panax ginseng* as the Key Enzyme Gene and Optimization of Fermentation Conditions

Xiaochen Yu [1], Jinghui Yu [1], Dinghui Wang [1], Sizhang Liu [1], Kangyu Wang [1,2], Mingzhu Zhao [1,2], Ping Chen [1,2], Yanfang Wang [3], Yi Wang [1,2,*] and Meiping Zhang [1,2,*]

[1] College of Life Science, Jilin Agricultural University, Changchun 130118, China; yu-xiaochen@foxmail.com (X.Y.); jinghui-yu@foxmail.com (J.Y.); dinghui-wang@foxmail.com (D.W.); lsz512411412@163.com (S.L.); kangyu.wang@jlau.edu.cn (K.W.); mingzhuzhao@jlau.edu.cn (M.Z.); chenping201407@126.com (P.C.)

[2] Jilin Engineering Research Center Ginseng Genetic Resources Development and Utilization, Jilin Agricultural University, Changchun 130118, China

[3] College of Chinese Medicinal Materials, Jilin Agricultural University, Changchun 130118, China; yfwang2014@163.com

* Correspondence: yi.wang@jlau.edu.cn (Y.W.); meiping.zhang@jlau.edu.cn (M.Z.)

Abstract: Ginsenoside Ro, as one of the few oleanane-type ginsenosides, is well known for its unique molecular structure and biological activities. Currently, research on the biosynthesis of ginsenoside Ro is still in its early stages. Therefore, the establishment of a new ginsenoside Ro cell factory is of great significance for the in-depth development and utilization of genes related to ginsenoside Ro synthesis, as well as for the exploration of pathways to obtain ginsenoside Ro. In this study, we cloned endogenous constitutive promoters, terminators, and other genetic elements from *S. cerevisiae* BY4741. These elements were then sequentially assembled with the uridine diphosphate glucuronic acid transferase gene identified in our previously study (*PgUGAT252645*) and several other reported key enzyme genes, to construct DNA fragments used for integration into the genome of *S. cerevisiae* BY4741. By sequentially transferring these DNA fragments into chemically competent cells of engineering strains and conducting screening and target product detection, we successfully constructed an engineered *S. cerevisiae* strain (BY-Ro) for ginsenoside Ro biosynthesis using *S. cerevisiae* BY4741 as the host cell. Strain BY-Ro produced 253.32 μg/L of ginsenoside Ro under optimal fermentation conditions. According to subsequent measurements and calculations, this equates to 0.033 mg/g DCW, corresponding to approximately 31% of the ginsenoside Ro content found in plant samples. This study not only included a deeper investigation into the function of *PgUGAT252645* but also provides a novel engineering platform for ginsenoside Ro biosynthesis.

Keywords: *Panax ginseng*; ginsenoside; biosynthesis; UDP-glucuronosyltransferase

1. Introduction

Ginsenosides are a class of secondary metabolites from plants of the *Panax* genus belonging to the triterpenoid class [1], generally with high medicinal value. To date, approximately 180 ginsenosides have been isolated and identified from 17 species of the *Panax* genus [2–4]. The differences in the molecular structures of different monomeric ginsenosides lead to general differences in their pharmacological activities. Among them, ginsenoside Ro is a highly specific ginsenoside. Ginsenoside Ro is one of the few reported oleanane-type ginsenosides to date [5]; its triterpene molecular skeleton is a pentacyclic triterpene compound, namely oleanolic acid [6]. In addition, among all the monomeric ginsenosides, ginsenoside Ro is one of the few ginsenosides known to contain a glucuronic acid group in its molecular structure.

Currently, the main method for obtaining ginsenosides is direct extraction from plant tissue. This production method is often limited by the efficiency of the extraction process, which, in turn, leads to a yield that struggles to meet market demand [7]. Although efficient systems for obtaining ginsenosides have been successfully established through biotechnological means such as cell engineering [8], metabolic engineering [9], and genetic engineering [10], these research results still face various limitations in practical applications, including with respect to cost and safety.

With the continuous development of synthetic biology, more and more active ingredients of medicinal plants have achieved heterologous biosynthesis and are expected to replace traditional planting and extraction methods and become a new way to obtain medicinal ingredients. Since the first study on the synthesis of dammarenediol-II and protopanaxadiol using *S. cerevisiae* was reported [11], research on the synthetic biology of ginsenosides using *E. coli* [12] and tobacco [13–15] as host organisms has also been carried out. Since the cell structure of *S. cerevisiae* is similar to that of plant cells, it can provide a suitable environment for cytochrome P450 and glycosyltransferase in the synthesis pathway of ginsenosides [16], and the inherent MVA pathway can provide sufficient precursor substrates [17]. In addition, the higher biosafety and higher tolerance to enzyme proteins encoded by genes from other species make provide advantages in the application of ginsenoside biosynthesis [18].

At present, exogenous key enzyme-encoding genes from species such as *P. notoginseng* [19], *G. uralensis* [20], *A. thaliana* [21], *T. subterraneum* [22], and *E. coli* [23] have been successively applied to the construction of the engineered *S. cerevisiae* strains of ginsenoside biosynthesis. Optimization strategies such as the development of ginsenoside synthesis compartments in engineered yeast strains [24] and modularization of ginsenoside synthesis pathways [25,26] have also been proposed. In addition, studies have shown that optimizing fermentation conditions including culture medium elements can effectively increase the yield of ginsenosides in engineered strains [27,28].

Although current synthetic biology research targeting ginsenoside Ro has made some progress [22,23], the number of known key genes involved in the biosynthesis of ginsenoside Ro is still limited, and the diversity of synthetic pathway construction schemes is insufficient. In the synthesis pathway of ginsenoside Ro, glucuronidation modification of the hydroxyl group at the C3 position of oleanolic acid is the first step in the glycosylation process. In a previous study, we identified uridine diphosphate glucuronic acid transferase gene *PgUGAT252645* [29], which catalyzes this step in *P. ginseng*. In this study, we used *PgUGAT252645* as one of the key enzyme gene elements to construct a new engineered *S. cerevisiae* strain for the biosynthesis of ginsenoside Ro. By optimizing fermentation conditions, we further improved the yield of ginsenoside Ro in the engineered *S. cerevisiae* strain.

2. Results

2.1. Construction of the Engineered S. cerevisiae Strain

By performing agarose gel electrophoresis and DNA sequencing, it was confirmed that all gene elements of the assembled key enzyme gene expression cassettes and genomic integration fragments were correctly connected in the order designed in the experiment. Competent *S. cerevisiae* cells were then transformed using the described methods, constructing and screening (Figure S1) the engineered *S. cerevisiae* strains with the genotypes shown in Table 1.

The results of liquid chromatography–mass spectrometric (LC-MS) detection of target products for each strain during the construction process are shown in Figure 1. The retention times of oleanolic acid (OA), calenduloside E (CE), and ginsenoside Ro (Ro) in the standard solution correspond to the chromatographic peaks detected in the extract solutions of different strain cultures. Mass spectrometry results showed ions with the same mass as the target products at the corresponding times, indicating that the extract solutions of each strain culture contained the respective target products. This confirms that all gene fragments were successfully integrated into the genome and correctly performed their catalytic functions. Based on the sample retention peak areas, the yields of oleanolic acid in strain BY-OA, calenduloside E in strain BY-CE, and ginsenoside Ro in strain BY-Ro were 56.31 mg/L, 3.8 mg/L, and 155 µg/L, respectively.

Table 1. Information on engineered brewing yeast strains.

Name	Description
BY4741	*MATα his3Δ1 leu2Δ0 met15Δ0 ura3Δ0*
BY-A	*BY4741, rDNA:: TPI 1P-tHMG 1-CYC 1T-KanR-TEF 1P-syn Gg.βAS-ADH 1T*
BY-OA	*BY-A, δDNA:: TEF 1P-syn Cr.OAS-FBA 1T-HygR-TDH 3P-syn Mt.CYPR-ENO 2T*
BY-CE	*BY-OA, tHIS 3:: TPI 1P-syn At.UGDH-CYC 1T-URA 3-TEF 1P-syn Pg.UGAT-ADH 1T*
BY-Ro	*BY-CE, tLEU 2:: TEF 1P-syn Mt.UGT-FBA 1T-HIS 3-TDH 3P-syn Ts.UGT-ENO 2T*

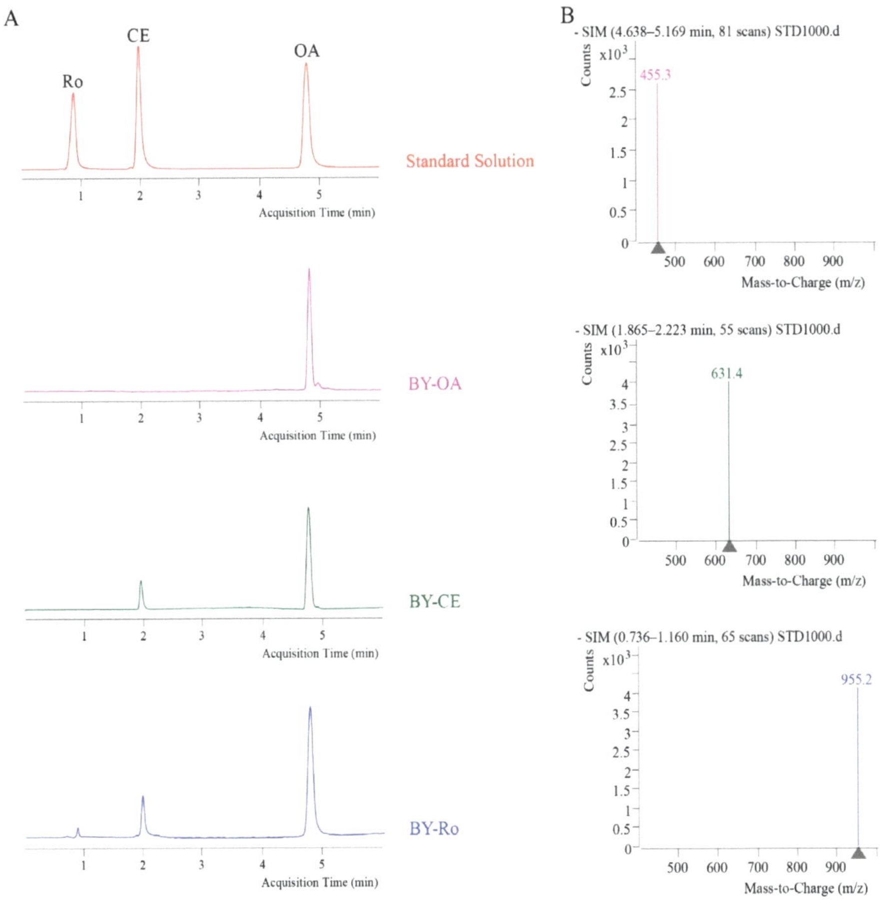

Figure 1. Liquid chromatography–mass spectrometry (LC-MS) detection of target products. (**A**) Liquid chromatographic analysis of the standard and extract samples. The red curve represents the standard solution containing oleanolic acid (OA), calenduloside E (CE), and ginsenoside Ro (OA). The pink, green, and blue curves represent the extract samples from the cultures of strains BY-OA, BY-CE, and BY-Ro, respectively. (**B**) Mass spectrometric detection of the standards.

2.2. Optimization of Fermentation Conditions: Single-Factor Experiment

As shown in Figure 2A, when the inoculum size is 1–3%, the yield of ginsenoside Ro in strain BY-Ro increases with the inoculum size and reaches a maximum of 183.60 μg/L at an inoculum size of 3%. When the inoculum size is greater than 3%, the yield of ginsenoside Ro decreases with increasing inoculum size. Therefore, an inoculum size range of 2–4% was selected for subsequent experiments.

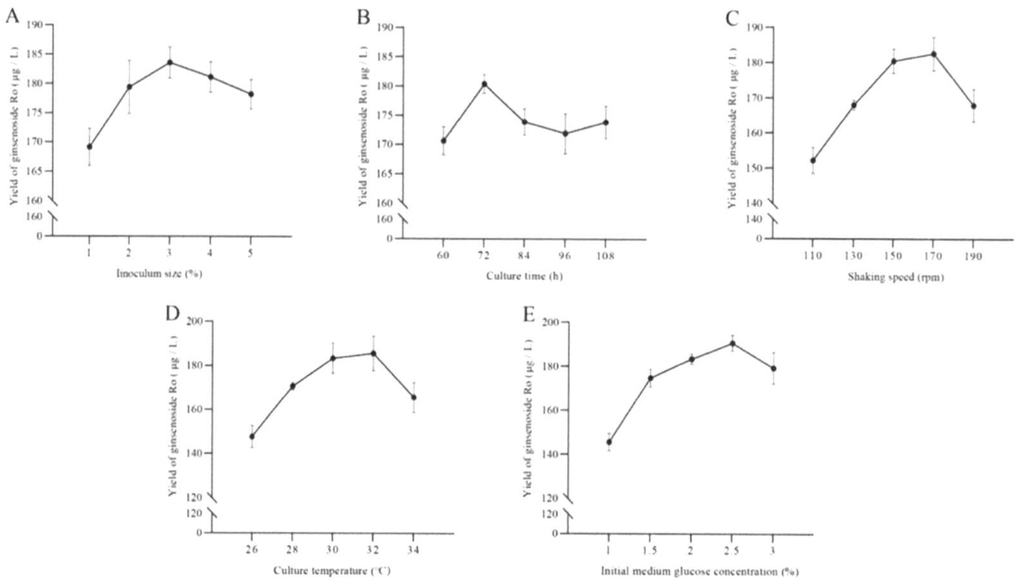

Figure 2. The effect of different cultivation conditions on the yield of ginsenoside Ro in strain BY-Ro. (**A**) Inoculum size. (**B**) Culture time. (**C**) Shaking speed. (**D**) Culture temperature. (**E**) Initial medium glucose concentration.

As shown in Figure 2B, the yield of ginsenoside Ro in strain BY-Ro reaches a maximum of 180.39 ng/L at a culture time of 72 h. When the culture time is greater than 72 h, the yield of ginsenoside Ro significantly decreases. Therefore, a culture time range of 60 h–84 h was selected for subsequent experiments.

As shown in Figure 2C, the yield of ginsenoside Ro in strain BY-Ro increases with shaking speed in the range of 110–170 rpm, reaching a maximum of 182.61 μg/L at 170 rpm. When the shaking speed is greater than 170 rpm, the yield of ginsenoside Ro significantly decreases. Therefore, a shaking speed range of 150–190 rpm was selected for subsequent experiments.

As shown in Figure 2D, the yield of ginsenoside Ro in strain BY-Ro reaches a maximum of 185.63 ng/L at a culture temperature at 32 °C. When the culture temperature is higher than 32 °C, the yield of ginsenoside Ro significantly decreases. Therefore, a culture temperature range of 30–34 °C was selected for subsequent experiments.

As shown in Figure 2E, the yield of ginsenoside Ro in strain BY-Ro increases with an initial medium glucose concentration in the range of 1–2.5%, reaching a maximum of 190.70 μg/L at 2.5%. When the initial medium glucose concentration is higher than 2.5%, the yield of ginsenoside Ro decreases. Therefore, an initial medium glucose concentration range of 2–3% in the medium was selected for subsequent experiments.

2.3. Optimization of Fermentation Conditions: Plackett–Burman Design

The design of relevant factors and levels in the Plackett–Burman design based on the results of the single-factor experiments is shown in Table 2. Using ginsenoside Ro yield as the response value, the design scheme and experimental data for single-factor significance screening are shown in Table 3. The results of the statistical analysis of the experimental data are shown in Table 4. The analysis results show that the established model has a p value of 0.0001 < 0.01, an F value of 104.22 > 1, $R^2 = 0.9886$, and $R^2_{adj} = 0.9791$, indicating that the model can effectively explain the experimental results and provide guidance for further optimization. Additionally, the multiple linear regression equation between the factors and the response value is expressed as follows:

$$Y = 188.21 - 0.36A + 1.29B + 3.25C - 28.14D + 10.37E$$

The results indicate that shaking speed, culture time, and initial medium glucose concentration have positive effects, while inoculum size and culture temperature have negative effects. The order of factors affecting ginsenoside Ro yield is culture temperature > initial medium glucose concentration > shaking speed > culture time > inoculum size. Among them, culture temperature, initial medium glucose concentration, and shaking speed have significant impacts on ginsenoside Ro yield. Therefore, these factors were selected for subsequent response surface optimization design experiments and analysis.

Table 2. Factors and levels of the Plackett–Burman design.

Factor	Levels	
	−1	**1**
A. Inoculum Size (%)	2	4
B. Culture Time (h)	60	84
C. Shaker Speed (rpm)	150	190
D. Culture Temperature (°C)	30	34
E. Medium Glucose Concentration (%)	2	3

Table 3. Plackett–Burman design and experimental results.

Experiment Number	Factors					Ginsenoside Ro Yield (μg/L)
	A	B	C	D	E	
1	−1	1	1	−1	−1	231.11
2	−1	1	1	1	1	166.88
3	1	−1	1	−1	1	214.16
4	−1	−1	−1	1	−1	145.55
5	1	1	−1	1	1	171.51
6	1	1	1	−1	−1	224.48
7	−1	−1	−1	−1	−1	199.32
8	−1	−1	1	1	1	145.55
9	1	−1	1	1	−1	154.82
10	−1	1	−1	−1	1	221.35
11	1	−1	−1	−1	1	207.67
12	1	1	−1	1	−1	176.16

Table 4. Statistical analysis of the Plackett–Burman design and experimental results.

Source of Variation	Sum of Squares	Degrees of Freedom	Mean Square	F Value	p Value	Significance
Model	10,937.83	5	2187.57	104.22	<0.0001	**
A	1.56	1	1.56	0.074	0.7945	
B	19.89	1	19.89	0.95	0.368	
C	126.82	1	126.82	6.04	0.0493	*
D	9499.83	1	9499.83	452.58	<0.0001	**
E	1289.75	1	1289.75	61.44	0.0002	**
Residual	125.94	6	20.99			
Total	11,063.78	11				

* for significance of $p \leq 0.05$ and ** for significance of $p \leq 0.01$.

2.4. Optimization of Fermentation Conditions: Box–Behnken Design

The relevant factors and levels for the Box-Behnken design based on the analysis of the Plackett–Burman design and experimental results are shown in Table 5. We selected culture temperature, initial medium glucose concentration, and shaking speed, which have significant impacts on ginsenoside Ro yield, as the factors for further optimization. A three-factor, three-level experimental design was implemented. The Box–Behnken design scheme and experimental data obtained using ginsenoside Ro yield as the response value are shown in Table 6. The results of statistical analysis of the experimental data are shown in Table 7. The analysis results show that the established model has a p value of 0.0001 < 0.01, an F value of 173.03 > 1, R^2 = 0.9955, and R^2adj = 0.9898, indicating that the model can provide adequate guidance for the optimization of fermentation conditions. Additionally, the multiple quadratic regression equation between the factors and the response value is expressed as follows:

$$Y = 255.98 + 13.24A + 7.09B - 15.73C - 3.98AB + 4.33AC - 6.01BC - 31.59A^2 - 15.24B^2 - 29.43C^2$$

The results indicate that the initial medium glucose concentration, shaking speed, and culture temperature have a highly significant impact on ginsenoside Ro yield ($p \leq 0.01$), and the interaction between shaking speed and culture temperature also has a highly significant impact on ginsenoside Ro yield ($p \leq 0.01$).

Table 5. Factors and levels of the Box–Behnken design.

Factor	Levels		
	−1	0	1
A. Medium Glucose Concentration (%)	2	2.5	3
B. Shaker Speed (rpm)	150	170	190
C. Culture Temperature (°C)	30	32	34

Table 6. Box–Behnken design and experimental results.

Experiment Number	Factors			Ginsenoside Ro Yield (μg/L)
	A	B	C	
1	0	0	0	257.44
2	−1	0	1	163.52
3	0	1	−1	242.06
4	−1	0	−1	200.13
5	0	0	0	254.92
6	0	−1	1	192.59
7	0	1	1	196.50
8	1	0	−1	217.74
9	0	−1	−1	214.11
10	1	1	0	224.73
11	0	0	0	253.52
12	0	0	0	253.24
13	−1	1	0	206.00
14	0	0	0	260.79
15	1	0	1	198.46
16	−1	−1	0	185.60
17	1	−1	0	220.26

As shown in Figure 3A, when the culture temperature is constant, the contour plot appears as a closed ellipse, and the response surface is convex, indicating a strong interaction between initial medium glucose concentration and shaking speed, with a maximum value. The response surface plot (Figure 3B) shows that with increases in initial medium glucose concentration and shaking speed, the ginsenoside Ro yield initially increases, then decreases. As shown in Figure 3C, when the shaking speed is constant, the contour plot indicates a certain interaction between initial medium glucose concentration and culture temperature, with a maximum value. The response surface plot (Figure 3D) shows that with increases in initial medium glucose concentration and culture temperature, the ginsenoside Ro yield also initially increases, then decreases. As shown in Figure 3E, when the initial medium glucose concentration is constant, the contour plot indicates a certain interaction between shaking speed and culture temperature, with a maximum value. The response surface plot (Figure 3F) shows that with an increase in shaking speed, the ginsenoside Ro yield shows an increasing trend, while with an increase in culture temperature, the ginsenoside Ro yield initially increases, then decreases.

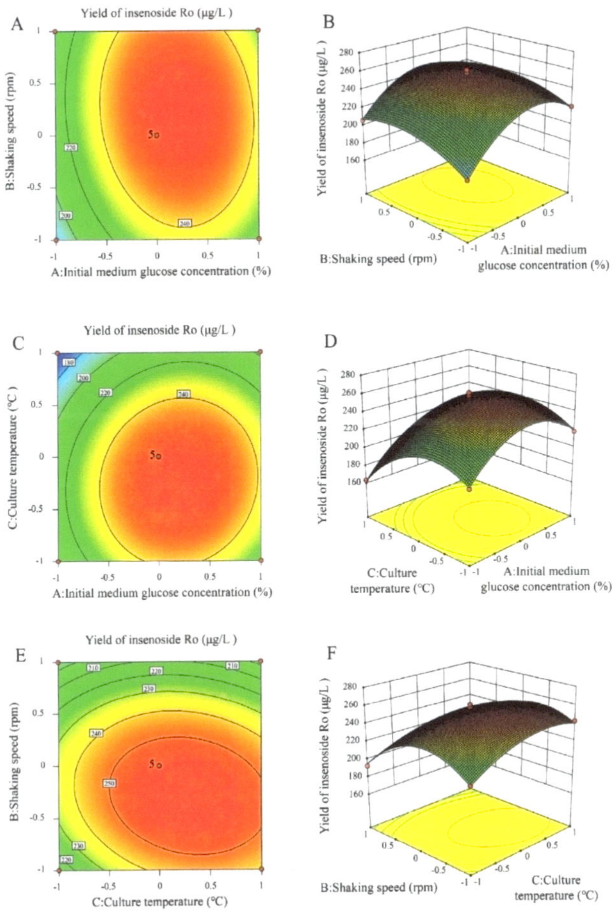

Figure 3. Contour plot and response surface plot of the effect of interaction between factors on the yield of ginsenoside Ro. (**A**,**B**) Shaking speed and initial medium glucose concentration. (**C**,**D**) Culture temperature and initial medium glucose concentration. (**E**,**F**) Shaking speed and culture temperature.

Based on the above data, the following optimal fermentation conditions are predicted for ginsenoside Ro production by strain BY-Ro: inoculum size, 2%; culture time, 84 h; initial medium glucose concentration, 2.31%; shaking speed, 175.27 rpm; and culture temperature, 31.15 °C. Under these conditions, the predicted ginsenoside Ro yield is 260.19 μg/L. Five validation experiments conducted under approximate levels of the predicted conditions showed an average ginsenoside Ro yield of 253.32 μg/L, with a deviation of 2.64% from the predicted value, indicating good data reproducibility. As shown in Figure 4, the results indicate that the ginsenoside Ro yield after optimization of fermentation conditions is significantly higher than that before optimization, corresponding to 1.63 times the pre-optimization yield.

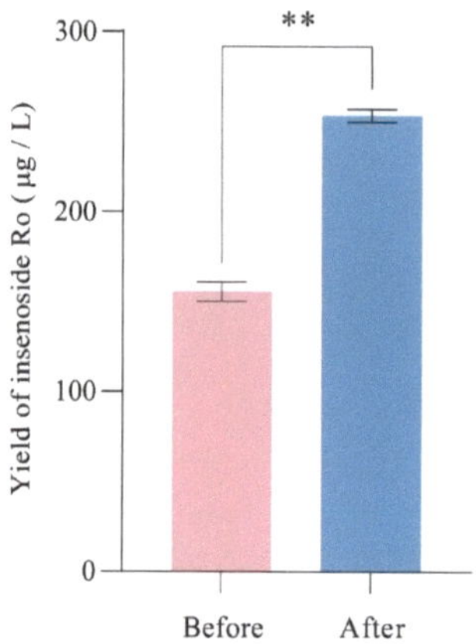

Figure 4. Comparison of the yield of ginsenoside Ro in strain BY-Ro before and after the optimization of fermentation conditions (** for significance of $p \leq 0.01$).

Table 7. Statistical analysis of the Box–Behnken design and experimental results.

Source of Variation	Sum of Squares	Degrees of Freedom	Mean Square	F Value	*p* Value	Significance
Model	13,732.47	9	1525.83	173.03	<0.0001	**
A	1402.84	1	1402.84	159.09	<0.0001	**
B	402.46	1	402.46	45.64	0.0003	**
C	1890.75	1	1890.75	214.42	<0.0001	**
AB	63.46	1	63.46	7.2	0.0314	*
AC	75.08	1	75.08	8.51	0.0224	*
BC	144.46	1	144.46	16.38	0.0049	**
A^2	4202.46	1	4202.46	476.57	<0.0001	**
B^2	978.02	1	978.02	110.91	<0.0001	**
C^2	3645.91	1	3645.91	413.46	<0.0001	**
Lack of fit	21.82	3	7.27	0.73	0.5861	
Residual	61.73	7	8.82			
Pure error	39.91	4	9.98			
Total	13,794.2	16				

* for significance of $p \leq 0.05$ and ** for significance of $p \leq 0.01$.

2.5. The Effect of Optimal Fermentation Conditions on the Expression of *PgUGAT252645*

The calculation results of the relative expression of *PgUGAT252645* in strain BY-Ro under different fermentation conditions are shown in Figure 5. The calculation results show that, compared to the strain cultured under pre-optimized conditions, the relative expression of *PgUGAT252645* in strain BY-Ro cultured under optimal fermentation conditions significantly increased to 1.45 times that of the control group. This increase is roughly consistent with the increase in ginsenoside Ro yield, indicating that *PgUGAT252645* is a key enzyme in the constructed heterologous synthesis pathway of ginsenoside Ro, responding to the regulation of culture conditions such as initial medium glucose concentration, culture temperature, and shaking speed.

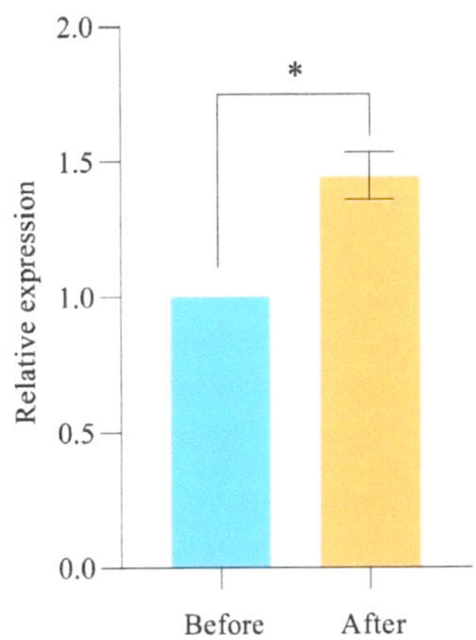

Figure 5. Comparison of the relative expression levels of *PgUGAT252645* in strain BY-Ro before and after the optimization of fermentation conditions (* for significance of $p \leq 0.05$).

3. Discussion

Synthetic biology research focused on improving product yield often maximizes the optimization of the synthetic pathways for target products at the genetic level. Optimized biosynthetic pathways are often constructed from genes spanning multiple species. For example, after several iterations of updates, some key enzyme genes in the heterologous synthesis pathway of β-carotene have been replaced with modified genes derived from *M. circinelloides* [30]. Although higher yields represent greater application value, such value-driven research may lead to the solidification of synthetic pathway construction schemes. When functional research on new genes is only used for the analysis of synthetic pathways rather than for synthetic applications, the actual value of such genes may be overlooked due to a lack of in-depth exploration.

With the continuous development of relevant technologies, the research focus on heterologous biosynthesis will inevitably expand beyond the key enzyme genes in target pathways. Instead, it will involve more complex studies conducted from a holistic perspective, including investigations of synthesis, metabolism, and regulatory mechanisms. At that time, chassis cells and genetic elements will undergo multidimensional evaluations. The in-depth development of new genes can effectively supplement current genetic resources and prepare for future specific needs. Accordingly, we identified ginsenoside Ro synthesis-related gene *PgUGAT252645* in previous research and used it as one of the key enzyme gene elements to construct ginsenoside Ro biosynthesis-engineered Saccharomyces cerevisiae strain BY-Ro. Preliminary explorations were conducted on the biosynthetic application of *PgUGAT252645*.

S. cerevisiae possesses an efficient homologous recombination mechanism based on the DNA double-strand break (DSB) repair model [31]. Only short, homologous arm fragments are needed between the foreign DNA fragments and the genomic DNA of *S. cerevisiae* to trigger homologous recombination, achieving genomic integration of the foreign DNA [32]. Compared to transformation using episomal plasmid vectors, this method provides higher genetic stability and expression efficiency for foreign genes. In this study, we used the pFA6a plasmid series. The target gene fragment was gradually constructed using seamless cloning and other methods and integrated into the *S. cerevisiae* genome. The integration of the target gene and its functional expression were verified by detecting specific products. The advantages of this approach, in addition to having a well-established technical system and highly efficient supporting tools, lie in the high modularity of the gene elements during the assembly of the target gene fragment. This allows for flexible replacement of gene elements to meet different requirements Additionally, since the methods used in this approach are based on fundamental molecular biology techniques, the operating costs can be maintained at a relatively low level.

However, limited integration sites remain a major constraint in constructing engineered *S. cerevisiae* strains using this method. In this study, besides the commonly used rDNA site [33] and δ site [34], we also employed the *tHIS 3* site, which is the residual sequence of the structural *HIS 3* gene present in a single copy in the genome of BY4741, and the *tLEU 2* site, which was artificially constructed and nested within foreign gene fragment C integrated into the *tHIS 3* site. To ensure the integrity of the ginsenoside Ro synthesis pathway, we had to allocate the limited positions to a minimal number of genes. However, product yield and other results indicate that there is still considerable room for optimization in balancing gene copy number and catalytic efficiency under this scheme.

Strain BY-Ro produced 253.32 μg/L of ginsenoside Ro under optimal fermentation conditions. According to subsequent measurements and calculations, this equates to 0.033 mg/g DCW, corresponding to approximately 31% of the ginsenoside Ro content found in plant samples. Although this yield is still significantly lower than the highest reported yield of heterologous biosynthesis of ginsenoside Ro, considering the cost of plant cultivation, including the production cycle, strain BY-Ro still has potential as a production tool for ginsenoside Ro. *PgUGAT252645* sourced from *P. ginseng* also has potential application value as a synthetic biology gene element. However, we have not yet implemented large-scale cultivation of strain BY-Ro in a bioreactor, so we have not achieved the fermentation method commonly used in the heterologous biosynthesis of ginsenosides by *S cerevisiae*, i.e., fed-batch fermentation. Therefore, further research on the fermentation scale and methods is crucial for the enhancement of the production yield of ginsenoside Ro by strain BY-Ro.

The promoters used to construct strain BY-Ro in this study are constitutive promoters known for their ability to maintain relatively stable gene expression under different conditions. However, under the optimized fermentation conditions, the expression level of uridine diphosphate glucose-aldehyde transferase gene *PgUGAT252645* in strain BY-Ro significantly increased. Based on this, we speculate that the introduced exogenous key enzymes have specific requirements for the catalytic environment, such as with respect to temperature, substrate concentration, and oxygen content. These unique conditions may activate regulatory mechanisms within *S cerevisiae*, thereby enhancing transcription levels or the stability of mRNA of downstream genes regulated by relevant promoters. Currently, research on regulatory mechanisms of heterologous biosynthetic systems in *S cerevisiae* is in its early stages. Further in-depth studies in this area could lead to the development of engineered strains with stronger adaptability, aimed at increasing the production yield of target products and further reducing production costs.

4. Materials and Methods

4.1. Data and Materials

E. coli DH5α containing plasmid pET-28a and the chemically competent cells of *E. coli* DH5α were purchased from Sangon Biotech Co., Ltd., Shanghai, China. *S. cerevisiae* S288C direct derivative strain BY4741 [35] was purchased from Huayueyang Biotech Co., Ltd., Beijing, China. The pFA6a-series plasmid vectors listed in Table S1 were purchased from Addgene plasmid repository, Watertown, MA, USA. Codon optimization of open reading frames of key enzyme encoding genes and synthesis of all DNA fragments were completed by Sangon Biotech Co., Ltd., Shanghai, China.

4.2. Cloning of Endogenous Gene Elements of BY4741

The BY4741 strain was inoculated into YPD liquid medium and cultured at 30 °C and 150 rpm until an OD_{600} of 1.2 was reached. Genomic DNA was then extracted using the TIANamp Yeast DNA Kit (TIANGEN, Beijing, China). Then, using the extracted genomic DNA of BY4741 as the template, endogenous constitutive promoters $TPI1_P$, $TEF1_P$, and $TDH3_P$; terminators $CYC1_T$, $ADH1_T$, $FBA1_T$, and $ENO2_T$; genomic integration sites;

rDNA-locus and δDNA-locus upstream and downstream homologous arms *RU*, *RD*, *DU*, and *DD*; and the C-terminal catalytic domain coding sequence of the rate-limiting enzyme *HMG1* in the MVA pathway, *tHMG1*, were cloned using PrimeSTAR® Max DNA Polymerase (TAKARA, Beijing, China). The sequences of primers, annealing temperatures, and extension times used in this study are listed in Table S2.

4.3. Assembly of Key Enzyme Gene Expression Cassettes

Using the Ready-to-Use Seamless Cloning Kit (SANGON, Shanghai, China), the cloned promoters were ligated into the pFA6a-KanMX6 plasmid vector at the *Eco*R I and *Eco*R V restriction sites; the cloned terminators were ligated into the pFA6a-KanMX6 plasmid vector at the *Spe* I and *Sac* II restriction sites; and the cloned *tHMG1* fragments and the exogenous key enzyme genes from different species listed in Table S3, which were codon-optimized for *S. cerevisiae*, were ligated into the pFA6a-KanMX6 plasmid vector at the *Eco*R V and *Spe* I restriction sites to construct the recombinant plasmid vectors shown in Table 8.

Table 8. Recombinant plasmid vectors containing the key enzyme gene expression cassette.

Name	Promoter	Key Enzyme Gene	Terminator
pK + I	*TPI 1P*	*tHMG 1*	*CYC 1T*
pK + II	*TEF 1P*	*Syn Gg.βAS*	*ADH 1T*
pK + III	*TEF 1P*	*Syn Cr.OAS*	*FBA 1T*
pK + IV	*TDH 3P*	*Syn Mt.CYPR*	*ENO 2T*
pK + V	*TPI 1P*	*Syn At.UGDH*	*CYC 1T*
pK + VI	*TEF 1P*	*Syn Pg.UGAT*	*ADH 1T*
pK + VII	*TEF 1P*	*Syn Mt.UGT*	*FBA 1T*
pK + VIII	*TDH 3P*	*Syn Ts.UGT*	*ENO 2T*

During the ligation process, the ligation products were transformed into chemically competent *E. coli* DH5α cells by heat shock and screened on LB solid medium containing ampicillin (final concentration of 100 μg/mL) to obtain positive clones carrying the recombinant plasmid vectors. The positive clones obtained through screening were cultured in LB liquid medium containing ampicillin (final concentration of 100 μg/mL) at 37 °C and 180 rpm until an OD600 of 0.8 was reached. Culturing was then stopped, and the cells were stored at 4 °C for plasmid extraction.

4.4. Assembly of Genomic Integration Fragments

Using the Ready-to-Use Seamless Cloning Kit, the cloned upstream homologous *RU* and *DU* arms of the genomic integration sites were ligated into the pET-28a plasmid vector at the *Not* I and *Xho* I restriction sites, and the cloned downstream homologous *RD* and *DD* arms were ligated into the same plasmid vector at the *Bam*H I and *Sac* I restriction sites.

Additionally, residual sequence *tHIS3* of the *HIS3* gene in the BY4741 genome was retrieved from the *Saccharomyces* Genome Database (https://yeastgenome.org/ (accessed on 16 October 2023)) [36], and upstream and downstream homologous *HU* and *HD* arms were designed. Similarly, the sequence of the *LEU2* gene in the *S. cerevisiae* S288C genome was retrieved from the Saccharomyces Genome Database, and a portion of sequence *LEU2* was selected (*tLEU2*) and placed after *HISU* as an additional genomic integration site. Based on the sequence of *tLEU2*, upstream and downstream homologous *LU* and *LD* arms were designed. After the design step, these DNA fragments were synthesized and ligated into the pET-28a plasmid vector at the *Not* I and *Xho* I and *Bam*H I and *Sac* I restriction sites, respectively.

Using recombinant plasmid vectors pK + III, pK + V, and pK + VII as templates, connected key enzyme gene expression cassettes III, V, and VII were cloned. The primer sequences, annealing temperatures, and extension times used in this study are listed in Table S2.

Then, using the Ready-to-Use Seamless Cloning Kit, cloned key enzyme gene expression cassette fragment III was ligated into the pFA6a-5FLAG-hphMX6 plasmid vector at the *Eco*R V and *Spe* I restriction sites. Cloned key enzyme gene expression cassette fragment V was ligated into the pFA6a-URA3 plasmid vector at the *Eco*R I and *Spe* I restriction sites. Cloned key enzyme gene expression cassette fragment VII was ligated into the pFA6a-His3MX6 plasmid vector at the *Eco*R I and *Eco*R V restriction sites, constructing recombinant plasmid vectors for the transfer of selection markers. Subsequently, using these recombinant plasmid vectors and recombinant plasmid vector pK + I as templates, PCR cloning was performed on the combination fragments of the key enzyme gene expression cassettes and selection markers. The primer sequences, annealing temperatures, and extension times are listed in Table S2. The Ready-to-Use Seamless Cloning Kit was used to ligate the cloned combination fragments of the key enzyme gene expression cassettes and selection markers into the *Sac* I and *Sal* I restriction sites of the aforementioned recombinant plasmid vectors containing homologous arm fragments.

Finally, using the constructed recombinant plasmid vectors (pK + II, pK + IV, pK + VI, and pK + VIII) as templates, PCR cloning was performed on the connected key enzyme gene expression cassettes (II, IV, VI, and VIII). The primer sequences, annealing temperatures, and extension times used in this study are listed in Table S2. Using the Ready-to-Use Seamless Cloning Kit, the cloned key enzyme gene expression cassette fragments were ligated into the constructed recombinant plasmid vectors at the *Not* I and *Sal* I restriction sites to create recombinant plasmid vectors with genomic integration fragments for the transformation of *S. cerevisiae*, as shown in Table 9.

Table 9. Recombinant plasmid vector containing integrated genomic integration genetic transformation fragments.

Name	Upstream Homologous Arm	Expression Cassette 1	Selection Marker	Expression Cassette 2	Downstream Homologous Arm
p28a-A	*RU*	I	*KanR*	II	*RD*
p28a-B	*DU*	III	*HygR*	IV	*DD*
p28a-C	*HU* + *tLEU 2*	V	*URA 3*	VI	*HD*
p28a-D	*LU*	VII	*HIS 3*	VIII	*LD*

4.5. Construction of the Engineered S. cerevisiae Strain

To integrate genomic integration fragment A containing the *tHMG1* and beta-amyrin synthase gene (*syn Gg.βAS*) expression cassettes (I, II) into the rDNA site of the genome of BY4741, we used recombinant plasmid vector p28a-A as a template to clone genomic integration fragment A and transformed the BY4741-competent cells using the standard lithium acetate method [37]. The transformed cells were plated on YPDA solid medium containing geneticin G418 (final concentration of 300 μg/mL) for selection to obtain positive strain BY-A.

Subsequently, using recombinant plasmid vector p28a-B as a template, genomic integration fragment B, containing the oleanolic acid synthase gene (*syn Cr.OAS*) and cytochrome P450 reductase gene (*syn Mt.CYPR*) expression cassettes (III, IV), was cloned and transformed into BY-A-competent cells, using the aforementioned method to integrate it into the δDNA site of the genome of BY-A. Then, strain BY-OA was obtained by selection on YPDA solid medium containing geneticin G418 (final concentration of 300 μg/mL) and hygromycin B (final concentration of 500 μg/mL).

The construction of strain BY-CE was achieved by using the aforementioned method to integrate genomic integration fragment C, which contains the uridine diphosphate glucose dehydrogenase gene (*syn At.UGDH*) and the uridine diphosphate glucuronic acid transferase gene (*syn Pg.UGAT*) expression cassette, into the *HIS3* residual sequence site of the genome of BY-OA. Strain BY-CE was obtained by selection on uracil-deficient SD solid medium containing geneticin G418 (final concentration of 300 μg/mL) and hygromycin B (final concentration of 500 μg/mL).

Strain BY-RO was constructed by using the aforementioned method to integrate genomic integration fragment D, which contains two uridine diphosphate glucose transferase gene (*Syn Mt.UGT* and *Syn Ts.UGT*) expression cassettes, into the artificially added *tLEU* sequence site of the genome of BY-CE. Strain BY-Ro was screened on SD solid medium deficient in uracil and histidine containing geneticin G418 (final concentration of 300 μg/mL) and hygromycin B (final concentration of 500 μg/mL).

4.6. Detection of Target Products

During the strain construction process, all screened positive strains, except for strain BY-A, were inoculated into YPD liquid medium and cultured at 30 °C and 150 rpm until an OD600 of 1.2 was reached. The cells were then disrupted using an ultrasonic cell disruptor. The cell lysate was then centrifuged at room temperature and 12,000 rpm for 45 min to collect the supernatant. The filtered supernatant was then loaded onto a separation column packed with octadecylsilane (ODS)-bonded silica gel. After the supernatant completely passed through the column, impurities were eluted with 5 column volumes of 40% methanol solution. Finally, the separated products were eluted with 5 column volumes of 100% methanol, the eluate was collected and evaporated to dryness, and the residue was dissolved in chromatographic methanol. The target products of each strain were then detected using Agilent 1290-6470B LC-MS/MS (Agilent, Santa Clara, CA, USA) under the conditions described in Table S4.

4.7. Single-Factor Experiment

Using 500 mL Erlenmeyer flasks with 250 mL standard YPD liquid medium, an inoculum size of 3%, a shaking speed of 150 rpm, a culture temperature of 30 °C, and a culture time of 84 h as the initial culture conditions, the effects of five major factors on the yield of ginsenoside Ro in strain BY-RO were studied in sequence as follows: inoculum size, 1%, 2%, 3%, 4%, and 5%; culture time, 60 h, 72 h, 84 h, 96 h, and 108 h; shaking speed, 110 rpm, 130 rpm, 150 rpm, 170 rpm, and 190 rpm; culture temperature, 26 °C, 28 °C, 30 °C, 32 °C, and 34 °C; and initial medium glucose concentration, 1%, 1.5%, 2%, 2.5%, and 3%. Three biological replicates were set up for each condition. After cultivation, the ginsenoside Ro in the cultures was extracted and detected using the aforementioned method.

4.8. Experiment and Analysis of the Plackett–Burman Design

By using the Plackett–Burman Design function in Design-Expert (Version 10.0.3), the Plackett–Burman experiment was designed based on the results of the aforementioned single-factor experiments and conducted according to the designed protocol. Subsequently, through statistical analysis of the results obtained from the Plackett-Burman experiment, factors significantly affecting the yield of ginsenoside Ro in strain BY-Ro were selected among inoculum size, culture time, shaking speed culture temperature, and initial medium glucose concentration to obtain fundamental data for subsequent response surface optimization experiments.

4.9. Experiment and Analysis of the Box—Behnken Design

Using the Box–Behnken Design function in Design-Expert, the response surface methodology was employed to design optimization experiments based on the results of the Plackett–Burman experiment, with subsequent experiments conducted accordingly. Ginsenoside Ro yield was used as the response value, with three levels set for each factor. Subsequently, the fermentation conditions for strain BY-Ro were optimized based on the statistical analysis of the experimental optimization results. Finally, verification experiments were conducted under the optimal fermentation conditions to validate the effectiveness of the established model.

4.10. The Effect of Optimal Fermentation Conditions on the Expression of *PgUGAT252645*

After fermenting strain BY-Ro under optimal fermentation conditions, total RNA was extracted from strain BY-Ro using TRNzol Universal total RNA extraction reagent (TIANGEN, Beijing, China). A PrimeScript™ FAST RT reagent Kit with gDNA Eraser (TAKARA, Beijing, China) was then used to further remove genomic DNA from the RNA samples, and the RNA was reverse-transcribed into cDNA. Subsequently, TB Green® Fast qPCR Mix (TAKARA, Beijing, China) was used with the primers listed in Table S2, using cDNA as the template and *S. cerevisiae* TATA-binding protein-associated factor gene *TAF10* [38] as the reference gene. Real-time quantitative PCR was performed using an Applied Biosystems 7500 Real-Time PCR System. Strain BY-Ro cells cultured under pre-optimization fermentation conditions were used as controls to detect changes in the expression levels of the key enzyme gene *PgUGAT252645* before and after optimization of fermentation conditions.

5. Conclusions

Using *PgUGAT252645* as identified in previous studies and several other reported key enzyme genes, as well as the endogenous promoters and terminators of *S. cerevisiae* BY4741, as elements, various plasmid vectors were employed to assemble genomic integration fragments. These genomic integration fragments were transferred into chemically competent *S. cerevisiae* cells in batches, successfully constructing strain BY-Ro for the biosynthesis of ginsenoside Ro. Among various fermentation conditions, the culture temperature, initial medium glucose concentration, and shaking speed significantly impacted the ginsenoside Ro yield in strain BY-Ro. The interaction between shaking speed and culture temperature also significantly affected the ginsenoside Ro yield. Under the initially optimized fermentation conditions, the ginsenoside Ro yield in strain BY-Ro was 253.32 μg/L, which is 1.63 times the yield before optimization. *PgUGAT252645*, as a key enzyme gene, responded to the regulation of culture conditions, with its relative expression level increasing to 1.45 times that before the optimization of fermentation conditions. This study not only deepened the understanding of the function of *PgUGAT252645* but also explored its potential application as a genetic element for the industrial production of ginsenoside Ro.

Supplementary Materials: The following supporting information can be downloaded at: https://www.mdpi.com/article/10.3390/ijms252011331/s1.

Author Contributions: Conceptualization, Y.W. (Yi Wang) and M.Z. (Meiping Zhang); methodology, X.Y. and K.W.; software, J.Y. and S.L; validation, X.Y., J.Y. and D.W.; formal analysis, S.L.; investigation, P.C.; resources, Y.W. (Yi Wang) and Y.W. (Yanfang Wang); data curation, K.W.; writing—original draft preparation, X.Y. and D.W.; writing—review and editing, X.Y.; visualization, K.W.; supervision, M.Z. (Mingzhu Zhao); project administration, M.Z. (Mingzhu Zhao); funding acquisition, Y.W. (Yi Wang) and M.Z. (Meiping Zhang). All authors have read and agreed to the published version of the manuscript.

Funding: This research was funded by awards from the Department of Science and Technology of Jilin Province (20240402046GH and 20210402043GH) and Department of Education of Jilin Province (JJKH20230378KJ).

Institutional Review Board Statement: Not applicable.

Informed Consent Statement: Not applicable.

Data Availability Statement: Data are contained within the article.

Acknowledgments: All authors extend their appreciation to the College of Life Science, Jilin Agricultural University and Jilin Engineering Research Center Ginseng Genetic Resources Development and Utilization.

Conflicts of Interest: The authors declare no conflicts of interest.

References

1. Wei, G.; Yang, F.; Wei, F.; Zhang, L.; Gao, Y.; Qian, J.; Chen, Z.; Jia, Z.; Wang, Y.; Su, H.; et al. Metabolomes and transcriptomes revealed the saponin distribution in root tissues of *Panax quinquefolius* and *Panax notoginseng*. *J. Ginseng Res.* **2020**, *44*, 757–769. [CrossRef] [PubMed]
2. Liang, Y.; Zhao, S. Progress in understanding of ginsenoside biosynthesis. *Plant Biol.* **2008**, *10*, 415–421. [CrossRef] [PubMed]
3. Kim, Y.J.; Zhang, D.; Yang, D.C. Biosynthesis and biotechnological production of ginsenosides. *Biotechnol. Adv.* **2015**, *33*, 717–735. [CrossRef] [PubMed]
4. Yang, J.L.; Hu, Z.F.; Zhang, T.T.; Gu, A.D.; Gong, T.; Zhu, P. Progress on the Studies of the Key Enzymes of Ginsenoside Biosynthesis. *Molecules* **2018**, *23*, 589. [CrossRef]
5. Han, J.Y.; Kim, H.J.; Kwon, Y.S.; Choi, Y.E. The Cyt P450 Enzyme CYP716A47 Catalyzes the Formation of Protopanaxadiol from Dammarenediol-II During Ginsenoside Biosynthesis in *Panax ginseng*. *Plant Cell Physiol.* **2011**, *52*, 2062–2073. [CrossRef]
6. Wu, W.; Lu, Z.; Teng, Y.; Guo, Y.; Liu, S. Structural Characterization of Ginsenosides from Flower Buds of *Panax ginseng* by RRLC-Q-TOF MS. *J. Chromatogr. Sci.* **2016**, *54*, 136–143.
7. Kim, J.S. Optimization of Accelerated Solvent Extraction of Ginsenosides from Cultivated Wild Ginseng Using Response Surface Methodology. *Prev. Nutr. Food Sci.* **2022**, *27*, 315–322. [CrossRef]
8. Roberts, S.C. Production and engineering of terpenoids in plant cell culture. *Nat. Chem. Biol.* **2007**, *3*, 387–395. [CrossRef]
9. Murthy, H.N.; Dandin, V.S.; Paek, K.Y. Tools for biotechnological production of useful phytochemicals from adventitious root cultures. *Phytochem. Rev.* **2016**, *15*, 129–145. [CrossRef]
10. Choi, H.S.; Koo, H.B.; Jeon, S.W.; Han, J.Y.; Kim, J.S.; Jun, K.M.; Choi, Y.E. Modification of ginsenoside saponin composition via the CRISPR/Cas9-mediated knockout of protopanaxadiol 6-hydroxylase gene in *Panax ginseng*. *J. Ginseng Res.* **2022**, *46*, 505–514. [CrossRef]
11. Dai, Z.; Liu, Y.; Zhang, X.; Shi, M.; Wang, B.; Wang, D.; Huang, L.; Zhang, X. Metabolic engineering of Saccharomyces cerevisiae for production of ginsenosides. *Metab. Eng.* **2013**, *20*, 146–156. [CrossRef] [PubMed]
12. Li, D.; Zhang, Q.; Zhou, Z.; Zhao, F.; Lu, W. Heterologous biosynthesis of triterpenoid dammarenediol-II in engineered *Escherichia coli*. *Biotechnol. Lett.* **2016**, *38*, 603–609. [CrossRef] [PubMed]
13. Gwak, Y.S.; Han, J.Y.; Adhikari, P.B.; Ahn, C.H.; Choi, Y.E. Heterologous production of a ginsenoside saponin (compound K) and its precursors in transgenic tobacco impairs the vegetative and reproductive growth. *Planta* **2017**, *245*, 1105–1119. [CrossRef]
14. Gwak, Y.S.; Han, J.Y.; Choi, Y.E. Production of ginsenoside aglycone (protopanaxatriol) and male sterility of transgenic tobacco co-overexpressing three *Panax ginseng* genes: PgDDS, CYP716A47, and CYP716A53v2. *J. Ginseng Res.* **2019**, *43*, 261–271. [CrossRef] [PubMed]
15. Chen, Q.; Liu, D.; Qu, Y.; Lei, J.; Zhang, J.; Cui, X.; Ge, F. Construction of plant cell factory for biosynthesis of ginsenoside Rh2 in tobacco. *Ind. Crops Prod.* **2023**, *192*, 116057. [CrossRef]
16. Pompon, D.; Louerat, B.; Bronine, A.; Urban, P. Yeast expression of animal and plant P450s in optimized redox environments. *Methods Enzymol.* **1996**, *272*, 51–64.
17. Ren, H.; Hu, P.; Zhao, H. A plug-and-play pathway refactoring workflow for natural product research in *Escherichia coli* and *Saccharomyces cerevisiae*. *Biotechnol. Bioeng.* **2017**, *114*, 1847–1854. [CrossRef]
18. Wang, C.L.; Liwei, M.; Park, J.B.; Jeong, S.H.; Wei, G.Y.; Wang, Y.J.; Kim, S.W. Microbial Platform for Terpenoid Production: *Escherichia coli* and Yeast. *Front. Microbiol.* **2018**, *9*, 2460. [CrossRef]
19. Wang, D.; Wang, J.; Shi, Y.; Li, R.; Fan, F.; Huang, Y.; Li, W.; Chen, N.; Huang, L.; Dai, Z.; et al. Elucidation of the complete biosynthetic pathway of the main triterpene glycosylation products of Panax notoginseng using a synthetic biology platform. *Metab. Eng.* **2020**, *61*, 131–140. [CrossRef]
20. Dai, Z.; Wang, B.; Liu, Y.; Shi, M.; Wang, D.; Zhang, X.; Liu, T.; Huang, L.; Zhang, X. Producing aglycons of ginsenosides in bakers' yeast. *Sci. Rep.* **2014**, *4*, 3698. [CrossRef]
21. Yan, X.; Fan, Y.; Wei, W.; Wang, P.; Liu, Q.; Wei, Y.; Zhang, L.; Zhao, G.; Yue, J.; Zhou, Z. Production of bioactive ginsenoside compound K in metabolically engineered yeast. *Cell Res.* **2014**, *24*, 770–773. [CrossRef]
22. Ren, S.; Sun, Q.; Zhang, L.; Sun, W.; Li, Y.; Feng, X.; Li, C. Sustainable production of rare oleanane-type ginsenoside Ro with an artificial glycosylation pathway in *Saccharomyces cerevisiae*. *Green Chem.* **2022**, *24*, 8302–8313. [CrossRef]
23. Zhang, H.; Hua, X.; Zheng, D.; Wu, H.; Li, C.; Rao, P.; Wen, M.; Choi, Y.E.; Xue, Z.; Wang, Y.; et al. De Novo Biosynthesis of Oleanane-Type Ginsenosides in *Saccharomyces cerevisiae* Using Two Types of Glycosyltransferases from *Panax ginseng*. *J. Agric. Food Chem.* **2022**, *70*, 2231–2240. [CrossRef]
24. Shi, Y.; Wang, D.; Li, R.; Huang, L.; Dai, Z.; Zhang, X. Engineering yeast subcellular compartments for increased production of the lipophilic natural products ginsenosides. *Metab. Eng.* **2021**, *67*, 104–111. [CrossRef] [PubMed]
25. Wang, P.; Wei, W.; Ye, W.; Li, X.; Zhao, W.; Yang, C.; Li, C.; Yan, X.; Zhou, Z. Synthesizing ginsenoside Rh2 in *Saccharomyces cerevisiae* cell factory at high-efficiency. *Cell Discov.* **2019**, *5*, 5. [CrossRef] [PubMed]
26. Zhao, C.; Gao, X.; Liu, X.; Wang, Y.; Yang, S.; Wang, F.; Ren, Y. Enhancing Biosynthesis of a Ginsenoside Precursor by Self-Assembly of Two Key Enzymes in Pichia pastoris. *J. Agric. Food Chem.* **2016**, *64*, 3380–3385. [CrossRef]
27. Wu, Y.; Xu, S.; Gao, X.; Li, M.; Li, D.; Lu, W. Enhanced protopanaxadiol production from xylose by engineered Yarrowia lipolytica. *Microb. Cell Factories* **2019**, *18*, 83. [CrossRef]

28. Wang, Y.; Zhang, H.; Ri, H.C.; An, Z.; Wang, X.; Zhou, J.-N.; Zheng, D.; Wu, H.; Wang, P.; Yang, J.; et al. Deletion and tandem duplications of biosynthetic genes drive the diversity of triterpenoids in *Aralia elata*. *Nat. Commun.* **2022**, *13*, 2224. [CrossRef]
29. Yu, X.; Yu, J.; Liu, S.; Liu, M.; Wang, K.; Zhao, M.; Wang, Y.; Chen, P.; Lei, J.; Wang, Y.; et al. Transcriptome-Wide Identification and Integrated Analysis of a UGT Gene Involved in Ginsenoside Ro Biosynthesis in *Panax ginseng*. *Plants* **2024**, *13*, 604. [CrossRef]
30. Varghese, R.; Buragohain, T.; Banerjee, I.; Mukherjee, R.; Penshanwar, S.N.; Agasti, S.; Ramamoorthy, S. The apocarotenoid production in microbial biofactories: An overview. *J. Biotechnol.* **2023**, *374*, 5–16. [CrossRef]
31. Lettier, G.; Feng, Q.; de Mayolo, A.A.; Erdeniz, N.; Reid, R.J.D.; Lisby, M.; Mortensen, U.H.; Rothstein, R. The role of DNA double-strand breaks in spontaneous homologous recombination in S. cerevisiae. *PLoS Genet.* **2006**, *2*, 1773–1786. [CrossRef] [PubMed]
32. Yoshida, J.; Umezu, K.; Maki, H. Positive and negative roles of homologous recombination in the maintenance of genome stability in *Saccharomyces cerevisiae*. *Genetics* **2003**, *164*, 31–46. [CrossRef]
33. Zheng, H.; Wang, K.; Xu, X.; Pan, J.; Sun, X.; Hou, J.; Liu, W.; Shen, Y. Highly efficient rDNA-mediated multicopy integration based on the dynamic balance of rDNA in *Saccharomyces cerevisiae*. *Microb. Biotechnol.* **2022**, *15*, 1511–1524. [CrossRef] [PubMed]
34. Yamada, R.; Tanaka, T.; Ogino, C.; Fukuda, H.; Kondo, A. Novel strategy for yeast construction using δ-integration and cell fusion to efficiently produce ethanol from raw starch. *Appl. Microbiol. Biotechnol.* **2010**, *85*, 1491–1498. [CrossRef] [PubMed]
35. Winston, F.; Dollard, C.; Ricupero-Hovasse, S.L. Construction of a set of convenient *Saccharomyces cerevisiae* strains that are isogenic to S288C. *Yeast* **1995**, *11*, 53–55. [CrossRef]
36. Wong, E.D.; Miyasato, S.R.; Aleksander, S.; Karra, K.; Nash, R.S.; Skrzypek, M.S.; Weng, S.; Engel, S.R.; Cherry, J.M. Saccharomyces genome database update: Server architecture, pan-genome nomenclature, and external resources. *Genetics* **2023**, *224*, iyac191. [CrossRef]
37. Amberg, D.C.; Burke, D.J.; Strathern, J.N. *Methods in Yeast Genetics: A Cold Spring Harbor Laboratory Course Manual*, 2005 ed.; David, C., Ed.; Cold Spring Harbor Laboratory Press: New York, NY, USA, 2005.
38. Teste, M.A.; Duquenne, M.; Francois, J.M.; Parrou, J.L. Validation of reference genes for quantitative expression analysis by real-time RT-PCR in *Saccharomyces cerevisiae*. *BMC Mol. Biol.* **2009**, *10*, 99. [CrossRef]

International Journal of
Molecular Sciences

MDPI

Article

Leveraging Therapeutic Proteins and Peptides from *Lumbricus* Earthworms: Targeting SOCS2 E3 Ligase for Cardiovascular Therapy through Molecular Dynamics Simulations

Nasser Alotaiq [1,*], Doni Dermawan [2] and Nasr Eldin Elwali [3]

1 Health Sciences Research Center, Imam Mohammad Ibn Saud Islamic University (IMSIU), Riyadh 11432, Saudi Arabia
2 Department of Applied Biotechnology, Faculty of Chemistry, Warsaw University of Technology, 00-661 Warsaw, Poland; doni.dermawan.stud@pw.edu.pl
3 Division of Biochemistry, Research Center for Health Sciences, Deanship of Scientific Research, Imam Mohammad Ibn Saud Islamic University, Riyadh 11432, Saudi Arabia; nmwali@imamu.edu.sa
* Correspondence: naalotaiq@imamu.edu.sa; Tel.: +966-112037109; Fax: +966-112037110

Abstract: Suppressor of cytokine signaling 2 (SOCS2), an E3 ubiquitin ligase, regulates the JAK/STAT signaling pathway, essential for cytokine signaling and immune responses. Its dysregulation contributes to cardiovascular diseases (CVDs) by promoting abnormal cell growth, inflammation, and resistance to cell death. This study aimed to elucidate the molecular mechanisms underlying the interactions between *Lumbricus*-derived proteins and peptides and SOCS2, with a focus on identifying potential therapeutic candidates for CVDs. Utilizing a multifaceted approach, advanced computational methodologies, including 3D structure modeling, protein–protein docking, 100 ns molecular dynamics (MD) simulations, and MM/PBSA calculations, were employed to assess the binding affinities and functional implications of *Lumbricus*-derived proteins on SOCS2 activity. The findings revealed that certain proteins, such as Lumbricin, Chemoattractive glycoprotein ES20, and Lumbrokinase-7T1, exhibited similar activities to standard antagonists in modulating SOCS2 activity. Furthermore, MM/PBSA calculations were employed to assess the binding free energies of these proteins with SOCS2. Specifically, Lumbricin exhibited an average $\Delta G_{binding}$ of −59.25 kcal/mol, Chemoattractive glycoprotein ES20 showed −55.02 kcal/mol, and Lumbrokinase-7T1 displayed −69.28 kcal/mol. These values suggest strong binding affinities between these proteins and SOCS2, reinforcing their potential therapeutic efficacy in cardiovascular diseases. Further in vitro and animal studies are recommended to validate these findings and explore broader applications of *Lumbricus*-derived proteins.

Keywords: cardiovascular disease; earthworm; lumbrokinase; molecular dynamics simulation; SOCS2; protein–protein interactions

1. Introduction

Cardiovascular diseases (CVDs) represent a major global health issue, profoundly affecting both health outcomes and healthcare systems. According to the World Health Organization (WHO), CVDs cause an alarming 17.9 million deaths each year, positioning them as the primary cause of mortality worldwide [1]. The range of CVDs includes a variety of conditions affecting the heart and blood vessels, each characterized by distinct pathophysiological mechanisms and clinical presentations [2,3]. Some of the most common CVDs include coronary artery disease, hypertension, stroke, heart failure, and peripheral artery disease. These conditions result from complex pathological mechanisms such as endothelial dysfunction, atherosclerosis, inflammation, thrombosis, and vascular remodeling, which collectively disturb cardiovascular homeostasis [4,5]. Despite advancements in the field of medical science and healthcare delivery, conventional treatment methods for cardiovascular diseases predominantly concentrate on addressing risk factors, relieving symptoms, and modifying the progression of the disease [6]. Medications like statins, antiplatelet agents, and drugs for controlling blood pressure have proven effective in lowering cardiovascular morbidity and mortality. Yet, these treatments have drawbacks, including potential side effects, interactions with other drugs, and variability in effectiveness among different groups of patients [7,8]. Additionally, the rise of drug resistance presents a substantial obstacle to the sustained efficacy of current therapies, underscoring the necessity for alternative treatment strategies [9,10].

A suppressor of cytokine signaling 2 (SOCS2), an E3 ubiquitin ligase, has emerged as a critical regulator within the Janus kinase/signal transducer and activator of transcription (JAK/STAT) signaling pathway, a key mediator of cytokine signaling involved in inflammation and immune responses [11,12]. Dysregulated SOCS2 expression has been implicated in various pathogenic processes underlying CVDs, including aberrant cellular proliferation, inflammatory cytokine production, and apoptotic resistance in cardiac and vascular tissues [13,14]. Research has highlighted SOCS2's multifaceted roles in cardiovascular pathophysiology. For instance, in experimental models of myocardial infarction, elevated SOCS2 levels correlate with increased myocardial fibrosis and impaired cardiac function, suggesting a contributory role in adverse remodeling post-infarction [15,16]. Moreover, SOCS2 influences vascular smooth muscle cell proliferation and migration, which are critical processes in atherosclerotic plaque development and stability. Targeting SOCS2 presents a promising therapeutic strategy for mitigating cardiovascular pathology. By modulating SOCS2 activity, either through direct or indirect inhibition, researchers aim to temper excessive inflammation, promote vascular health, and potentially attenuate the progression of atherosclerosis and other CVDs [17,18].

Besides focusing on SOCS2, natural products have attracted growing interest as reservoirs of bioactive compounds with therapeutic promise against cardiovascular diseases. Earthworms, specifically from the *Lumbricus* genus, have become particularly fascinating due to their abundant biochemical composition and pharmacological properties [19]. Earthworms have been utilized across different cultures and over centuries in traditional medicine for treating a range of conditions such as inflammation, wounds, and gastrointestinal disorders [20]. This historical background highlights the promise of bioactive compounds derived from earthworms in tackling modern healthcare issues like CVDs. Specifically, earthworms contain lumbrokinase, a bioactive component known for its potent fibrinolytic effects. Lumbrokinase has demonstrated thrombolytic properties, aiding in the dissolution of fibrin clots, which could potentially mitigate or reverse cardiac fibrosis, a characteristic feature of various cardiovascular disorders [21,22]. Bioactive compounds (like lumbrokinase and fibrinolytic enzymes) sourced from earthworms contribute positively to cardiovascular health by employing diverse mechanisms, encompassing anti-inflammatory, antioxidant, and vasodilatory activities [23–26]. Through regulation of these pathways, bioactive compounds from earthworms show promise in attenuating the fundamental pathological mechanisms behind CVDs, thereby enhancing patient outcomes. Moreover, investigating these compounds from earthworms presents a fresh and hopeful avenue in drug research and advancement, holding the potential for discovering novel therapeutic agents that boast enhanced effectiveness and safety profiles in combating CVDs. Molecular simulation techniques such as molecular docking and molecular dynamics simulations are pivotal in modern drug discovery. They are treasured for uncovering intricate protein–protein interactions and aiding in the discovery of promising therapeutic candidates [27,28]. In the context of CVDs, where the molecular mechanisms underlying pathogenesis are often intricate and multifactorial [29,30], molecular simulation techniques offer valuable insights into the interactions between bioactive molecules and their target proteins.

The goals of this study encompassed a thorough exploration of the potential therapeutic implications of targeting SOCS2 in CVDs using bioactive compounds derived from earthworms of the *Lumbricus* genus. Initially, our objective was to investigate the intricate protein–protein interactions between earthworm-derived bioactive compounds and SOCS2 through molecular docking techniques. By simulating how these compounds bind to specific regions within the SOCS2 protein structure, we aimed to identify lead molecules with strong binding affinity and favorable interactions, guiding the selection and optimization of potential therapeutic candidates. Subsequently, our study aimed to predict the binding affinities and interaction modes between the identified earthworm-derived compounds and distinct binding sites on the SOCS2 protein. Through molecular docking simulations, we aimed to characterize the molecular events responsible for the formation of stable protein–protein complexes, providing insights into the structural basis of compound–target interactions. Moreover, we sought to explore the dynamic behavior and structural integrity of protein–protein complexes formed between earthworm-derived compounds and SOCS2 using molecular dynamics simulations. By modeling the movements and interactions of atoms within these complexes over time, we aimed to elucidate the conformational dynamics and energetic principles governing the binding process, thereby deepening our understanding of the underlying mechanisms of action. Furthermore, our study aimed to assess the therapeutic potential of the identified earthworm-derived compounds in CVDs by targeting SOCS2. Through computational analyses and in silico modeling, we aimed to evaluate the effectiveness and specificity of these compounds in modulating SOCS2 function and alleviating the pathological mechanisms associated with cardiovascular diseases. This research lays the groundwork for future preclinical and clinical investigations into potential therapeutic interventions.

2. Results

2.1. Three-Dimensional Structure Modeling

Three-dimensional structure modeling was performed for the selected proteins and peptides derived from the *Lumbricus* earthworm. Table 1 shows 10 out of 78 selected proteins and peptides. The complete database can be seen in Supplementary Data S1.

Table 1. Selected proteins and peptides derived from *Lumbricus* earthworm.

Protein/Peptide	UniProt ID	Sequence	Size (kDa)
Actin-1	P92182	MCDEEVTALVVDNGSGMCKAGFAGDDAPRAVFPSIVGRPRH QGVMVGMGQKDSYVGDEAQSKRGILTLKYPIEHGIVTNWD DMEKIWHHTFYNELRVAPEEHPVLLTEAPLNPKANREKMT QIMFETFNSPAMYVAIQAVLSLYASGRTTGIVLDSGDGVTHT VPIYEGYALPHAILRLDLAGRDLTDYLMKILTERGYSFTTT AEREIVRDIKEKLCYVALDFDQEMGTAASSSSLEKSYELPDG QVITIGNERFRCPESMFQPAFLGMESAGIHETTFNSIMKCDVD IRKDLYANTVMSGGTTMFPGIADRMQKEITSMAPSTMKIKIIA PPERKYSVWIGGSILASLSTFQQMWISKQEYDESGPSIVHRKCF	41.85
Chemoattractive glycoprotein ES20	O44335	MKTYLLLVFLVGAHALVCPPGFTYLPAGESCYKVIFESHDW HSATERCRQESRGLAAISTPEESIAVKEFIDTEISKDSAGAAVC HPTGQSGIRFWTSGLQTKDTCTKTSFLLKITNTFEVPFDFTN WADGEPTLPRKTEKFSALIVGSSERTPSGTTMTATSSCVHSAN ISNDTLKRISVLPHLYVGFCDEIWLYLNFCLISIQILI	22.96
Cytochrome b	Q34945	MFKPIRTTHPAIKIINSTLIDLPAPNNISIWWNYGSLLGLCLV IQVLTGLFLSMHYVPNIEMAFSSVALISRDVNYGWLLRSIHA NGASMFFLFIYLHAGRGLYYGSYNLSETWNIGVILFLLTMA TAFMGYVLPWGQMSFWGATVITNLFSAIPYIGKTLVEWIWG GFAVDNATLNRFFAFHFILPFAIMGATILHIMFLHESGSNNPIG LNADSDRIPFHPYYSIKDTLGYTLAISALSLMVLFEPNLFTDP ENFLMANPLVTPIHIKPEWYFLWMYAILRSIPNKLGGVMALF AAIVILFIPPLTSVMNKRSLSFYPLNKTMFWGLVASWAILTWI GGRPVEDPFIIIGQVFTSLYFIYFISSPTISKLWDDSIII	42.88

Table 1. *Cont.*

Protein/Peptide	UniProt ID	Sequence	Size (kDa)
Fibrinolytic enzyme	P83298	VIGGTNASPGEFPWQLSQQRQSGSWSHSCGASLLSSTSALS ASHCVDGVLPNNIRVIAGLWQQSDTSGTQTANVDSYTMHE NYGAGTASYSNDIAILHLATSISLGGNIQAAVLPANNNNDYA GTTCVISGWGRTDGTNNLPDILQKSSIPVITTAQCTAAMVG VGGANIWDNHICVQDPAGNTGACNGDSGGPLNCPDGGTR VVGVTSWVVSSGLGTCLPDYPSVYTRVSAYLGWIGDNSR	24.84
Histone H3	A0A1C9UP21	GGKAPRKQLATKAARKSAPATGGVKKPHRYRPGTVALRE IRRYQKSTELLIRKLPFQRLVREIAQDFKTDLRFQSSAVMA LQEASEAYLVGLFEDTNLCAIH	11.47
Lumbricin	O96447	MSLCISDYLYLTLTFSKYERQKDKRPYSERKNQYTGPQFL YPPERIPPQKVIKWNEEGLPIYEIPGEGGHAEPAAA	8.85
Lumbrokinase-7T1	B8ZZ01	MRSFVAFLAALSLCQARPQKFLDGARPSFRMGGEQYIIG GSNASPGEFPWQLSQTRGGSHSCGASLLNALNGLSAFHC VDGAAPGTITVIAGLHDRSGTPGSQEVDITGYTMHENYN QGTNTYANDIAILHFASAINIGGNGQAALLPANNDNDYSG LTCVISGWGRKGSSNVLPDTLQKASIQVIGTTQCQSLMGS IGHIWDNHICLYNNTNNVGSCNGDSGGPLNCPDGGTRVA GVTSWGVSSGAGNCLQTYPTVYTRTSAYLSWIANNS	28.35
Ribosomal protein S27	Q9U5N5	MPLTRDLLHPTLKDEKRKCKLKRLVQSPNSFFMDVKCP GCYKITTVFSHAQTVVLCVGCNTVLCQPTGGKARLTEG CSFRRKQH	9.50
SCBP3 protein	Q7YWL4	VWEQYLKGVVSDGTRLTQAVFVEAVKKQLGDPNFKKV LAGPLPLFFSAVDGNGDGLIQKDEFQLFFKLLGIPESAEK SFEAIDTNKDGDISKEEFVIAGTDFFTSTDESSPSKYFWGPLV	13.25
Ubiquitin	P84589	TITLEVEPSDTIENVKAKIQDKEGIPPDQQRLIFAGKQLED GRTLSDYNIQKESTLHLVLRLR	7.20

The 3D structures of these proteins and peptides were generated using specialized computational tools designed for their specific structural attributes. For proteins and peptides that had suitable templates in the PDB, the I-TASSER algorithm was utilized. I-TASSER is recognized for its capability to predict precise 3D structures by utilizing template structures from the PDB. This method ensures accurate predictions of tertiary structures through analysis of sequence similarity and structural homology [31]. For proteins and peptides without existing templates in the PDB or newly introduced sequences in UniProt release 2024_01, AlphaFold v2.3.0. was employed. AlphaFold represents a cutting-edge deep learning technique for predicting protein structures, revolutionizing this field with its high accuracy. It is particularly adept at predicting 3D structures for sequences lacking homologous structures in the PDB. By leveraging sophisticated machine learning algorithms trained on extensive protein databases, AlphaFold can predict precise 3D structures based solely on the amino acid sequences of proteins [32]. The 3D structure modeling phase of this research employed advanced computational techniques to create precise and dependable structural models for the proteins and peptides sourced from *Lumbricus* earthworms (Figure 1). These models form the basis for subsequent analyses, such as protein–protein docking simulations and molecular dynamics simulations, facilitating a thorough exploration of the interactions between bioactive compounds from earthworms and the target protein SOCS2. This investigation aims to identify potential therapeutic strategies for treating CVDs.

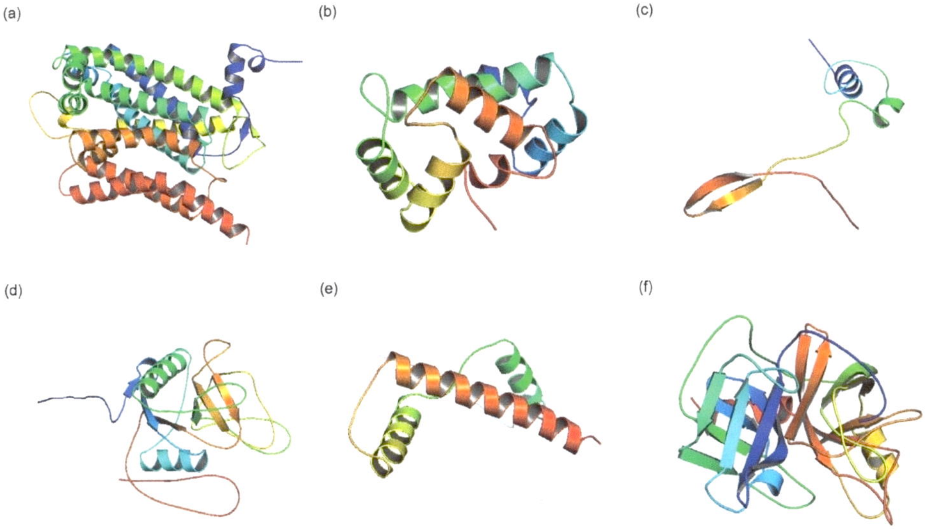

Figure 1. Three-dimensional structural modeling results illustrating proteins and peptides derived from earthworms of the *Lumbricus* genus. (**a**) Cytochrome b. (**b**) SCBP3 protein. (**c**) Lumbricin. (**d**) Chemoattractive glycoprotein ES20. (**e**) Histone H3. (**f**) Lumbrokinase-7T1. Three proteins exhibit well-folded, globular structures, while another three display large, unfolded regions or loops, indicating potential structural fluctuations that may influence their interactions with SOCS2 during docking.

2.2. Protein–Protein Docking Simulation

Figure 2 illustrates the optimal binding poses of a standard agonist, a standard antagonist, and the top-performing *Lumbricus*-derived protein. The molecular docking results provide a comprehensive overview of the binding affinity, structural parameters, and clustering characteristics of the various SOCS2 complexes formed with proteins and peptides from earthworms. Specifically, comparisons were made with standard agonist (SOCS2: EpoR peptide) and antagonist (SOCS2: N4BP1) complexes, which served as benchmarks for assessing the effectiveness of the *Lumbricus*-derived compounds. A critical measure of the interaction strength between molecules is the binding affinity, quantified by the ΔG (Gibbs free energy) value. Lower ΔG values indicate stronger binding, suggesting more stable and favorable interactions between the proteins or peptides and SOCS2. The comparison between the top-performing proteins and peptides derived from *Lumbricus* and the standard agonist (SOCS2: EpoR peptide) and antagonist (SOCS2: N4BP1) complexes based on binding affinity provides valuable insights into the potential efficacy of *Lumbricus*-derived bioactive compounds in modulating SOCS2 activity compared to established proteins.

Firstly, the standard agonist, SOCS2: EpoR peptide, demonstrated a binding affinity with a ΔG value of −8.80 kcal/mol and a dissociation constant (Kd) of 6.50×10^{-7} M. This indicates a strong interaction between SOCS2 and EpoR peptide, highlighting its efficacy as an agonist for SOCS2 activity modulation. In contrast, the standard antagonist, SOCS2: N4BP1, exhibited a slightly lower binding affinity, with a ΔG value of −8.30 kcal/mol and a Kd of 1.50×10^{-6} M. Despite its lower binding affinity compared to the agonist, SOCS2: N4BP1 still displayed a notable interaction with SOCS2, indicative of its efficacy as an antagonist for SOCS2 activity inhibition. Among the *Lumbricus*-derived complexes, cytochrome b, SCBP3 protein, Lumbricin, Chemoattractive glycoprotein ES20, Histone H3, and Lumbrokinase-7T1 emerged as top-performing candidates based on their low values of binding affinity ΔG (kcal/mol). Cytochrome b exhibited an ΔG value of −15.1 kcal/mol, surpassing the binding affinity of the standard agonist, SOCS2: EpoR peptide. This suggests that cytochrome b may possess a stronger interaction with SOCS2, potentially surpassing the efficacy of EpoR peptide and N4BP1 in modulating SOCS2 activity. Similarly, SCBP3 protein, Lumbricin, and Chemoattractive glycoprotein ES20 displayed notable binding affinities with ΔG values of −12.6 kcal/mol, −12.3 kcal/mol, and −12.2 kcal/mol, respectively (Table 2 and Figure 3c). These values compare favorably with the binding affinities of the standard agonist and antagonist, underscoring the potential of these *Lumbricus*-derived compounds as effective modulators of SOCS2 activity for addressing cardiovascular diseases. Complete docking results are available in Supplementary Data S2. This dataset includes complex identifiers, HADDOCK scores, binding affinities (ΔG), dissociation constants (Kd), cluster sizes, RMSD values, van der Waals energies, electrostatic energies, desolvation energies, restraints violation energies, buried surface areas, and Z-scores.

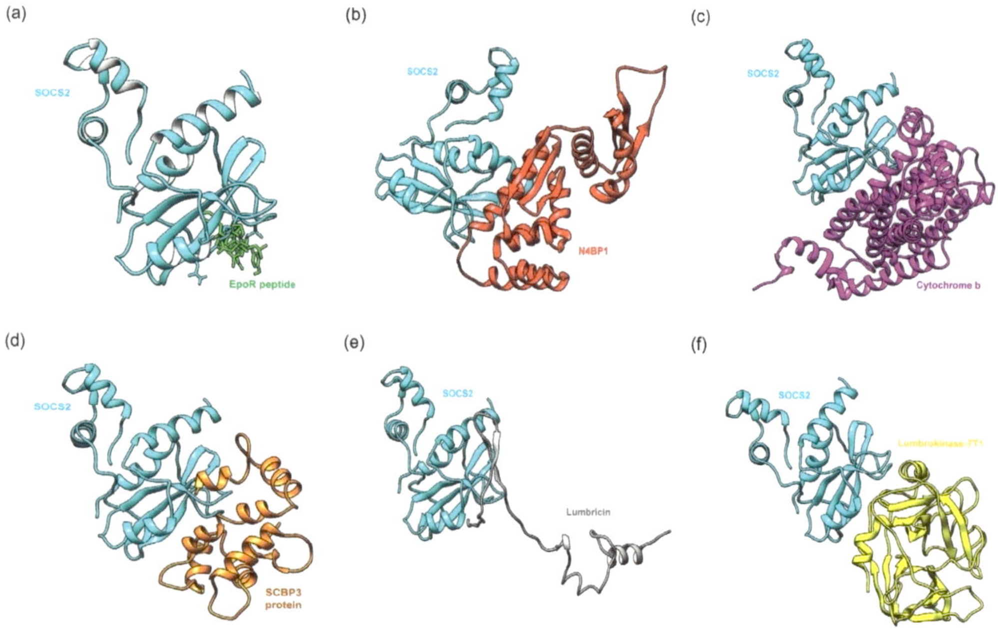

Figure 2. Molecular docking simulations illustrate the optimal binding orientations for interactions between proteins derived from the earthworm (*Lumbricus* genus) and SOCS2. (**a**) SOCS2: EpoR peptide (agonist) complex. (**b**) SOCS2: N4BP1 (antagonist) complex. (**c**) SOCS2: Cytochrome b complex. (**d**) SOCS2: SCBP3 protein complex. (**e**) SOCS2: Lumbricin complex. (**f**) SOCS2: Lumbrokinase-7T1 complex.

Table 2. Molecular docking outcomes: binding affinities and structural attributes of SOCS2 interactions with proteins and peptides obtained from *Lumbricus* earthworms.

Complex	HADDOCK Score (a.u.)	Binding Affinity ΔG (kcal/mol)	Kd (nM)	Cluster Size	RMSD (Å)
	Standard				
SOCS2: EpoR peptide (standard agonist)	−81.0 ± 3.5	−8.8	650	22	1.7 ± 0.1
SOCS2: N4BP1 (standard antagonist)	−94.3 ± 11.4	−8.3	1500	17	1.2 ± 0.3

Table 2. *Cont.*

Complex	HADDOCK Score (a.u.)	Binding Affinity ΔG (kcal/mol)	Kd (nM)	Cluster Size	RMSD (Å)
Protein and peptide derived from earthworm (*Lumbricus* genus)					
SOCS2: Cytochrome b	−99.0 ± 8.2	−15.1	0.02	27	1.0 ± 0.6
SOCS2: Cytochrome c oxidase subunit 3	−104.7 ± 6.3	−12.7	1.09	15	1.3 ± 0.5
SOCS2: SCBP3 protein	−96.2 ± 6.2	−12.6	1.30	41	1.2 ± 0.9
SOCS2: Lumbricin	−72.1 ± 3.7	−12.3	2.10	43	2.3 ± 0.3
SOCS2: Chemoattractive glycoprotein ES20	−72.4 ± 5.3	−12.2	2.30	20	1.5 ± 1.0
SOCS2: NADH-ubiquinone oxidoreductase chain 6	−109.0 ± 7.1	−12.2	2.30	14	1.1 ± 0.7
SOCS2: Histone H3	−96.6 ± 4.0	−12.1	3.09	8	1.2 ± 0.2
SOCS2: Peroxidasin	−74.6 ± 7.6	−12.1	2.70	8	2.8 ± 0.0
SOCS2: Lumbrokinase-7T1	−84.3 ± 2.8	−12.0	3.69	43	2.1 ± 0.2
SOCS2: Lysosomal membrane glycoprotein	−70.3 ± 2.3	−11.9	4.29	5	2.6 ± 0.9
SOCS2: NADH-ubiquinone oxidoreductase chain 4L	−90.4 ± 5.4	−11.9	4.10	12	1.6 ± 1.0
SOCS2: NADH-ubiquinone oxidoreductase chain 5	−102.9 ± 11.6	−11.8	4.69	7	1.7 ± 0.3
SOCS2: Preprocarboxypeptidase	−102.7 ± 9.1	−11.4	9.70	19	0.6 ± 0.4
SOCS2: Extracellular globin-4	−79.4 ± 10.8	−11.3	12.00	6	1.0 ± 0.6
SOCS2: Lumbrokinase-7T2	−68.5 ± 2.5	−11.3	10.99	35	1.6 ± 0.5

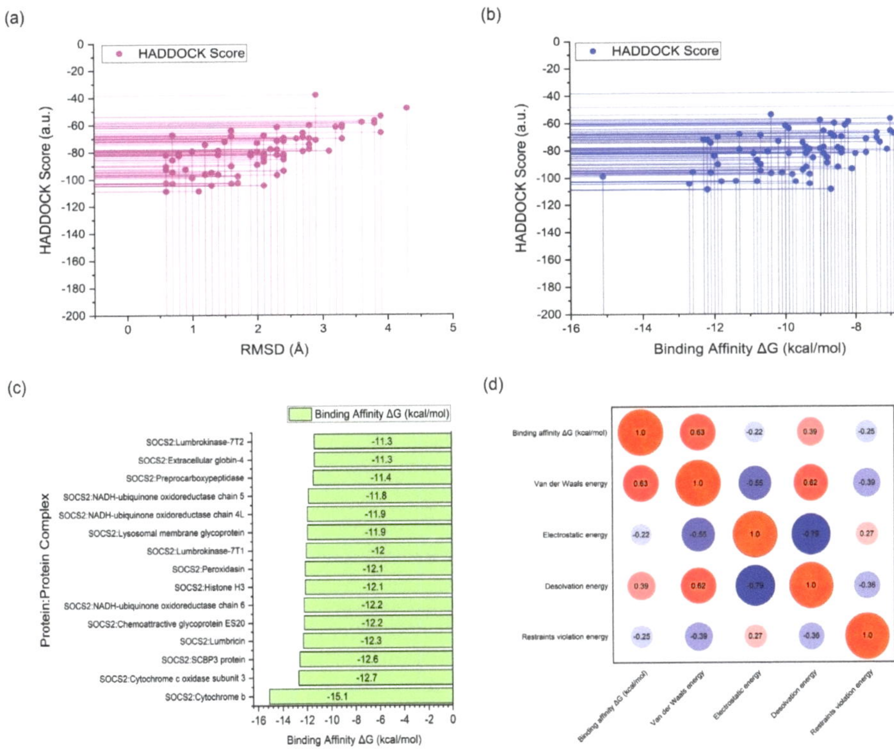

Figure 3. Summary of molecular docking outcomes. (**a**) Correlation between HADDOCK score and RMSD. (**b**) Correlation between HADDOCK score and binding affinity. (**c**) Binding affinity values of the top-performing proteins derived from *Lumbricus* earthworms, with a threshold of −11.0 kcal/mol. (**d**) Correlation matrix illustrating the relationship between binding energy (kcal/mol) and individual energy components.

Furthermore, the cluster size and RMSD values provide crucial structural insights into the protein–protein complexes. Cluster size refers to the number of distinct conformations or poses observed within the ensemble of generated protein–protein complexes [33]. A larger cluster size indicates greater structural diversity, suggesting multiple binding modes or orientations between the proteins and peptides from *Lumbricus* and SOCS2. This diversity within the complexes reflects potential variations in interaction configurations, influencing their functional properties and biological impacts [34]. For instance, the standard agonist SOCS2: EpoR peptide complex exhibits a notably large cluster size of 22, indicating the presence of diverse conformations or binding modes between SOCS2 and EpoR peptide. This diversity suggests the potential for versatile interactions, which could play crucial roles in various cellular processes regulated by ubiquitination. Conversely, the standard antagonist SOCS2: N4BP1 complex has a smaller cluster size of 19, implying fewer distinct binding configurations compared to the agonist complex. This difference in cluster size may reflect the specific nature of the antagonist interaction and its regulatory role in modulating SOCS2 activity. Analyzing the *Lumbricus*-derived protein/peptide complexes, it is evident that the cluster sizes vary across different interactions. For instance, complexes involving proteins like the SCBP3 protein and Lumbricin exhibit relatively larger cluster sizes (41 and 43, respectively), suggesting structural diversity and potential functional versatility in their interactions with SOCS2. On the other hand, complexes such as SOCS2: Lysosomal membrane glycoprotein and SOCS2: Extracellular globin-4 have smaller cluster sizes (five and six, respectively), indicating a more limited range of binding configurations.

Conversely, RMSD values measure the variations or discrepancies in structure among distinct conformations within a given cluster. Reduced RMSD values indicate limited deviation or greater structural similarity among various conformations, implying heightened stability and consistency in the binding interactions observed throughout the simulation period [35]. This stability demonstrates the resilience of the protein–protein complexes, ensuring they maintain their specific structural arrangements despite variations or disturbances in their environment [36]. For instance, the standard agonist SOCS2: EpoR peptide complex exhibits an acceptable RMSD value of 1.7 Å, indicating minimal structural deviation among its conformations and suggesting a stable and well-defined binding mode. Similarly, *Lumbricus*-derived complexes like SOCS2: Cytochrome b (1.0 Å) and SOCS2: SCBP3 protein show low RMSD values (1.2 Å), indicating stable binding interactions and consistent structural configurations. In contrast, complexes with higher RMSD values, such as SOCS2: Intermediate filament protein (RMSD = 4.3 Å), suggest greater structural variability among their conformations, potentially reflecting dynamic binding interactions with SOCS2. This higher RMSD value suggests potential conformational flexibility or transient interactions, which could impact the functional significance of these complexes in cellular processes [37].

The analysis of intermolecular contacts (ICs) and non-interacting surfaces (NIS) for the protein–protein complexes provides valuable insights into the interaction dynamics between SOCS2 and various earthworm-derived proteins and peptides. The comparison between these complexes and the standard agonist (SOCS2: EpoR peptide) and antagonist (SOCS2: N4BP1) reveals significant differences in their interaction profiles, which can be linked to their potential efficacy in modulating SOCS2 activity (Table 3). The SOCS2: EpoR peptide complex, serving as the standard agonist, exhibited 3 charged-charged, 8 charged-polar, 12 charged-apolar, 2 polar-polar, 11 polar-apolar, and 12 apolar-apolar contacts. The NIS values for charged and apolar interactions were 25.55 and 38.69, respectively. In comparison, the SOCS2: N4BP1 complex, which functions as the standard antagonist, displayed a higher number of charged-charged (5), charged-polar (11), charged-apolar (20), and polar-polar (6) contacts, with slightly lower polar-apolar (11) and apolar-apolar (8) contacts. The NIS values were 29.71 for charged and 39.49 for apolar interactions.

Table 3. Intermolecular contacts and non-interacting surface areas for SOCS2 complexes with standard agonist, antagonist, and earthworm-derived proteins and peptides.

Complex	ICs Charged-Charged	ICs Charged-Polar	ICs Charged-Apolar	ICs Polar-Polar	ICs Polar-Apolar	ICs Apolar-Apolar	NIS Charged	NIS Apolar
Standard								
SOCS2: EpoR peptide (standard agonist)	3	8	12	2	11	12	25.55	38.69
SOCS2: N4BP1 (standard antagonist)	5	11	20	6	11	8	29.71	39.49
Protein and peptide derived from earthworm (*Lumbricus* genus)								
SOCS2: Cytochrome b	0	0	25	0	38	33	15.13	52.85
SOCS2: Cytochrome c oxidase subunit 3	2	7	22	6	30	26	14.71	49.41
SOCS2: SCBP3 protein	2	3	24	2	27	16	27.83	41.74
SOCS2: Lumbricin	7	9	20	0	22	12	26.34	40.98
SOCS2: Chemoattractive glycoprotein ES20	8	15	24	5	23	16	22.98	42.39
SOCS2: NADH-ubiquinone oxidoreductase chain 6	0	3	21	7	32	29	14.29	52.01
SOCS2: Histone H3	7	6	27	0	20	14	27.56	42.67
SOCS2: Peroxidasin	13	20	24	5	22	6	26.23	41.86
SOCS2: Lumbrokinase-7T1	3	15	12	4	24	10	17.72	41.14
SOCS2: Lysosomal membrane glycoprotein	12	17	18	5	18	11	20.49	39.02
SOCS2: NADH-ubiquinone oxidoreductase chain 4L	0	8	21	2	24	14	17.73	46.82
SOCS2: NADH-ubiquinone oxidoreductase chain 5	0	3	17	1	25	29	14.12	50.10
SOCS2: Preprocarboxypeptidase	7	17	23	5	20	28	26.18	40.05
SOCS2: Extracellular globin-4	15	18	28	2	12	6	31.28	37.04
SOCS2: Lumbrokinase-7T2	5	10	16	9	24	11	19.03	41.69

Note: ICs: Number of intermolecular contacts; NIS: Non-interacting surface.

Several proteins and peptides derived from *Lumbricus* showed distinct interaction profiles with SOCS2, suggesting varying efficacy in modulating its activity. The SOCS2: Cytochrome b complex, for instance, displayed high charged-apolar (25), polar-apolar (38), and apolar-apolar (33) contacts, with non-interacting surface (NIS) values of 15.13 for charged and 52.85 for apolar interactions. This indicates a potentially strong binding affinity. Cytochrome c oxidase subunit 3 showed balanced interactions, with significant polar-apolar and apolar-apolar contacts and NIS values of 14.71 (charged) and 49.41 (apolar), suggesting strong stability and efficacy. The SCBP3 protein complex, with 27.83 NIS (charged) and 41.74 (apolar), also displayed robust binding capability. Lumbricin exhibited higher numbers of charged interactions, indicating strong potential as a SOCS2 modulator. Chemoattractive glycoprotein ES20, with high numbers of charged and polar interactions and NIS values of 22.98 (charged) and 42.39 (apolar), also showed promise.

The statistical analysis aimed at exploring the relationship between the HADDOCK score and root mean square deviation (RMSD) unveiled a Pearson correlation coefficient (r) of 0.687. This coefficient quantifies the strength and direction of the linear relationship between these two variables. A value of 0.687 signifies a medium positive correlation, suggesting that there is a tendency for the HADDOCK score to decrease as the RMSD decreases. In the context of molecular docking, a lower HADDOCK score indicates a better docking quality, and a lower RMSD reflects a closer match to the reference structure. Thus, the positive correlation implies that improvements in docking quality, as indicated by lower HADDOCK scores, are associated with a more accurate docking pose, as represented by lower RMSD values. This relationship underscores the consistency between these two metrics in evaluating the quality of protein–protein docking simulations. Further analysis revealed that approximately 46.75% of the proteins and peptides derived from *Lumbricus* demonstrated favorable RMSD values, defined as RMSD values equal to or less than 2.00 Å (Figure 3a). The favorable RMSD values indicate that many of the predicted protein–protein complexes have structural conformations closely matching experimental or reference structures, highlighting the overall success of the docking simulations in accurately predicting these arrangements. However, the analysis also revealed three outliers with very large RMSD values. These outliers suggest instances where the predicted structures significantly deviate from the experimental or reference structures. Such deviations could be due to inaccuracies in the docking algorithm, limitations in the input experimental data, or the inherent complexities of the protein–protein interactions being studied [38].

The statistical analysis to examine the relationship between the HADDOCK score and binding affinity (ΔG) revealed a Pearson correlation coefficient (r) of 0.491 (Figure 3b). This coefficient reflects the strength and direction of the linear relationship between these two variables. A value of 0.491 suggests a moderate positive correlation, indicating that as the HADDOCK score decreases (indicating better docking quality), the binding affinity also tends to decrease. This means that improvements in docking quality, as indicated by lower HADDOCK scores, are associated with stronger binding between the proteins, as shown by lower binding affinity values. Conversely, higher HADDOCK scores correlate with weaker binding affinity, underscoring a discernible trend between docking quality and binding strength.

The HADDOCK scoring system effectively captures key aspects of protein–protein interactions that influence binding affinity, including molecular surface complementarity, electrostatic interactions, and van der Waals forces. This indicates that HADDOCK's scoring metrics align well with the physical principles governing these interactions, enhancing the reliability of its predictions. The correlation matrix depicted in Figure 3d provides deeper insights into the relationship between the binding energy (kcal/mol) and its individual energy components. Specifically, the positive correlation scores between the binding affinity and both the van der Waals energy (correlation score: 0.63) and the desolvation energy (correlation score: 0.39) are particularly noteworthy. Van der Waals forces are crucial in stabilizing the complex by facilitating close contact between the protein surfaces, while the desolvation energy reflects the energetic cost of removing water molecules from the binding interface. A positive correlation with the van der Waals energy implies that, as these interactions become stronger, the overall binding affinity increases, indicating a more stable protein–protein complex [39,40]. Similarly, the positive correlation with the desolvation energy suggests that, as the system pays the energetic cost to remove water molecules, the resulting interactions between the proteins are more favorable, thus increasing the binding affinity [41]. Conversely, the correlation between binding affinity and electrostatic energy revealed a negative value (−0.22), signifying an inverse relationship between these factors. This implies that, as electrostatic energy increases, the binding affinity tends to decrease. Electrostatic interactions involve the attraction or repulsion between charged residues on the protein surfaces, influencing the stability of protein–protein complexes [42]. However, interpreting this correlation requires caution due to the multifaceted nature of protein–protein interactions. While electrostatic energy plays a significant role, other variables beyond the HADDOCK score can also impact binding affinity. Factors such as specific amino acid residues at the binding interface, which may facilitate or hinder interactions, post-translational modifications that alter protein structure and function, and environmental conditions such as pH and ion concentrations can all influence the overall stability and strength of protein–protein complexes [43,44].

The thorough examination of hydrogen bond interactions between SOCS2 and the highest-performing proteins sourced from the earthworm (*Lumbricus* genus) provides insights into the molecular mechanisms that govern their binding interactions (Table 4). These interactions are pivotal for stabilizing protein–protein complexes and facilitating precise recognition between the involved proteins [45]. The interactions predominantly involve specific residues and atoms crucial for stabilizing the protein–protein complexes, focusing particularly on the ligand binding sites of SOCS2, namely, Arg73, Ser75, Ser76, and Arg96. The interactions observed between SOCS2 and its standard agonist (EpoR peptide) and antagonist (N4BP1) molecules served as benchmarks for evaluating the binding efficacy of *Lumbricus*-derived proteins and peptides. The EpoR peptide exhibited multiple hydrogen bonds with key residues such as Val55, Ser76, Arg96, Lys113, and others, with interaction distances ranging from 2.69 Å to 3.33 Å. Conversely, N4BP1 demonstrated interactions involving Gln32, Arg41, Tyr49, Asp74, and others, emphasizing its distinct binding profile, characterized by shorter interaction distances (2.57 Å to 2.78 Å). These results underscore the specificity and strength of the interactions necessary for SOCS2 modulation by both agonist and antagonist molecules. The proteins and peptides from the *Lumbricus* genus, including Cytochrome b, SCBP3 protein, Lumbricin, Lumbrokinase-7T1, and others, displayed diverse binding interactions with SOCS2, highlighting their potential as novel modulators of SOCS2 activity. Cytochrome b, for instance, engaged in hydrogen bonds with His77 and Arg96, indicating a stable binding interface, critical for functional interaction. Similarly, the SCBP3 protein exhibited interactions involving Lys59 and Arg96, demonstrating specificity in binding residues essential for complex stability. Lumbricin and Lumbrokinase-7T1 revealed extensive hydrogen bonding networks with residues such as Arg41, Tyr49, Asp74, and Arg96, underscoring their potential to competitively bind with SOCS2 and influence its regulatory functions. These interactions were characterized by moderate to strong interaction distances, indicative of a robust binding affinity, crucial for therapeutic efficacy.

Table 4. A detailed examination of hydrogen bond interactions occurring between SOCS2 at its binding sites and the most effective proteins originating from earthworms (*Lumbricus* genus), focusing on identifying specific residues and atoms crucial for the binding mechanism.

Complex	Residue (Receptor)	Protein Atom (Receptor)	Residue (Interacting Protein/Peptide)	Protein Atom (Interacting Protein/Peptide)	Interaction Distance (Å)
SOCS2: EpoR peptide (standard agonist)	Val55	N	Asp8	OD1	2.76
	Ser76	OG	Asp8	OD1	3.09
	Arg96	NH1	Glu3	O	2.69
	Arg96	NH2	Glu3	O	3.33
	Lys113	NZ	Ser1	OG	2.73
SOCS2: N4BP1 (standard antagonist)	Gln32	NE2	Glu178	OE1	2.77
	Arg41	NH1	Glu118	OE1	2.65
	Arg41	NH2	Glu118	OE2	2.57
	Tyr49	OH	Lys132	NZ	2.78
	Asp74	OD2	Lys132	NZ	2.55
	Ser75	O	Asn136	ND2	3.04
	Ser78	O	Ser174	OG	2.63
	Asp79	OD2	Lys145	NZ	2.62
	Arg96	NH2	Glu138	OE1	2.67
SOCS2: Cytochrome b	His77	NE2	Leu122	O	2.74
	Arg96	NH2	Ile119	O	2.72
SOCS2: SCBP3 protein	Lys59	NZ	Asp50	O	2.58
	Arg96	NH1	Leu68	O	2.66
SOCS2: Lumbricin	Arg41	NH1	Glu56	O	3.08
	Arg41	NH1	Glu56	OE1	2.60
	Arg41	NH2	Glu56	O	2.98
	Tyr49	OH	Lys53	NZ	2.83
	Ser52	OG	Glu63	OE1	2.59
	Asp74	OD2	Lys53	NZ	2.65
	Ser78	OG	Ile46	O	2.67
	Asp79	OD1	Arg45	NH1	2.65
	Thr93	OG1	Glu72	OE2	2.70
	Asn94	N	Glu72	OE2	2.73
	Asn94	ND2	Glu72	OE1	2.64
	Arg96	NE	Glu72	O	2.74
	Arg96	NH2	Glu72	O	2.87
SOCS2: Lumbrokinase-7T1	Glu57	OE2	Lys20	NZ	2.55
	Asp74	OD2	Arg17	NH1	2.69
	Asp74	OD2	Arg17	NH2	2.72
	Ser76	O	Gln19	NE2	3.20
	Asp101	OD1	Asn249	ND2	2.66
	Asp101	OD2	Ser170	OG	2.66
	Cys111	O	Lys168	NZ	2.91
	Lys113	NZ	Gly32	O	2.87
	Lys113	NZ	Gln35	OE1	2.66
	Leu116	O	Asn214	ND2	2.98

2.3. Molecular Dynamics (MD) Simulation

The MD simulation results offer a comprehensive understanding of the behavior of protein–protein complexes formed between *Lumbricus*-derived proteins and SOCS2. Throughout the 100 ns simulation, SOCS2 maintained a relatively stable conformation, as evidenced by the average RMSD values ranging from 2.413 to 2.599 Å without significant spikes (Figure 4a). This stability suggests that SOCS2 interactions with both standard agonist and antagonist, as well as *Lumbricus*-derived proteins, were dynamically consistent over the simulation period. When analyzing the RMSD values, which indicate the deviation of protein structures from their initial conformations, notable differences emerge among the complexes. For instance, the SOCS2: EpoR peptide (standard agonist) complex displays a slightly higher average RMSD (2.423 Å) compared to apo-protein SOCS2 (2.417 Å). This observation suggests a moderate increase in structural flexibility upon the binding of the standard agonist, indicating potential conformational adjustments required for effective binding. Conversely, the RMSD value for the SOCS2: N4BP1 (standard antagonist) complex (2.496 Å) remains similar to that of the apo-protein, indicating that the binding of the antagonist may not significantly perturb the structural stability of SOCS2. Furthermore, the *Lumbricus*-derived protein complexes, including cytochrome b, SCBP3 protein, Lumbricin, Chemoattractive glycoprotein ES20, and Lumbrokinase-7T1, exhibit slightly higher average RMSD values ranging from 2.467 to 2.587 Å compared to the standard complexes. These differences suggest potential variations in the dynamic behavior and conformational changes induced by the binding of *Lumbricus*-derived proteins. The higher RMSD values imply that these *Lumbricus*-derived proteins may interact with SOCS2 in a manner that elicits different structural adjustments or conformational dynamics compared to the standard agonist and antagonist.

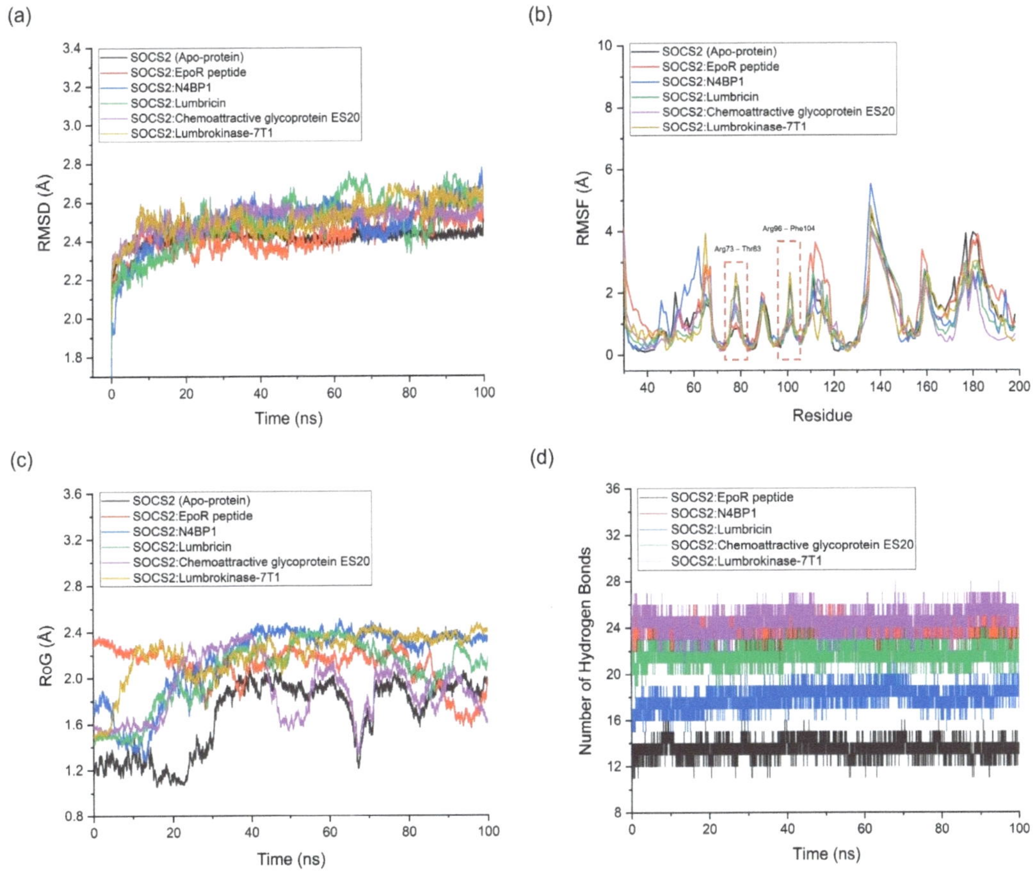

Figure 4. An analysis of molecular dynamics (MD) simulations for complexes formed between proteins sourced from the *Lumbricus* genus earthworm and SOCS2, including several key parameters: (**a**) the root mean square deviation (RMSD) assessed structural stability, (**b**) the root mean square fluctuation (RMSF) depicted residue flexibility, (**c**) the radius of gyration (RoG) illustrated structural compactness, and (**d**) the number of hydrogen bonds highlighted intermolecular interactions.

During MD simulations, RMSF analysis was performed to assess the flexibility of individual amino acid residues within SOCS2. The average RMSF values obtained ranged from 0.876 to 1.397 Å, reflecting moderate flexibility in various regions of the protein. This analysis provides valuable insights into the mobility and flexibility of specific residues, enhancing our understanding of their functional roles within the protein structure [46]. Upon examining the interaction between the top-performing *Lumbricus*-derived protein complex and SOCS2, it was apparent that this interaction disrupted hydrogen bonds with specific residues, especially within the regions encompassing amino acid residues Arg73 to Thr83 and Arg96 to Phe104 (Figure 4b). These regions are notably critical as they constitute the active binding site of SOCS2 as a target receptor [47]. The disruption of hydrogen bonds in critical areas led to greater residue fluctuations compared to the SOCS2 agonist and apo-protein complex. This increase in residue fluctuations suggests enhanced mobility and flexibility of these residues when the top-performing *Lumbricus*-derived protein binds to SOCS2. Notably, this pattern of increased flexibility resembles that of N4BP1, a known standard antagonist. The similarity in flexibility patterns at these specific residues indicates that the top-performing proteins from *Lumbricus* have the potential to act as inhibitors, much like standard antagonists. This finding is significant, as it suggests that *Lumbricus*-derived proteins could modulate SOCS2 activity by acting as inhibitors, similar to known antagonists. By disrupting hydrogen bonds and increasing the flexibility of key binding site residues, these proteins may interfere with SOCS2′s functional interactions, presenting promising avenues for developing therapeutic interventions targeting SOCS2-mediated cellular processes. Further investigation into the structural and functional implications of these interactions is necessary to fully understand their therapeutic potential and to develop new treatments for SOCS2-related diseases.

The radius of gyration (RoG) offered insights into the compactness or degree of expansion of protein structures during MD simulations [48]. For the SOCS2 complexes, the average RoG values ranged from 1.674 to 2.678 Å. The RoG values for the standard SOCS2 agonist (EpoR peptide) and antagonist (N4BP1) complexes were 2.101 Å and 2.162 Å, respectively. For the *Lumbricus*-derived protein complexes, RoG values ranged from 2.176 to 2.678 Å, slightly higher than the standard agonist but comparable to the antagonist. The number of hydrogen bonds formed between the two interacting proteins was indicative of the strength and specificity of their interaction. For the standard complexes, the SOCS2: EpoR peptide agonist complex formed 11 hydrogen bonds, while the SOCS2: N4BP1 antagonist complex formed 24 hydrogen bonds. Comparatively, the *Lumbricus*-derived protein complexes displayed similar or slightly higher numbers of hydrogen bonds, with values ranging from 16 to 39. Notably, the SOCS2: Cytochrome b complex formed the highest number of hydrogen bonds (39), followed closely by the SOCS2: Peroxidasin complex (35) (Table 5). The findings indicated that the interactions between SOCS2 and *Lumbricus*-derived proteins featured a comparable or slightly higher number of hydrogen bonds relative to the standard interactions. This suggests robust and specific binding between these proteins, which could enhance their functional relevance. Furthermore, the RoG values indicated that the *Lumbricus*-derived protein complexes displayed similar levels of compactness or dispersion compared to the standard complexes, suggesting they maintained stable structural conformations throughout the simulation period.

Table 5. Time-averaged structural properties obtained from the MD simulations of SOCS2 protein–protein complexes.

Complex	Average RMSD (Å)	Average RMSF (Å)	Average RoG (Å)	Number of Hydrogen Bonds between the Two Proteins	Potential Energy (kcal/mol)
Standard					
SOCS2 (apo-protein)	2.417	1.089	1.674	N/A	−158,603.87
SOCS2: EpoR peptide (standard agonist)	2.423	1.397	2.101	11	−159,708.72
SOCS2: N4BP1 (standard antagonist)	2.496	1.179	2.162	24	−459,214.66
Protein and peptide derived from earthworm (*Lumbricus* genus)					
SOCS2: Cytochrome b	2.467	0.876	2.182	39	−471,304.36
SOCS2: Cytochrome c oxidase subunit 3	2.504	1.122	2.176	30	−450,869.63
SOCS2: SCBP3 protein	2.587	1.301	2.198	22	−316,369.57
SOCS2: Lumbricin	2.495	1.031	2.148	18	−694,628.89
SOCS2: Chemoattractive glycoprotein ES20	2.512	0.941	2.176	22	−621,068.14
SOCS2: NADH-ubiquinone oxidoreductase chain 6	2.487	1.001	2.178	20	−466,577.25
SOCS2: Histone H3	2.413	0.917	2.287	16	−629,402.13
SOCS2: Peroxidasin	2.599	1.106	2.678	35	−2,208,424.10
SOCS2: Lumbrokinase-7T1	2.523	1.123	2.213	27	−1,065,029.39
SOCS2: Lysosomal membrane glycoprotein	2.511	1.115	2.298	31	−1,645,741.31
SOCS2: NADH-ubiquinone oxidoreductase chain 4L	2.498	1.178	2.199	26	−406,049.87
SOCS2: NADH-ubiquinone oxidoreductase chain 5	2.487	1.101	2.190	21	−1,187,770.16
SOCS2: Preprocarboxypeptidase	2.596	1.259	2.188	30	−560,689.83
SOCS2: Extracellular globin-4	2.543	1.028	2.287	23	−473,808.36
SOCS2: Lumbrokinase-7T2	2.524	1.060	2.214	24	−764,436.84

2.4. Molecular Mechanics/Poisson–Boltzmann Surface Area (MM/PBSA) Calculations

We assessed the strength of the protein–protein interactions by calculating the average $\Delta G_{binding}$ values, which indicate the binding free energy, for each complex. In the case of the standard complexes, SOCS2 bound to EpoR peptide (standard agonist) demonstrated a mean $\Delta G_{binding}$ of −42.60 kcal/mol, indicating a substantial level of stability in the interaction. Conversely, when bound to N4BP1 (standard antagonist), SOCS2 exhibited a higher mean $\Delta G_{binding}$ of −42.51 kcal/mol, suggesting a comparatively weaker interaction. However, the *Lumbricus*-derived protein complexes presented intriguing findings, showcasing similar or even more negative mean $\Delta G_{binding}$ values compared to the standard complexes. Notably, SOCS2 complexed with Lumbrokinase-7T1 displayed a notably high mean $\Delta G_{binding}$ of −69.28 kcal/mol, indicative of a robust and energetically favorable interaction. This suggests that the *Lumbricus*-derived protein has a strong affinity for SOCS2, potentially surpassing the binding strength observed with the standard agonist and antagonist. Furthermore, the complexes formed between SOCS2 and Lumbricin and Chemoattractive glycoprotein ES20 exhibited mean $\Delta G_{binding}$ values of −59.25 kcal/mol and −55.02 kcal/mol, respectively (Table 6). These values suggest strong binding affinities comparable to or even exceeding those observed in the standard complexes. This implies that the *Lumbricus*-derived proteins possess significant potential in modulating SOCS2 activity, potentially rivaling or surpassing the efficacy of standard proteins.

Table 6. The average binding free energy ($\Delta G_{binding}$) for the SOCS2 protein–protein complexes is reported with standard deviation in kcal/mol units, as determined by MM/PBSA calculations.

Complex	MM/PBSA Calculation Results $\Delta G_{binding}$ (kcal/mol)			Average (kcal/mol)
	I	II	III	
	Standard			
SOCS2: EpoR peptide (standard agonist)	−42.84	−43.48	−41.48	−42.60
SOCS2: N4BP1 (standard antagonist)	−42.34	−42.34	−42.85	−42.51
Protein and peptide derived from earthworm (*Lumbricus* genus)				
SOCS2: Cytochrome b	−50.57	−49.93	−49.82	−50.11
SOCS2: Cytochrome c oxidase subunit 3	−44.93	−44.93	−44.87	−44.91
SOCS2: SCBP3 protein	−29.33	−29.45	−29.45	−29.41
SOCS2: Lumbricin	−59.22	−59.26	−59.26	−59.25
SOCS2: Chemoattractive glycoprotein ES20	−53.69	−57.76	−53.60	−55.02
SOCS2: NADH-ubiquinone oxidoreductase chain 6	−52.81	−51.48	−53.15	−52.48
SOCS2: Histone H3	−44.05	−44.84	−45.82	−44.90
SOCS2: Peroxidasin	−42.91	−42.91	−42.91	−42.91
SOCS2: Lumbrokinase-7T1	−69.22	−69.35	−69.27	−69.28
SOCS2: Lysosomal membrane glycoprotein	−34.42	−34.89	−34.75	−34.69
SOCS2: NADH-ubiquinone oxidoreductase chain 4L	−48.14	−48.35	−48.35	−48.28
SOCS2: NADH-ubiquinone oxidoreductase chain 5	−37.74	−36.11	−36.11	−36.65
SOCS2: Preprocarboxypeptidase	−39.55	−38.03	−39.58	−39.05
SOCS2: Extracellular globin-4	−50.09	−49.68	−50.30	−50.02
SOCS2: Lumbrokinase-7T2	−46.18	−45.80	−47.11	−46.36

Furthermore, a detailed examination of the binding free energy of individual amino acid residues within SOCS2 provided profound insights into the molecular mechanisms governing binding specificity and affinity. Notably, Arg96 and Ser78 were identified as crucial contributors to the observed antagonistic activity within the complexes. In interactions between SOCS2 and the top-performing protein from *Lumbricus*, these specific residues demonstrated significantly higher binding affinity values compared to the standard agonist complex. The elevated binding affinity of Arg96 and Ser78 underscores their pivotal roles in mediating the observed antagonistic effects (Figure 5). These residues are likely involved in essential interactions that govern complex stability and specificity, influencing overall functional outcomes. The heightened affinity observed in the complexes involving *Lumbricus*-derived proteins highlights the importance of these interactions in modulating SOCS2 activity and emphasizes their potential as key factors in therapeutic efficacy. By elucidating the roles of individual amino acid residues in binding energetics, this analysis offers critical insights into the structural basis of protein–protein interactions. The identification of Arg96 and Ser78 as significant contributors to antagonistic activity enhances our understanding of the molecular determinants underlying the intricate interplay between SOCS2 and its interacting partners. These findings pave the way for the targeted manipulation of specific residues to effectively modulate SOCS2 function, presenting promising opportunities for developing therapeutic interventions aimed at SOCS2-associated pathways.

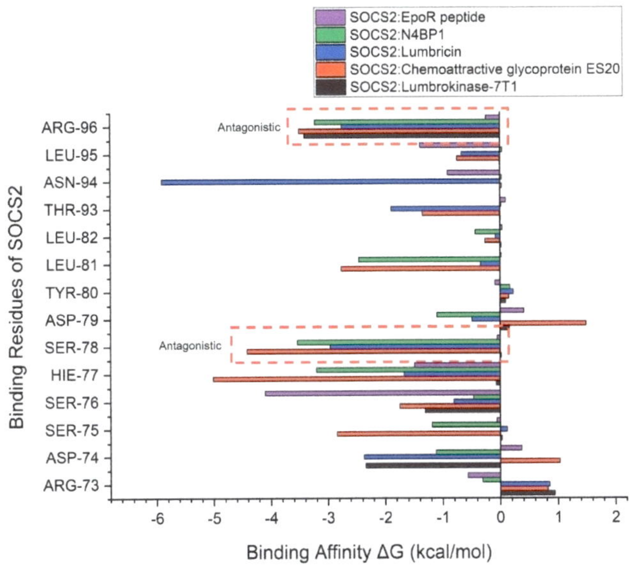

Figure 5. An analysis of the binding free energy for individual amino acids in SOCS2 interactions with the EpoR peptide (standard agonist), N4BP1 (standard antagonist), and the top three proteins derived from earthworms (*Lumbricus* genus) using MM/PBSA calculations.

3. Discussion

This research employed a comprehensive approach to explore the interactions between SOCS2 and proteins and peptides sourced from *Lumbricus* earthworms, focusing particularly on their potential therapeutic implications in cardiovascular diseases. Emphasizing meticulous data collection and stringent quality control measures, this study utilized advanced computational techniques and methodologies for tasks such as 3D structure modeling, protein–protein docking simulations, molecular dynamics (MD) simulations, and binding free energy calculations. During the 3D structure modeling phase, this study utilized bioinformatics and computational biology tools to generate precise and dependable structural models of the *Lumbricus*-derived proteins and peptides. By systematically gathering sequences from the UniProt database and applying rigorous selection criteria, this study ensured the dataset's integrity and relevance. This approach aligns with previous research efforts that have successfully predicted protein structures with high precision using bioinformatics tools [49]. Furthermore, employing both I-TASSER and AlphaFold algorithms enabled thorough modeling of proteins and peptides, regardless of whether homologous structures were available in the PDB. This approach significantly expanded the breadth and depth of our structural modeling endeavors [32]. The flexibility of proteins is an important factor that can significantly influence the results of docking studies. As shown in Figure 1, while some proteins appear well-folded and globular, others exhibit large, unfolded regions or loops, indicating potential structural fluctuations. Although our initial docking studies were performed using rigid models, we recognized the limitations of this approach in capturing the dynamic nature of protein interactions. To address this, we conducted MD simulations for the proteins with identified flexible regions. The MD simulations provided valuable insights into conformational changes and the dynamic behavior of these proteins, helping us understand the potential impact of flexibility on their interactions with SOCS2. Future docking studies could benefit from incorporating flexible docking methods to account for the dynamic nature of these proteins.

This research explored the intricate interactions between SOCS2 and proteins derived from *Lumbricus* earthworms using protein–protein docking simulations, aiming to identify potential therapeutic strategies for cardiovascular diseases. By meticulously comparing the binding affinities of these *Lumbricus*-derived proteins with those of standard agonist and antagonist complexes, this study identified promising candidates that exhibited comparable or enhanced effectiveness in modulating SOCS2 activity. These findings align with earlier studies emphasizing the therapeutic potential of natural compounds in managing cardiovascular diseases. Previous research has particularly highlighted the medicinal properties of *Lumbricus* earthworms, including their primary component, lumbrokinase, which has shown promise in treating cardiovascular conditions [25]. By elucidating the molecular intricacies of lumbrokinase's interaction with the SOCS2 pathway, this study contributes novel insights into the therapeutic potential of lumbrokinase for cardiovascular treatment. Additionally, the analysis of the cluster size and RMSD values provided insights into the structural diversity and stability of protein–protein complexes, offering valuable information for rational drug design and optimization [50]. MD simulations further elucidated the dynamic behavior of protein–protein complexes over time, uncovering potential mechanisms for modulating SOCS2 activity. This study provided detailed insights into the structural dynamics and functional implications of SOCS2–therapeutic protein interactions by examining residue flexibility and hydrogen bond interactions. This aligned with previous studies that used MD simulations to investigate protein–protein interactions and elucidate their dynamic behavior [51]. This study observed that Lumbricin, Chemoattractive glycoprotein ES20, and Lumbrokinase-7T1 demonstrated sustained interactions with the SOCS2 protein throughout the simulation period, akin to the behavior exhibited by the standard antagonist complex. However, it is important to note that, while these proteins showed promising interactions with SOCS2, specific experimental studies directly linking them to SOCS2′s E3 ligase activity are currently lacking in the literature. Furthermore, this study discussed the structural features and key interaction residues within these protein–protein complexes, elucidating the molecular basis underlying their binding affinity and activity. By analyzing the conformational changes and intermolecular forces at play during the MD simulations, this study highlighted the structural motifs and binding pockets crucial for stabilizing the interactions between the *Lumbricus*-derived proteins and SOCS2. The resemblance in activity between these *Lumbricus*-derived proteins and the standard antagonist suggests that they may function through similar mechanisms or binding modes, warranting further investigation into their therapeutic potential.

Several compounds have been identified as SOCS2 inhibitors, each exhibiting unique binding affinities and mechanisms of action. For instance, small-molecule inhibitors such as the KIAA0317 protein have shown promising binding affinities in the nanomolar range, effectively disrupting SOCS2′s interactions with its targets [52]. The mechanism of inhibition for these small molecules typically involves blocking the binding site of SOCS2, thereby preventing its E3 ligase activity and subsequent ubiquitination of target proteins [53]. In contrast, proteins derived from natural sources often present complex interactions. For example, interferons, which are known to modulate SOCS2 activity, bind to the receptor complex and induce SOCS2 expression, creating a negative feedback loop that can lead to enhanced SOCS2 activity rather than inhibition [54]. This underscores the importance of understanding the nuanced interactions of each inhibitor. Our study demonstrates that the proteins and peptides from *Lumbricus* earthworms exhibit unique binding patterns with SOCS2, driven by their distinct amino acid compositions and structural conformations. The binding affinities of these earthworm-derived compounds, while not yet quantified in the nanomolar range, exhibit potential based on our docking studies and molecular dynamics simulations. Specifically, the presence of specific residue substitutions, such as lysine and threonine in cytochrome b, results in altered hydrogen bonding patterns, which may lead to different binding affinities compared to established inhibitors.

The proteins from *Lumbricus* earthworms exhibit several distinguishing features compared to their human counterparts, significantly impacting their interactions with SOCS2 and potential therapeutic applications. A key example is cytochrome b, which demonstrates notable structural and functional differences. The amino acid sequence of earthworm cytochrome b shares less than 60% identity with its human variant, resulting in altered binding interfaces and unique interaction patterns with SOCS2. While the human variant exhibits a more conserved sequence, the earthworm version contains specific residue substitutions, such as lysine and threonine, which can lead to different hydrogen bonding patterns and potentially affect binding affinity. Structurally, earthworm cytochrome b displays greater flexibility due to less compact folding in some regions, contrasting with the more stable and globular structure seen in humans. This increased flexibility may influence interaction dynamics with SOCS2, either enhancing or weakening binding depending on the context. Additionally, the molecular weight of earthworm cytochrome b (approximately 42.88 kDa) [55] is much lower than that of the human variant (approximately 91 kDa) [56], which may affect its structural stability during binding. The hydrophobicity in the transmembrane regions is also lower in the earthworm version, potentially impacting its membrane integration and positioning during docking. Functionally, while both cytochrome b proteins are involved in electron transport, the earthworm variant is adapted to specific environmental conditions, which could influence its interaction with SOCS2 under oxidative stress.

In addition to cytochrome b, proteins like lumbrokinase, a serine protease, illustrate significant differences in enzymatic activity and specificity compared to human proteases. Lumbrokinase's unique mechanism of action enhances fibrinolysis, suggesting a tailored adaptation for promoting blood flow and preventing thrombosis in the earthworm's natural habitat, with potential cardiovascular benefits for humans [57]. Furthermore, *Lumbricus*-derived proteins often possess variations in key residues that influence structural stability and interaction profiles. Differences in glycosylation patterns may affect how these proteins interact with receptors or other proteins within the human body, potentially enhancing their bioactivity or altering their pharmacokinetics. Notably, some *Lumbricus* proteins, such as the Chemoattractive glycoprotein ES20, exhibit distinct structural motifs compared to similar human proteins, leading to different binding affinities and mechanisms of action. These proteins are often more hydrophilic and possess unique surface charges that influence their solubility and interactions in biological systems.

Considering the administration of whole proteins for therapeutic use, it is crucial to identify and study the active portions of these molecules. Isolating specific peptide sequences that exhibit bioactivity may enhance therapeutic efficacy while minimizing potential side effects associated with whole protein administration. Future studies should focus on proteolytic digestion or recombinant techniques to generate and characterize these active peptides derived from *Lumbricus* proteins. By elucidating the structure–function relationships of these peptides, we can better understand their mechanisms of action, improve their pharmacokinetic properties, and optimize their therapeutic applications in cardiovascular therapy.

4. Limitations, Clinical Implications, and Future Works

While this study provides valuable insights into the potential therapeutic interactions between SOCS2 and proteins derived from *Lumbricus* earthworms, several limitations should be acknowledged. Firstly, the findings are based on computational protein–protein docking simulations, which, while robust, rely on predictive models and may not fully capture the complexities of real biological systems. Experimental validation is essential to confirm the observed interactions and their functional implications in vitro and in vivo. Secondly, this study primarily focused on binding affinities and structural dynamics without directly measuring biological outcomes such as enzymatic inhibition or cellular responses. Future research should incorporate functional assays to validate the therapeutic effects of these interactions on SOCS2 activity and explore their broader implications in disease models. Thirdly, the specific mechanisms through which *Lumbricus*-derived proteins modulate SOCS2 function, particularly regarding E3 ligase activity, remain speculative. Experimental studies investigating these mechanisms are crucial to elucidate their precise therapeutic potential and optimize their application in clinical settings. Additionally, we recognize the limitation of our search for proteins and peptides derived from *Lumbricus* earthworms in the UniProt database, which yielded a total of 978 entries filtered down to 78 non-redundant proteins/peptides. While this number appears low, it is important to note that our laboratory has actively worked with *Lumbricus* bioactive protein extracts and performed LC-MS/MS protein sequencing. The majority of the identified proteins from *Lumbricus* earthworms correspond to entries covered in the UniProt database. Although additional sequences may be available through NCBI databases, our focus on proteins characterized in established databases enhances the reliability of our computational analysis.

This study's findings have significant clinical implications for cardiovascular disease management and potentially other SOCS2-related conditions. Identifying natural compounds from *Lumbricus* earthworms that exhibit comparable efficacy to standard agonists and antagonists highlights their potential as alternative or adjunctive therapies. These compounds may offer novel strategies for modulating SOCS2 activity, potentially leading to more targeted and effective treatments with fewer side effects compared to conventional therapies. Furthermore, understanding the molecular interactions between *Lumbricus*-derived proteins and SOCS2 can pave the way for the development of tailored therapeutic approaches. This knowledge could inform the design of novel medications that specifically target SOCS2-related pathways, potentially improving patient outcomes and quality of life in various disease contexts.

Future research directions should aim to address the identified limitations and expand upon the current findings. Experimental studies are crucial to validating the computational predictions and confirming the biological relevance of the observed interactions. Specifically, employing biochemical assays and cellular models will provide deeper insights into how *Lumbricus*-derived proteins affect SOCS2 function and E3 ligase activity. Moreover, investigating the therapeutic efficacy of these proteins in preclinical disease models will be essential to evaluating their potential clinical translation. This includes assessing their pharmacokinetics, bioavailability, and safety profiles to establish their feasibility as therapeutic agents. Furthermore, exploring the synergy between *Lumbricus*-derived proteins and existing therapies could uncover combination strategies that enhance therapeutic outcomes. This approach could potentially lead to personalized treatment regimens tailored to individual patient needs, maximizing therapeutic efficacy while minimizing adverse effects.

5. Materials and Methods

5.1. Materials

Protein and peptide sequences derived from earthworms of the *Lumbricus* genus were obtained through a systematic search approach using UniProt, a comprehensive database housing protein sequences and functional information. The search strategy utilized the MeSH term "*Lumbricus*" within UniProtKB to specifically target proteins and peptides associated with this genus. Subsequently, retrieved data underwent meticulous curation to remove duplicate entries, ensuring the dataset's integrity and accuracy for subsequent analyses. This curation process was essential to minimizing redundancy and potential biases, thereby enhancing the reliability and validity of this study's findings. By rigorously retrieving and curating data, a comprehensive and high-quality dataset was established, forming the basis for investigating the bioactive constituents derived from earthworms and their potential therapeutic roles in cardiovascular diseases.

5.2. Computing Power

For this study, computational analyses were conducted using a workstation equipped with the following specifications: an Intel® Core™ i9-12900KF CPU (Intel, Santa Clara, CA, USA) running at 3.90 GHz with 16 cores, an NVIDIA GeForce RTX 4090 graphics card featuring 24 GB GDDR6X, 64 GB DDR5 RAM (NVIDIA, Santa Clara, CA, USA), an 8TB hard disk drive, and an Ubuntu operating system.

5.3. Three-Dimensional Structure Modeling

In the initial phase of 3D structure modeling in this study, we systematically gathered peptide and protein sequences sourced from earthworms of the *Lumbricus* genus using the UniProt database. This process yielded a substantial dataset comprising 979 records. To maintain the dataset's quality and reliability for subsequent analyses, stringent criteria were applied during the selection of peptide and protein sequences from *Lumbricus* earthworms. Several key criteria were employed to filter and refine the dataset obtained from UniProt. Initially, sequences were assessed for completeness to ensure only those with sufficient information for accurate 3D modeling were included, thereby reducing the risk of generating incomplete or erroneous structural models [58]. Furthermore, rigorous quality control protocols were implemented to identify and rectify sequencing errors, ambiguous residues, or any irregularities that might undermine the integrity of subsequent analyses. Sequences that did not meet these quality standards were excluded to ensure the overall accuracy and reliability of the dataset [59]. Additionally, duplicate entries were identified and eliminated from the dataset to avoid redundancy and potential biases in subsequent analyses. A thorough comparison and validation process of sequence data ensured that each unique protein or peptide appeared only once in the final dataset. Moreover, maintaining the taxonomic accuracy of the earthworm species represented in the dataset was essential to ensuring specificity and relevance to the *Lumbricus* genus. Taxonomic verification procedures [60] were utilized to confirm the origin of all sequences from earthworms belonging to the *Lumbricus* genus, thereby preventing the inclusion of irrelevant or misclassified data. The database of protein sequences sourced from earthworms (*Lumbricus* genus) is available in Supplementary Data S1. This database includes detailed information on protein/peptide names, UniProt IDs, sequences, the size (kDa), identified protein binding sites (number of residues), the prediction method, the assessment method, and the quality score. Several computational tools were utilized to generate the three-dimensional structures of the protein and peptide sequences acquired. Initially, I-TASSER (Iterative Threading ASSEmbly Refinement) was applied to proteins and peptides that had corresponding templates in the Protein Data Bank (PDB). I-TASSER excels in constructing 3D structures using available templates from the PDB, utilizing these templates to predict the tertiary structure of the queried sequences [61]. Proteins and peptides without templates in the PDB were modeled using AlphaFold. AlphaFold is a cutting-edge deep learning approach for protein structure prediction, renowned for its ability to accurately forecast the 3D structures of proteins, including cases where homologous structures are unavailable [32]. Next, the active sites of proteins and peptides sourced from *Lumbricus* were examined using CASTp 3.0. CASTp is a tool designed to detect and analyze active sites and binding pockets within protein structures, offering crucial insights into their functional characteristics and potential sites for ligand binding [62]. Additionally, the SOCS2 protein, which was selected as the target for subsequent docking studies, was obtained from the PDB under the accession code 6I5N with a resolution of 1.98 Å [63]. This step ensured that the target receptor structure was readily available for the following molecular docking simulations, enabling the study of protein–protein interactions between the bioactive compounds derived from *Lumbricus* earthworm and SOCS2.

5.4. Protein–Protein Docking Simulation

In this section, we thoroughly investigated the interactions between the proteins and peptides obtained from *Lumbricus* and the SOCS2 target receptor. Employing protein–protein docking simulations, we aimed to uncover critical aspects of the binding mechanism. This included identifying pivotal residues crucial for forming protein–protein complexes, understanding the types of intermolecular interactions at play, assessing binding affinities, examining binding modes, and analyzing orientations. To delineate the binding sites of the target receptor, we utilized PDBSum, a computational tool tailored for detailed summaries of protein structures and their interactions [64]. This examination provided us with insights into how essential residues are spatially arranged within SOCS2′s binding site and their potential interactions with *Lumbricus*-derived proteins and peptides. To maintain the precision of our analysis, we refined the receptor using Swiss-PdbViewer v4.1.1 [65] before proceeding with protein–protein docking analysis, ensuring a reliable foundation for our subsequent investigations. To delve deeper into understanding whether the proteins and peptides derived from *Lumbricus* earthworms exhibit agonistic or antagonistic effects toward SOCS2, we utilized well-established molecules for comparison. EpoR peptide, identified by its PDB ID, 6I5N [63], serves as a standard agonist known to bind to SOCS2 within the PDB complex. In contrast, N4BP1, with a PDB ID of 8T48 [66], acts as a potent suppressor of E3 ligase activity. By comparing *Lumbricus*-derived compounds to these known standards, we aimed to evaluate their interactions with SOCS2, particularly focusing on binding pocket residues (Arg73, Ser75, Ser76, and Arg96) [67]. These residues are pivotal in determining the specificity and strength of interactions between SOCS2 and its binding partners. Following this, protein–protein docking calculations were conducted using the advanced interface option within the standalone version of High Ambiguity Driven Protein–protein Docking (HADDOCK). HADDOCK is a well-established computational tool for modeling protein–protein interactions, utilizing ambiguous interaction restraints derived from experimental or computational sources [68,69]. Using this method, we evaluated the interactions between *Lumbricus* proteins and peptides and SOCS2, predicting potential binding modes and affinities. Optimal protein–protein docking results for each complex were selected based on two key criteria: the highest number of observed clusters or populations, indicating the reliability of predicted interactions, and the highest docking score (HADDOCK score), which measures the strength of the binding affinity between *Lumbricus* proteins and peptides and SOCS2 within the protein–protein complex.

5.5. Molecular Dynamics (MD) Simulation

Molecular dynamics (MD) simulations were used to study the dynamics and stability of the protein–protein complexes involving *Lumbricus* proteins and peptides with SOCS2. These simulations were performed using GROMACS 2022.5, a well-established molecular dynamics simulation software renowned for its efficiency and accuracy in modeling biomolecular systems [70]. For the simulations, we utilized the Optimized Potentials for Liquid Simulations (OPLS-AA/L) force field, which accurately models molecular interactions within the system [71]. The dimensions of the simulation box were set using default cubic box parameters to accommodate the biomolecular complexes effectively. Standard protocols were followed to prepare simulation input files, which included adding water molecules using the Single Point Charge Extended (SPCE) model and incorporating counterions to maintain system neutrality [72]. Energy minimization was performed using the steepest-descent method to eliminate steric clashes and achieve system relaxation to a stable state. Following this, a two-phase equilibration process was executed. Initially, in phase 1, the system underwent equilibration in the NVT ensemble to regulate temperature fluctuations and stabilize conditions. Subsequently, in phase 2, equilibration was conducted in the NPT ensemble to maintain constant pressure and temperature. Once equilibrated, production MD simulations were conducted over 100 nanoseconds to observe the dynamics of the protein–protein complexes over an extended period. Throughout the simulations, parameters such as the root mean square deviation (RMSD), the root mean square fluctuation (RMSF), the radius of gyration (RoG), potential energies, and intermolecular hydrogen bonding interactions were monitored and analyzed to assess complex stability and conformational dynamics. The visualization of critical residues and intermolecular interactions within the predicted protein–protein complexes was conducted through manual inspection using molecular visualization software like PyMOL [73] and UCSF Chimera [74]. These tools enabled the examination and interpretation of essential structural characteristics and interactions within the simulated complexes, offering valuable insights into the mechanisms that govern the binding and stability of the proteins and peptides from *Lumbricus* with SOCS2.

5.6. Molecular Mechanics/Poisson–Boltzmann Surface Area (MM/PBSA) Calculations

The Molecular Mechanics/Poisson–Boltzmann Surface Area (MM/PBSA) method, employing MD simulations, was employed to investigate the protein–protein interactions involving proteins and peptides derived from *Lumbricus* and SOCS2. MD simulations produced a range of protein conformations, and representative snapshots were chosen for detailed analysis [75]. Each snapshot was subjected to thorough energy computations, encompassing calculations of gas-phase energy, the estimation of solvation energy using a continuum solvent model, and entropy calculations. These energy components were integrated to determine the binding free energy of the protein–protein complex [76,77]. To conduct these computations, we utilized the gmx_MMPBSA module available within the GROMACS simulation package [78,79]. This module facilitates the precise and efficient calculation of binding free energies for biomolecular complexes. The MM/PBSA method is renowned for its ability to predict the binding free energy of protein–protein interactions, establishing it as a valuable tool for understanding the energetic aspects of biomolecular interactions [80]. The MM/PBSA binding free energy calculation is derived from the following equation:

$$\Delta G\_binding = \Delta G\_complex - \Delta G\_proteinX_1 - \Delta G\_proteinX_2$$

where the variables are defined as follows:

$\Delta G\_binding$: the binding free energy associated with the formation of the protein–protein complex.

$\Delta G\_complex$: the free energy of the fully solvated protein–protein complex.

$\Delta G\_proteinX_1$: the free energy of protein 1 in its solvated state when unbound.

$\Delta G\_proteinX_2$: the free energy of protein 2 in its solvated state when unbound.

The binding free energy was determined by calculating the difference between the free energy of the complex and the combined free energies of the unbound proteins. This calculation provided insights into the energetic alterations that occurred during the formation of the protein–protein complex, thereby elucidating the strength and stability of the interaction.

5.7. Statistical Analysis

In this study, statistical analysis was performed to investigate the relationships among all parameters derived from both molecular docking and molecular dynamics simulations. The data from these computational experiments underwent thorough statistical analysis and interpretation using SPSS 25 statistical software [81] and OriginLab Pro 2022 [82]. This analytical approach allowed for exploring correlations, trends, and patterns within the dataset, offering valuable insights into the relationships between various variables and parameters. Moreover, statistical analysis helped validate computational outcomes and identify significant findings, enhancing the understanding of the molecular interactions under scrutiny.

6. Conclusions

In conclusion, this study conducted a thorough investigation into the interactions between *Lumbricus*-derived proteins and SOCS2, revealing their potential therapeutic implications for cardiovascular diseases. Employing a comprehensive approach integrating bioinformatics, computational modeling, and molecular dynamics simulations, this research yielded valuable insights into the structural characteristics and binding affinities of *Lumbricus*-derived proteins and peptides with SOCS2. The findings indicated that certain proteins such as Lumbricin, Chemoattractive glycoprotein ES20, and Lumbrokinase-7T1 demonstrated activity comparable to standard antagonists in regulating SOCS2 function, suggesting their candidacy for therapeutic applications in cardiovascular diseases. Moreover, this study underscored the significance of computational methodologies in drug discovery, particularly for natural products possessing diverse chemical structures and biological activities. By leveraging advanced computational tools, this study efficiently screened a wide array of candidate compounds for their potential interactions with SOCS2, overcoming the challenges associated with traditional drug discovery methods. This approach not only accelerated the discovery process but also provided mechanistic insights into how these compounds operate at the molecular level. However, it is essential to recognize this study's limitations, primarily its reliance on computational predictions that necessitate validation through experimental assays. Future research efforts should focus on validating these findings using biochemical and structural biology techniques, as well as conducting preclinical studies in relevant disease models. Furthermore, exploring the broader therapeutic potential of *Lumbricus*-derived proteins beyond cardiovascular diseases and investigating their synergistic effects with existing pharmacological agents could open new avenues for drug development and personalized medicine initiatives.

Supplementary Materials: The following supporting information can be downloaded at: https://www.mdpi.com/article/10.3390/ijms251910818/s1.

Author Contributions: Conceptualization, N.A. and D.D.; Data curation, N.A.; Formal analysis, D.D. and N.E.E.; Funding acquisition, N.A. and N.E.E.; Investigation, D.D.; Methodology, N.A., D.D. and N.E.E.; Project administration, N.A.; Resources, N.A.; Software, D.D.; Supervision, N.A. and N.E.E.; Validation, N.A. and D.D.; Visualization, D.D.; Writing—original draft, N.A. and D.D.; Writing—review and editing, N.A., D.D. and N.E.E. All authors have read and agreed to the published version of the manuscript.

Funding: The authors extend their appreciation to the Deanship of Scientific Research, Imam Mohammad Ibn Saud Islamic University (IMSIU), Saudi Arabia, for funding this research work through Grant No. (221413005).

Institutional Review Board Statement: Not applicable.

Informed Consent Statement: Not applicable.

Data Availability Statement: The original contributions presented in this study are included in the article/Supplementary Material, and further inquiries can be directed to the corresponding author.

Conflicts of Interest: The authors declare no conflicts of interest.

References

1. WHO. Cardiovascular Diseases (CVDs): World Health Organization. 2021. Available online: https://www.who.int/news-room/fact-sheets/detail/cardiovascular-diseases-(cvds) (accessed on 1 April 2024).
2. Roth Gregory, A.; Mensah George, A.; Johnson Catherine, O.; Addolorato, G.; Ammirati, E.; Baddour Larry, M.; Barengo Noël, C.; Beaton Andrea, Z.; Benjamin Emelia, J.; Benziger Catherine, P.; et al. Global Burden of Cardiovascular Diseases and Risk Factors, 1990–2019. *J. Am. Coll. Cardiol.* **2020**, *76*, 2982–3021. [CrossRef] [PubMed]
3. Del Buono, M.G.; Montone, R.A.; Camilli, M.; Carbone, S.; Narula, J.; Lavie, C.J.; Niccoli, G.; Crea, F. Coronary Microvascular Dysfunction Across the Spectrum of Cardiovascular Diseases: JACC State-of-the-Art Review. *J. Am. Coll. Cardiol.* **2021**, *78*, 1352–1371. [CrossRef] [PubMed] [PubMed Central]
4. Gimbrone, M.A.; García-Cardeña, G. Endothelial Cell Dysfunction and the Pathobiology of Atherosclerosis. *Circ. Res.* **2016**, *118*, 620–636. [CrossRef] [PubMed]
5. Marc, T. Cardiovascular Disease: An Introduction. *Vasculopathies* **2018**, *8*, 1–90.
6. Ullah, A.; Kumar, M.; Sayyar, M.; Sapna, F.; John, C.; Memon, S.; Qureshi, K.; Agbo, E.C.; Ariri, H.I.; Chukwu, E.J.; et al. Revolutionizing Cardiac Care: A Comprehensive Narrative Review of Cardiac Rehabilitation and the Evolution of Cardiovascular Medicine. *Cureus* **2023**, *15*, e46469. [CrossRef] [PubMed] [PubMed Central]
7. Jansen-Chaparro, S.; López-Carmona, M.D.; Cobos-Palacios, L.; Sanz-Cánovas, J.; Bernal-López, M.R.; Gómez-Huelgas, R. Statins and Peripheral Arterial Disease: A Narrative Review. *Front. Cardiovasc. Med.* **2021**, *8*, 1–18. [CrossRef] [PubMed] [PubMed Central]
8. Rousan, T.A.; Thadani, U. Stable Angina Medical Therapy Management Guidelines: A Critical Review of Guidelines from the European Society of Cardiology and National Institute for Health and Care Excellence. *Eur. Cardiol.* **2019**, *14*, 18–22. [CrossRef] [PubMed] [PubMed Central]
9. Kario, K.; Hoshide, S.; Narita, K.; Okawara, Y.; Kanegae, H.; Aoki, K.; Kihara, H.; Koga, T.; Nakata, T.; Oku, K.; et al. Cardiovascular Prognosis in Drug-Resistant Hypertension Stratified by 24-Hour Ambulatory Blood Pressure: The JAMP Study. *Hypertension* **2021**, *78*, 1781–1790. [CrossRef]

10. Wijkman, M.O.; Malachias, M.V.B.; Claggett, B.L.; Cheng, S.; Matsushita, K.; Shah, A.M.; Jhund, P.S.; Coresh, J.; Solomon, S.D.; Vardeny, O. Resistance to antihypertensive treatment and long-term risk: The Atherosclerosis Risk in Communities study. *J. Clin. Hypertens.* **2021**, *23*, 1887–1896. [CrossRef] [PubMed] [PubMed Central]
11. Hu, X.; Li, J.; Fu, M.; Zhao, X.; Wang, W. The JAK/STAT signaling pathway: From bench to clinic. *Signal Transduct. Target. Ther.* **2021**, *6*, 402. [CrossRef]
12. Morris, R.; Kershaw, N.J.; Babon, J.J. The molecular details of cytokine signaling via the JAK/STAT pathway. *Protein Sci.* **2018**, *27*, 1984–2009. [CrossRef] [PubMed] [PubMed Central]
13. Liang, C.; Zhang, L.; Lian, X.; Zhu, T.; Zhang, Y.; Gu, N. Circulating Exosomal SOCS2-AS1 Acts as a Novel Biomarker in Predicting the Diagnosis of Coronary Artery Disease. *BioMed Res. Int.* **2020**, *2020*, 9182091. [CrossRef]
14. Zhang, H.; Dhalla, N.S. The Role of Pro-Inflammatory Cytokines in the Pathogenesis of Cardiovascular Disease. *Int. J. Mol. Sci.* **2024**, *25*, 1082. [CrossRef]
15. Shook, P.L.; Singh, M.; Singh, K. Macrophages in the Inflammatory Phase following Myocardial Infarction: Role of Exogenous Ubiquitin. *Biology* **2023**, *12*, 1258. [CrossRef] [PubMed] [PubMed Central]
16. Klaeske, K.; Dix, M.; Adams, V.; Jawad, K.; Eifert, S.; Etz, C.; Saeed, D.; Borger, M.A.; Dieterlen, M.T. Differential Regulation of Myocardial E3 Ligases and Deubiquitinases in Ischemic Heart Failure. *Life* **2021**, *11*, 1430. [CrossRef] [PubMed] [PubMed Central]
17. Rico-Bautista, E.; Flores-Morales, A.; Fernández-Pérez, L. Suppressor of cytokine signaling (SOCS) 2, a protein with multiple functions. *Cytokine Growth Factor Rev.* **2006**, *17*, 431–439. [CrossRef] [PubMed]
18. Durham, G.; Williams, J.; Nasim, T.; Palmer, T. Targeting SOCS Proteins to Control JAK-STAT Signalling in Disease. *Trends Pharmacol. Sci.* **2019**, *40*, 298–308. [CrossRef]
19. Mustafa, R.; Saiqa, A.; Domínguez, J.; Jamil, M.; Manzoor, S.; Wazir, S.; Shaheen, B.; Parveen, A.; Khan, R.; Ali, S. Therapeutic Values of Earthworm Species Extract from Azad Kashmir as Anticoagulant, Antibacterial, and Antioxidant Agents. *Can. J. Infect. Dis. Med. Microbiol.* **2022**, *2022*, 1–20. [CrossRef]
20. Cooper, E.L.; Balamurugan, M.; Huang, C.Y.; Tsao, C.R.; Heredia, J.; Tommaseo-Ponzetta, M.; Paoletti, M.G. Earthworms dilong: Ancient, inexpensive, noncontroversial models may help clarify approaches to integrated medicine emphasizing neuroimmune systems. *Evid. Based Complement. Altern. Med.* **2012**, *2012*, 164152. [CrossRef] [PubMed]
21. Dong, Y.; Woo, Y.M.; Lee, Y.H.; Ahn, M.Y.; Lee, D.G.; Lee, S.H.; Ha, J.M.; Park, C.I.; Kim, A. Data on the potent fibrinolytic effects of the Lumbricus rubellus earthworm and the Perinereis linea lugworm. *Data Brief* **2019**, *26*, 104484. [CrossRef] [PubMed]
22. Lai, C.H.; Han, C.K.; Shibu, M.A.; Pai, P.Y.; Ho, T.J.; Day, C.H.; Tsai, F.J.; Tsai, C.H.; Yao, C.H.; Huang, C.Y. Lumbrokinase from earthworm extract ameliorates second-hand smoke-induced cardiac fibrosis. *Environ. Toxicol.* **2015**, *30*, 1216–1225. [CrossRef] [PubMed]
23. Chen, M.; Wang, L.; Zheng, C.; Ma, A.; Hu, K.; Xiang, A.; Sun, Z.; Xie, B.; Xiong, G.; Shi, L.; et al. Novel ACE inhibitory peptides derived from bighead carp (Aristichthys nobilis) hydrolysates: Screening, inhibition mechanisms and the bioconjugation effect with graphene oxide. *Food Biosci.* **2023**, *52*, 102399. [CrossRef]
24. Prakash, M.; Balamurugan, M.; Parthasarathi, K.; Gunasekaran, G.; Cooper, E.; Ranganathan, L. Anti-ulceral and anti-oxidative properties of "earthworm paste" of Lampito mauritii (Kinberg) on Rattus Norvegicus. *Eur. Rev. Med. Pharmacol. Sci.* **2006**, *11*, 9–15.
25. Wang, Y.-H.; Chen, K.-M.; Chiu, P.-S.; Lai, S.-C.; Su, H.-H.; Jan, M.-S.; Lin, C.-W.; Lu, D.-Y.; Fu, Y.-T.; Liao, J.-M.; et al. Lumbrokinase attenuates myocardial ischemia-reperfusion injury by inhibiting TLR4 signaling. *J. Mol. Cell. Cardiol.* **2016**, *99*, 113–122. [CrossRef] [PubMed]
26. Yu, Q.; Li, P.; Yang, Q. Improving the absorption of earthworm fibrinolytic enzymes with mucosal enhancers. *Pharm. Biol.* **2010**, *48*, 816–821. [CrossRef] [PubMed]
27. Dermawan, D.; Prabowo, B.A.; Rakhmadina, C.A. In silico study of medicinal plants with cyclodextrin inclusion complex as the potential inhibitors against SARS-CoV-2 main protease (Mpro) and spike (S) receptor. *Inform. Med. Unlocked* **2021**, *25*, 100645. [CrossRef] [PubMed]
28. Challapa-Mamani, M.R.; Tomás-Alvarado, E.; Espinoza-Baigorria, A.; León-Figueroa, D.A.; Sah, R.; Rodriguez-Morales, A.J.; Barboza, J.J. Molecular Docking and Molecular Dynamics Simulations in Related to Leishmania donovani: An Update and Literature Review. *Trop. Med. Infect. Dis.* **2023**, *8*, 457. [CrossRef] [PubMed] [PubMed Central]
29. Sánchez-Gloria, J.L.; Arellano-Buendía, A.S.; Juárez-Rojas, J.G.; García-Arroyo, F.E.; Argüello-García, R.; Sánchez-Muñoz, F.; Sánchez-Lozada, L.G.; Osorio-Alonso, H. Cellular Mechanisms Underlying the Cardioprotective Role of Allicin on Cardiovascular Diseases. *Int. J. Mol. Sci.* **2022**, *23*, 9082. [CrossRef] [PubMed] [PubMed Central]
30. Choudhury, T.Z.; Garg, V. Molecular genetic mechanisms of congenital heart disease. *Curr. Opin. Genet. Dev.* **2022**, *75*, 101949. [CrossRef]
31. Zhang, Y. I-TASSER: Fully automated protein structure prediction in CASP8. *Proteins* **2009**, *77* (Suppl. S9), 100–113. [CrossRef]
32. Jumper, J.; Evans, R.; Pritzel, A.; Green, T.; Figurnov, M.; Ronneberger, O.; Tunyasuvunakool, K.; Bates, R.; Žídek, A.; Potapenko, A.; et al. Highly accurate protein structure prediction with AlphaFold. *Nature* **2021**, *596*, 583–589. [CrossRef] [PubMed]
33. Mohammadi, S.; Narimani, Z.; Ashouri, M.; Firouzi, R.; Karimi-Jafari, M.H. Ensemble learning from ensemble docking: Revisiting the optimum ensemble size problem. *Sci. Rep.* **2022**, *12*, 410. [CrossRef] [PubMed] [PubMed Central]
34. Shyamal, M.; Mandal, T.K.; Panja, A.; Saha, A. Influence of anionic co-ligands on the structural diversity and catecholase activity of copper(II) complexes with 2-methoxy-6-(8-iminoquinolinylmethyl)phenol. *RSC Adv.* **2014**, *4*, 53520–53530. [CrossRef]

35. Kufareva, I.; Abagyan, R. Methods of Protein Structure Comparison. *Methods Mol. Biol.* **2012**, *857*, 231–257. [CrossRef]
36. Russell, R.; Alber, F.; Aloy, P.; Davis, F.; Korkin, D.; Pichaud, M.; Topf, M.; Sali, A. A structural perspective on protein-protein interactions. *Curr. Opin. Struct. Biol.* **2004**, *14*, 313–324. [CrossRef]
37. Dagliyan, O.; Proctor, E.A.; D'Auria, K.M.; Ding, F.; Dokholyan, N.V. Structural and dynamic determinants of protein-peptide recognition. *Structure* **2011**, *19*, 1837–1845. [CrossRef] [PubMed] [PubMed Central]
38. Guedes, I.A.; Pereira, F.S.S.; Dardenne, L.E. Empirical Scoring Functions for Structure-Based Virtual Screening: Applications, Critical Aspects, and Challenges. *Front. Pharmacol.* **2018**, *9*, 1089. [CrossRef] [PubMed] [PubMed Central]
39. Desantis, F.; Miotto, M.; Di Rienzo, L.; Milanetti, E.; Ruocco, G. Spatial organization of hydrophobic and charged residues affects protein thermal stability and binding affinity. *Sci. Rep.* **2022**, *12*, 12087. [CrossRef] [PubMed] [PubMed Central]
40. Bitencourt-Ferreira, G.; Veit-Acosta, M.; De Azevedo, W., Jr. Van der Waals Potential in Protein Complexes. *Methods Mol. Biol.* **2019**, *2053*, 79–91.
41. Cramer, J.; Krimmer, S.; Heine, A.; Klebe, G. Paying the Price of Desolvation in Solvent-Exposed Protein Pockets: Impact of Distal Solubilizing Groups on Affinity and Binding Thermodynamics in a Series of Thermolysin Inhibitors. *J. Med. Chem.* **2017**, *60*, 5791–5799. [CrossRef]
42. Zhou, H.-X.; Pang, X. Electrostatic Interactions in Protein Structure, Folding, Binding, and Condensation. *Chem. Rev.* **2018**, *118*, 1691–1741. [CrossRef] [PubMed]
43. Kastritis, P.L.; Rodrigues, J.P.G.L.M.; Bonvin, A.M.J.J. HADDOCK2P2I: A Biophysical Model for Predicting the Binding Affinity of Protein–Protein Interaction Inhibitors. *J. Chem. Inf. Model.* **2014**, *54*, 826–836. [CrossRef] [PubMed]
44. Kastritis, P.L.; Bonvin, A.M. On the binding affinity of macromolecular interactions: Daring to ask why proteins interact. *J. R. Soc. Interface* **2013**, *10*, 20120835. [CrossRef] [PubMed] [PubMed Central]
45. Dermawan, D.; Sumirtanurdin, R.; Dewantisari, D. Simulasi dinamika molekular reseptor estrogen alfa dengan andrografolid sebagai anti kanker payudara. *Indones J. Pharm. Sci. Technol.* **2019**, *6*, 65–76. [CrossRef]
46. Craveur, P.; Joseph, A.P.; Esque, J.; Narwani, T.J.; Noël, F.; Shinada, N.; Goguet, M.; Leonard, S.; Poulain, P.; Bertrand, O.; et al. Protein flexibility in the light of structural alphabets. *Front. Mol. Biosci.* **2015**, *2*, 20. [CrossRef] [PubMed] [PubMed Central]
47. Maspero, E.; Mari, S.; Valentini, E.; Musacchio, A.; Fish, A.; Pasqualato, S.; Polo, S. Structure of the HECT:ubiquitin complex and its role in ubiquitin chain elongation. *EMBO Rep.* **2011**, *12*, 342–349. [CrossRef] [PubMed] [PubMed Central]
48. Sanusi, Z.K.; Lobb, K.A. Insights into the Dynamics and Binding of Two Polyprotein Substrate Cleavage Points in the Context of the SARS-CoV-2 Main and Papain-like Proteases. *Molecules* **2022**, *27*, 8251. [CrossRef] [PubMed] [PubMed Central]
49. Kryshtafovych, A.; Schwede, T.; Topf, M.; Fidelis, K.; Moult, J. Critical assessment of methods of protein structure prediction (CASP)-Round XIII. *Proteins* **2019**, *87*, 1011–1020. [CrossRef] [PubMed] [PubMed Central]
50. Lenselink, E.B.; Louvel, J.; Forti, A.F.; van Veldhoven, J.P.D.; de Vries, H.; Mulder-Krieger, T.; McRobb, F.M.; Negri, A.; Goose, J.; Abel, R.; et al. Predicting Binding Affinities for GPCR Ligands Using Free-Energy Perturbation. *ACS Omega* **2016**, *1*, 293–304. [CrossRef] [PubMed] [PubMed Central]
51. Salsbury, F.R., Jr. Molecular dynamics simulations of protein dynamics and their relevance to drug discovery. *Curr. Opin. Pharmacol.* **2010**, *10*, 738–744. [CrossRef] [PubMed] [PubMed Central]
52. Lear, T.B.; McKelvey, A.C.; Evankovich, J.W.; Rajbhandari, S.; Coon, T.A.; Dunn, S.R.; Londino, J.D.; McVerry, B.J.; Zhang, Y.; Valenzi, E.; et al. KIAA0317 regulates pulmonary inflammation through SOCS2 degradation. *JCI Insight* **2019**, *4*, e129110. [CrossRef] [PubMed] [PubMed Central]
53. Bulatov, E.; Ciulli, A. Targeting Cullin–RING E3 ubiquitin ligases for drug discovery: Structure, assembly and small-molecule modulation. *Biochem. J.* **2015**, *467*, 365–386. [CrossRef] [PubMed]
54. Piessevaux, J.; Lavens, D.; Montoye, T.; Wauman, J.; Catteeuw, D.; Vandekerckhove, J.; Belsham, D.; Peelman, F.; Tavernier, J. Functional Cross-modulation between SOCS Proteins Can Stimulate Cytokine Signaling. *J. Biol. Chem.* **2006**, *281*, 32953–32966. [CrossRef] [PubMed]
55. Boore, J.L.; Brown, W.M. Complete sequence of the mitochondrial DNA of the annelid worm Lumbricus terrestris. *Genetics* **1995**, *141*, 305–319. [CrossRef] [PubMed] [PubMed Central]
56. Parkos, C.A.; Dinauer, M.C.; Walker, L.E.; Allen, R.A.; Jesaitis, A.J.; Orkin, S.H. Primary structure and unique expression of the 22-kilodalton light chain of human neutrophil cytochrome b. *Proc. Natl. Acad. Sci. USA* **1988**, *85*, 3319–3323. [CrossRef] [PubMed] [PubMed Central]
57. Nguyen, Q.T.T.; Rhee, H.; Kim, M.; Lee, M.Y.; Lee, E.J. Lumbrokinase, a Fibrinolytic Enzyme, Prevents Intra-Abdominal Adhesion by Inhibiting the Migrative and Adhesive Activities of Fibroblast via Attenuation of the AP-1/ICAM-1 Signaling Pathway. *Biomed. Res. Int.* **2023**, *2023*, 4050730. [CrossRef] [PubMed] [PubMed Central]
58. Chen, S.; Fan, G.; Li, J. Improving completeness and accuracy of 3D point clouds by using deep learning for applications of digital twins to civil structures. *Adv. Eng. Inform.* **2023**, *58*, 102196. [CrossRef]
59. Davis, E.M.; Sun, Y.; Liu, Y.; Kolekar, P.; Shao, Y.; Szlachta, K.; Mulder, H.L.; Ren, D.; Rice, S.V.; Wang, Z.; et al. SequencErr: Measuring and suppressing sequencer errors in next-generation sequencing data. *Genome Biol.* **2021**, *22*, 37. [CrossRef] [PubMed]
60. Faits, T.; Odom, A.; Castro-Nallar, E.; Crandall, K.; Johnson, W. Metagenomic profiling pipelines improve taxonomic classification for 16S amplicon sequencing data2022. *Sci. Rep.* **2023**, *13*, 13957.
61. Zhang, Y. I-TASSER server for protein 3D structure prediction. *BMC Bioinform.* **2008**, *9*, 40. [CrossRef]

62. Tian, W.; Chen, C.; Lei, X.; Zhao, J.; Liang, J. CASTp 3.0: Computed atlas of surface topography of proteins. *Nucleic Acids Res.* **2018**, *46*, 363–367. [CrossRef] [PubMed]
63. Kung, W.-W.; Ramachandran, S.; Makukhin, N.; Bruno, E.; Ciulli, A. Structural insights into substrate recognition by the SOCS2 E3 ubiquitin ligase. *Nat. Commun.* **2019**, *10*, 2534. [CrossRef] [PubMed]
64. Laskowski, R.A.; Jabłońska, J.; Pravda, L.; Vařeková, R.S.; Thornton, J.M. PDBsum: Structural summaries of PDB entries. *Protein Sci.* **2018**, *27*, 129–134. [CrossRef] [PubMed] [PubMed Central]
65. Johansson, M.U.; Zoete, V.; Michielin, O.; Guex, N. Defining and searching for structural motifs using DeepView/Swiss-PdbViewer. *BMC Bioinform.* **2012**, *13*, 173. [CrossRef] [PubMed] [PubMed Central]
66. Gitlin, A.D.; Maltzman, A.; Kanno, Y.; Heger, K.; Reja, R.; Schubert, A.F.; Wierciszewski, L.J.; Pantua, H.; Kapadia, S.B.; Harris, S.F.; et al. N4BP1 coordinates ubiquitin-dependent crosstalk within the IκB kinase family to limit Toll-like receptor signaling and inflammation. *Immunity* **2024**, *57*, 973–986. [CrossRef]
67. Ramachandran, S.; Makukhin, N.; Haubrich, K.; Nagala, M.; Forrester, B.; Lynch, D.M.; Casement, R.; Testa, A.; Bruno, E.; Gitto, R.; et al. Structure-based design of a phosphotyrosine-masked covalent ligand targeting the E3 ligase SOCS2. *Nat. Commun.* **2023**, *14*, 6345. [CrossRef]
68. Dominguez, C.; Boelens, R.; Bonvin, A.M.J.J. HADDOCK: A Protein−Protein Docking Approach Based on Biochemical or Biophysical Information. *J. Am. Chem. Soc.* **2003**, *125*, 1731–1737. [CrossRef]
69. van Zundert, G.C.P.; Rodrigues, J.; Trellet, M.; Schmitz, C.; Kastritis, P.L.; Karaca, E.; Melquiond, A.S.J.; van Dijk, M.; de Vries, S.J.; Bonvin, A.; et al. The HADDOCK2.2 Web Server: User-Friendly Integrative Modeling of Biomolecular Complexes. *J. Mol. Biol.* **2016**, *428*, 720–725. [CrossRef] [PubMed]
70. Pronk, S.; Páll, S.; Schulz, R.; Larsson, P.; Bjelkmar, P.; Apostolov, R.; Shirts, M.R.; Smith, J.C.; Kasson, P.M.; van der Spoel, D.; et al. GROMACS 4.5: A high-throughput and highly parallel open source molecular simulation toolkit. *Bioinformatics* **2013**, *29*, 845–854. [CrossRef]
71. Robertson, M.J.; Tirado-Rives, J.; Jorgensen, W.L. Improved Peptide and Protein Torsional Energetics with the OPLSAA Force Field. *J. Chem. Theory Comput.* **2015**, *11*, 3499–3509. [CrossRef] [PubMed] [PubMed Central]
72. Yuet, P.; Blankschtein, D. Molecular Dynamics Simulation Study of Water Surfaces: Comparison of Flexible Water Models. *J. Phys. Chem. B.* **2010**, *114*, 13786–13795. [CrossRef] [PubMed]
73. Yuan, S.; Chan, H.C.S.; Hu, Z. Using PyMOL as a platform for computational drug design. *WIREs Comput Mol Sci.* **2017**, *7*, e1298. [CrossRef]
74. Pettersen, E.F.; Goddard, T.D.; Huang, C.C.; Couch, G.S.; Greenblatt, D.M.; Meng, E.C.; Ferrin, T.E. UCSF Chimera—A visualization system for exploratory research and analysis. *J. Comput. Chem.* **2004**, *25*, 1605–1612. [CrossRef] [PubMed]
75. Tian, S.; Sun, H.; Pan, P.; Li, D.; Zhen, X.; Li, Y.; Hou, T. Assessing an ensemble docking-based virtual screening strategy for kinase targets by considering protein flexibility. *J. Chem. Inf. Model.* **2014**, *54*, 2664–2679. [CrossRef] [PubMed]
76. Yuan, Z.; Chen, X.; Fan, S.; Chang, L.; Chu, L.; Zhang, Y.; Wang, J.; Li, S.; Xie, J.; Hu, J.; et al. Binding Free Energy Calculation Based on the Fragment Molecular Orbital Method and Its Application in Designing Novel SHP-2 Allosteric Inhibitors. *Int. J. Mol. Sci.* **2024**, *25*, 671. [CrossRef]
77. Rifai, E.A.; Ferrario, V.; Pleiss, J.; Geerke, D.P. Combined Linear Interaction Energy and Alchemical Solvation Free-Energy Approach for Protein-Binding Affinity Computation. *J. Chem. Theory Comput.* **2020**, *16*, 1300–1310. [CrossRef]
78. Valdés-Tresanco, M.S.; Valdés-Tresanco, M.E.; Valiente, P.A. Moreno E gmx_MMPBSA: ANew Tool to Perform End-State Free Energy Calculations with, G.R.O.M.A.C.S. *J. Chem. Theory Comput.* **2021**, *17*, 6281–6291. [CrossRef]
79. Miller, B.R., 3rd; McGee, T.D., Jr.; Swails, J.M.; Homeyer, N.; Gohlke, H.; Roitberg, A.E. MMPBSA.py: Effic. Program End-State Free Energy Calculations. *J. Chem. Theory Comput.* **2012**, *8*, 3314–3321. [CrossRef]
80. Panday, S.K.; Alexov, E. Protein-Protein Binding Free Energy Predictions with the MM/PBSA Approach Complemented with the Gaussian-Based Method for Entropy Estimation. *ACS Omega* **2022**, *7*, 11057–11067. [CrossRef] [PubMed] [PubMed Central]
81. IBM. *IBM SPSS Statistics for Windows*, Version 25.0 ed.; IBM Corp: New York, NY, USA, 2017.
82. OriginLab. *Origin(Pro)*; 2022 ed.; OriginLab Corporation: Northampton, MA, USA, 2022.

International Journal of
Molecular Sciences

MDPI

Article

In Vitro Study of the Differential Anti-Inflammatory Activity of Dietary Phytochemicals upon Human Macrophage-like Cells as a Previous Step for Dietary Intervention

Antonio J. Ruiz-Alcaraz [1,*], Lorena Baquero [1], Paula Martínez Pérez-Munar [1], Alba Oliva-Bolarín [2], María A. Sánchez-Martínez [1,2], Bruno Ramos-Molina [2], María A. Núñez-Sánchez [2,*] and Diego A. Moreno [3]

[1] Department of Biochemistry, Molecular Biology B and Immunology, School of Medicine, University of Murcia, Regional Campus of International Excellence, 30120 Murcia, Spain; lorena.baqueroa@um.es (L.B.); paula.martinezp1@um.es (P.M.P.-M.); mariaantonia.martinez1@um.es (M.A.S.-M.)

[2] Obesity and Metabolism Laboratory, Biomedical Research Institute of Murcia (IMIB), 30120 Murcia, Spain; alba.oliva@imib.es (A.O.-B.); bruno.ramos@imib.es (B.R.-M.)

[3] Grupo Laboratorio de Fitoquímica y Alimentos Saludables (LabFAS), CEBAS-CSIC, Campus Universitario de Espinardo-25, 30100 Murcia, Spain; dmoreno@cebas.csic.es

* Correspondence: ajruiz@um.es (A.J.R.-A.); mariaa.nunez@imib.es (M.A.N.-S.); Tel.: +34-868889521 (A.J.R.-A.); +34-868881839 (M.A.N.-S.)

Abstract: Chronic inflammatory diseases pose a substantial health challenge globally, significantly contributing to morbidity and mortality. Addressing this issue requires the use of effective anti-inflammatory strategies with fewer side effects than those provoked by currently used drugs. In this study, a range of phytochemicals (phenolic di-caffeoylquinic acid (Di-CQA), flavonoid cyanidin-3,5-diglucoside (Cy3,5DiG), aromatic isothiocyanate sinalbin (SNB) and aliphatic isothiocyanate sulforaphane (SFN)) sourced from vegetables and fruits underwent assessment for their potential anti-inflammatory activity. An in vitro model of human macrophage-like cells treated with a low dose of LPS to obtain a low degree of inflammation that emulates a chronic inflammation scenario revealed promising results. Cell viability and production of the key pro-inflammatory cytokines were assessed in the presence of various phytochemicals. The compounds Di-CQA and Cy-3,5-DiG, within low physiologically relevant doses, demonstrated notable anti-inflammatory effects by significantly reducing the production of key pro-inflammatory cytokines TNF-α and IL-6 without affecting cell viability. These findings underscore the potential of plant-derived bioactive compounds as valuable contributors to the prevention or treatment of chronic inflammatory diseases. These results suggest that these compounds, whether used individually or as part of natural mixtures, hold promise for their inclusion in nutritional interventions designed to mitigate inflammation in associated pathologies.

Keywords: bioactive phytochemicals; anti-inflammatory agents; inflammatory diseases; glucosinolates; isothiocyanate; anthocyanins; phenolic acids

1. Introduction

Human pathologies related to chronic inflammation represent relevant social and medical burdens, causing high morbidity and mortality worldwide [1]. The inflammatory process is part of the natural response of the organism to maintain homeostasis against external insults and tissue damage, playing a major role in tissue repair. However, if this acute process does not resolve properly, it may evolve into a chronic inflammatory scenario, which in turn may cause the appearance of different related diseases, such as rheumatoid arthritis and diabetes mellitus, among others [2]. In this aggravation of the pathophysiological situation, a series of cells, such as macrophages, and soluble mediators produced by immune system cells, including pro-inflammatory cytokines tumor necrosis factor alpha (TNF-α) and interleukin 6 (IL-6), play a key role [3,4].

Currently, non-steroidal anti-inflammatory drugs, or NSAIDs, represent the preferred treatment for inflammatory diseases, since they are able to inhibit the production of important proinflammatory mediators such as prostaglandins [5]. However, there is a great effort to develop alternatives to NSAIDs, as many of these agents present a variety of adverse effects, including cardiovascular, gastrointestinal, and renal toxicity [6,7]. This fact encourages the search for new natural therapeutic approaches with fewer side effects that may be of use as coadjuvants in the treatment of inflammatory diseases. In this regard, bioactive phytochemicals and ingredients derived from them, obtained from fruits and vegetables from the diet, have great anti-inflammatory potential and may contribute to preventing or even treating these pathologies [8].

Glucosinolates (GSLs) are amino acid-derived compounds classified according to their precursor amino acid (aliphatic-Met, Ala, Leu, Ile, or Val; aromatic–Phe or Tyr; and indolics, from Trp). Aliphatic GSLs are predominant in broccoli and cabbages (glucoraphanin, the parental GSL of sulforaphane (SFN)). On the other hand, mustards, especially oil-rich varieties, are rich in aromatic GSL, mainly glucosinalbin, and its hydrolysis yields, 4-hydroxybenzyl isothiocyanate, the compound responsible for its sharp flavor [9,10]. Another relevant mustard GSL is sinalbin (SNB), which, to the best of our knowledge, has not been mentioned in any publication related to human inflammation to the present date. The evaluation of the potential bioactivity of SNB as an anti-inflammatory agent would be useful in discriminating the specific bioactivity of isothiocyanates and GSL.

On the other hand, one of the most important representatives of bioactive isothiocyanates in human health is SFN, which has been described to exert its main function through the nuclear factor erythroid 2–related factor 2 (NRF2) intracellular pathway. SFN binds to KEAP1 (Kelch-like ECH-associated protein 1), the natural suppressor of NRF2, allowing the transcription of Phase II metabolism enzymes [11]. Furthermore, SFN is also able to interfere with the activity of the nuclear factor kappa-light-chain-enhancer of activated B cells (NFκB), decreasing its capability to bind target sequences of genes related to inflammatory processes, such as those codifying for TNF-α [12]. Thus, this ingredient can be taken as a gold standard of a bioactive anti-inflammatory natural compound to which compare the potency of novel potentially interesting phytochemicals.

Chlorogenic acids are very ubiquitous in nature and in plant-origin foods (vegetables and fruits, e.g., artichokes, berries), classically studied for their antioxidant capacity [13,14], and, more recently, evaluated for their potential anti-inflammatory effects mediated by the inhibition of related key enzymes (e.g., COX-2) and the release of pro-inflammatory cytokines (e.g., TNF-α, IL-6, and IL-1β) [15,16]. Similarly, colored flavonoids (anthocyanins), including acylated cyanidins such as those found in Brassicas and berries (e.g., cyanidin-3,5-diglucoside, Cy3,5DiG) have also been found to exert antioxidant, anti-inflammatory, anticancer, anti-diabetic, and cardiovascular protective activity [17]. Di-glycosylated anthocyanins diminished the induction of pro-inflammatory cytokines such as IL-1β, TNF-α, and IL-6, which awakened the interest in evaluating structure–activity relationships between these substances and their potential as agents for preventing inflammation and related conditions [16].

The hypothesis of this study is that several phytochemicals present in vegetables and fruits (e.g., green leafy vegetables, tropical fruits, etc.) may present relevant bioactivity of interest for human health as anti-inflammatory agents, and to prove this hypothesis, we tested and verified the capacity of single compounds to inhibit the production of key pro-inflammatory cytokines TNF-α and IL-6 by human macrophage-like cells in vitro within low physiologically-relevant dosing and without any cytotoxic effects. This preliminary evaluation supposes a basic requirement for additional research on the potential synergistic effects of combinations of the bioactive compounds for the design of further investigations at the preclinical and clinical levels.

2. Results

2.1. Effect of Phytochemicals Sulforaphane, Sinalbin, Di-Caffeoylquinic Acid and Cyanidin-3,5-Diglucoside on Human Macrophage-like Cell Viability

None of the tested compounds showed a significant effect on the viability of the human macrophage-like HL-60 differentiated cells in this in vitro model of low-degree inflammation at any of the tested doses (10 μM, 25 μM, and 50 μM). The only exception was SFN, which produced a significant reduction of about 30.4% in cell viability , but only at its highest dose (50 μM). This result was not unexpected as it was similar to what was observed in our previous studies [17] (Table 1).

Table 1. Effect of phytochemicals Sulforaphane, Sinalbin, Di-caffeoylquinic acid and Cyanidin-3,5-diglucoside on human macrophage-like HL-60 cell viability under low degree inflammatory conditions.

	Dose (μM)		
Compound	**10**	**25**	**50**
SFN	117.0 ± 13.1	101.8 ± 11.4	69.6 ± 9.51 *
SNB	123.9 ± 13.5	122.9 ± 8.8	126.6 ± 10.2
Di-CQA	138.5 ± 4.4	130.9 ± 8.4	120.2 ± 8.7
Cy3,5DiG	117.9 ± 5.7	106.5 ± 7.1	95.5 ± 6.8

Results represent viability (%) as the mean ± SEM following 24 h exposure of HL-60 differentiated cells to the tested compounds at the different doses indicated plus LPS (0.1 μg/mL). Data were obtained from three different assays (N = 3). * $p < 0.05$. Cy3,5DiG: cyanidin-3,5-diglucoside; Di-CQA: di-caffeoylquinic acid; SFN: sulforaphane; SNB: sinalbin.

2.2. Effect of Phytochemicals Sulforaphane, Sinalbin, Di-Caffeoylquinic Acid, and Cyanidin-3,5-Diglucoside on TNF-α Production

To evaluate the anti-inflammatory potential of the phytochemicals selected, we performed an ELISA analysis to determine the effect of the different compounds on the production of TNF-α and IL-6 in an LPS-induced inflammatory model of HL-60 cells. Our results showed that after 24 h, the treatment of macrophage-like differentiated HL-60 cells with SFN significantly reduced the LPS-induced production of TNF-α in a dose-dependent manner (*), lowering TNF-α production to similar levels as those observed in cells not exposed to LPS (negative control, cytokine normalized level of 100% represented as a dotted line) (Figure 1A). Furthermore, significant differences were also observed in TNF-α levels between the lowest (10 μM) and the highest (50 μM) doses (§ $p = 0.03$) (Figure 1A). On the other hand, the exposure of the cells to the aliphatic GSL SNB significantly reduced the production of TNF-α induced by LPS only with the highest tested dose of 50 μM down to basal levels (* $p = 0.02$) (Figure 1B). Regarding the hydroxycinnamic acid Di-CQA, the results showed a significant reduction in the LPS-induced TNF-α production both at 25 μM and 50 μM doses after 24 h treatment (down to 97%, ** $p < 0.01$; and 24%, *** $p < 0.001$; respectively) following a dose-response trend with significant differences between the higher dose (50 μM) and the lower doses tested (§ $p < 0.05$ compared to 25 μM, §§§ $p < 0.001$ compared to 10 μM) (Figure 1C). Remarkably, the reduction in TNF-α levels in cells treated with 50 μM Di-CQA was also significant compared to the negative control of untreated cells ($^{\#}$ $p < 0.05$) (Figure 1C). Similarly, the cells exposed to the anthocyanin Cy3,5DiG showed a similar pattern as those treated with Di-CQA, with significant reductions in TNF-α levels after 24 h of treatments with 25 and 50 μM, showing an increasing dose-response effect with significant differences between all the tested doses (§ $p < 0.05$ for 10 μM vs. 25 μM; §§ $p < 0.01$ for 25 μM vs. 50 μM; and, §§§ $p < 0.001$ for 10 μM vs. 50 μM), and a further significant reduction with the highest dose of 50 μM compared to control non-inflammatory conditions ($^{\#}$ $p < 0.05$) (Figure 1D).

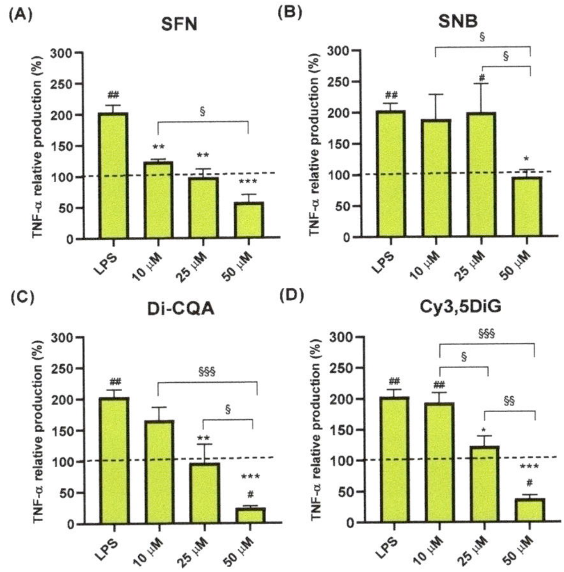

Figure 1. Effect of Sulforaphane, Sinalbin, Di-caffeoylquinic acid, and Cyanidin-3,5-diglucoside on TNF-α production. Data are presented as percentage compared to the negative control (no treatment) valued as 100% from three in vitro assays with different doses (10 μM, 25 μM and 50 μM) of SFN (**A**), SNB (**B**), Di-CQA (**C**) and Cy3,5DiG (**D**) for 24 h in the presence of LPS (0.1 μg/mL) as a pro-inflammatory stimulus (N = 3). Values of p were calculated using the ANOVA test with a post hoc Tukey's HSD test. Statistically significant differences are represented as: * $p < 0.05$, ** $p < 0.01$ and *** $p < 0.001$, compared to LPS treatment; # $p < 0.05$ and ## $p < 0.01$, compared to negative control; § $p < 0.05$, §§ $p < 0.01$ and §§§ $p < 0.001$, for comparisons between different doses of compounds. Dot lines represent cytokine normalized level of 100% corresponding to negative control. SFN, Sulforaphane; SNB, Sinalbin; Cy3,5DiG, Cyanidin-3,5-diglucoside; Di-CQA, Di-caffeoylquinic acid; LPS, lipopolysaccharide; TNF-α, tumor necrosis factor alpha.

2.3. Effect of Phytochemicals Sulforaphane, Sinalbin, Di-Caffeoylquinic Acid, and Cyanidin-3,5-Diglucoside on IL-6 Production

The exposure of macrophage-like cells to SFN significantly and drastically reduced the LPS-induced levels of IL-6 at all the tested concentrations (Figure 2A). However, contrary to the dose-dependent effect observed in SFN-treated cells in TNF-α production, even the exposure to the lower dose of 10 μM was already sufficient to reach the maximum reduction in IL-6 production after 24 h (Figure 2A). Furthermore, such reduction was not only significantly lower than that observed in LPS-stimulated proinflammatory *conditions* (*** $p < 0.001$) but also compared to basal conditions of non-stimulated cells (### $p < 0.001$, dotted line) (Figure 2A). On the other hand, the treatment of cells with SNB did not significantly affect the LPS-induced production of IL-6 at any of the tested doses (Figure 2B). In the case of Di-CQA, the production of IL-6 was only affected by the higher dose of 50 μM, lowering this cytokine levels by about 50% when compared to the negative control of unstimulated cells (## $p < 0.01$) and 75% compared to LPS-stimulated cells (** $p < 0.01$) (Figure 2C). On the other hand, no effect was observed after exposure to the 10 μM and 25 μM doses, and, furthermore, the effect of the 50 μM dose was also significantly lower compared to the 10 μM and 25 μM doses (§§§ $p < 0.001$ for 10 μM vs. 50 μM; and, § $p < 0.05$ for 25 μM vs. 50 μM) (Figure 2C). Finally, the treatment of cells with the anthocyanin Cy3,5DiG showed a similar pattern to that of Di-CQA since only the treatment with the dose of 50 μM significantly reduced the production of IL-6 compared to both LPS-stimulated (** $p < 0.01$) and basal conditions (# $p < 0.05$), while no effect was observed with minor doses of Cy3,5DiG. Furthermore, such an effect of the 50 μM dose of Cy3,5-CQA was also significantly lower when compared to the treatments with 10 μM and 25 μM (§§ $p < 0.01$ for 10 μM vs. 50 μM; and § $p < 0.05$ for 25 μM vs. 50 μM) (Figure 2D).

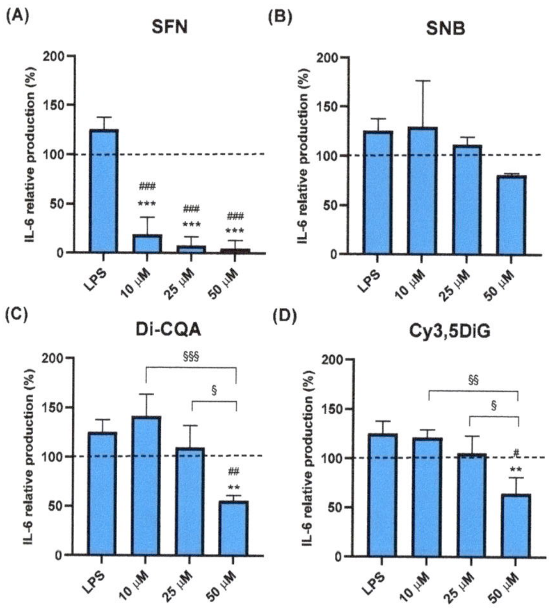

Figure 2. Effect of Sulforaphane, Sinalbin, Di-caffeoylquinic acid, and Cyanidin-3,5-diglucoside on IL-6 production. Data are presented as percentages compared to the negative control (no treatment) valued as 100% from three in vitro assays with different doses (10 μM, 25 μM, and 50 μM) of SFN (**A**), SNB (**B**), Di-CQA (**C**), and Cy3,5DiG (**D**) for 24 h in the presence of LPS (0.1 μg/mL) as a pro-inflammatory stimulus (N = 3). Values of p were calculated using the ANOVA test with a post hoc Tukey's HSD test. Statistically significant differences are represented as ** $p < 0.01$ and *** $p < 0.001$ compared to LPS treatment; # $p < 0.05$, ## $p < 0.01$ and ### $p < 0.001$ compared to the negative control; and § $p < 0.05$, §§ $p < 0.01$ and §§§ $p < 0.001$ for comparisons between different doses of compounds. Dotted lines represent normalized cytokine levels of 100% corresponding to the negative control. SFN, sulforaphane; SNB, sinalbin; Cy3,5DiG, cyanidin-3,5-diglucoside; Di-CQA, di-caffeoylquinic acid; IL-6, interleukin 6; LPS, lipopolysaccharide.

3. Discussion

Herein, we have assessed, for the first time, the potential anti-inflammatory bioactivity of a series of phytochemicals in a well-established in vitro model of human macrophage-like cells obtained from the differentiation of HL-60 cells maintained in a state of a low degree of inflammation after stimulation with a low dose of LPS for 24 hours that emulates chronic inflammation scenario. Our results show that two of those compounds, Di-CQA and Cy-3,5-DiG, promote remarkable anti-inflammatory activity, since both were capable of reducing the production of the systemic pro-inflammatory cytokines TNF-α and IL-6. These results support the use of plant-derived bioactive compounds (e.g., nutraceuticals, functional ingredients) to help prevent or treat chronic human inflammatory diseases [8]. Furthermore, this might help design natural mixtures of bioactive phytochemicals or determine single bioactive compounds that could be used as a new phytotherapeutic approach to treat inflammation-related pathologies.

The in vitro model used in this study, which was previously established and optimized by us, provides a reliable method to study the anti-inflammatory potential of compounds from vegetal origin in a scenario of low-degree inflammation resembling that found in several chronic inflammatory diseases [18]. Thus, we used this model to study the anti-inflammatory bioactivity of potentially active compounds present not only in cruciferous foods (e.g., sprouts, baby leaf varieties, cabbages, etc.) but also in a myriad of fruits and vegetables as sources of dichlorogenic acids (e.g., berries, artichoke, coffee, apples, etc.) in order to assess their potential to reduce the production and release of the systemic key pro-inflammatory cytokines TNF-α and IL-6. In addition, we included in this study a treatment with SFN as a positive control, as it is considered a gold standard of anti-inflammatory bioactive compounds from plant origin [19,20], and whose inflammatory activity has been confirmed here.

Our results showed that the new compounds tested had no effect on cell viability at any dose, thus showing no cytotoxicity whatsoever, while exerting high anti-inflammatory activity at the higher concentration tested, which was reflected in their ability to significantly reduce the production of TNF-α and IL-6, reducing the levels of these pro-inflammatory cytokines below the levels observed in control conditions of untreated cells. Furthermore, the hydroxycinnamic acid Di-CQA and the anthocyanin Cy3,5DiG were able to block the production of both TNF-α and IL-6 at basal levels when cells were treated with lower doses. It was also noteworthy that, at its highest dose (50 μM), Di-CQA was able to further reduce the production of TNF-α to a lower level than that observed with the gold standard SFN. Overall, the capacity of these compounds to efficiently inhibit the production of TNF-α and IL-6, especially Di-CQA and Cy3,5DiG, highlights their potential as anti-inflammatory agents capable of reducing the production of these pro-inflammatory factors by human macrophages that play a major function as inductors and maintainers of chronic inflammation in many related pathologies [21].

To the best of our knowledge, this is the first study reporting the anti-inflammatory effect of Di-CQA as a single agent capable of reducing the production and release of the key pro-inflammatory cytokines TNF-α and IL-6 by human macrophages in an in vitro model of differentiated HL-60 cells. Our data are in agreement with those reported by others who have shown the potential of different plant foods and derived products rich in chlorogenic acid derivatives, including Di-CQA, as anti-inflammatory agents and analgesics used in traditional medicine [22,23]. Furthermore, the results are also in line with previous data showing that complex leaf extracts rich in Di-CQA obtained from *Artemisia selengensis Turcz* could alleviate gout inflammation in an in vitro model of THP-1 cells by inhibiting the NLRP3 inflammasome and activating the NRF2 signaling pathway and thus the production of several pro-inflammatory cytokines [24]. This anti-inflammatory effect may be produced not only by the Di-CQA present in the leaf extracts but also by the many other phytochemicals present in the mix. Thus, our results now confirm that Di-CQA has a potent anti-inflammatory activity, per se, when used as a single agent.

Anthocyanins have been widely studied for their relevant bioactivity. The available literature supports the multiple functions of different natural products containing these types of compounds [25,26], with particular interest in the already proven bioactivity of their major representative, cyanidin-3-O-glucoside (C3G) [27]. This compound has not only been described as an anti-inflammatory agent [18,28], but it can also act as a potential anti-diabetic [29,30], anti-cancer [31], and antioxidant compound, among other functions [32–34]. In this study, we have focused on the analysis of a less-known anthocyanin, Cy3,5DiG, which has been previously predicted in silico to be one of the bioactive agents responsible for the reduction in the levels of IL-6, and, therefore, of the anti-inflammatory activity, of *Ficus carica* [35]. Here, we have confirmed the capacity of Cy3,5DiG to reduce not only the production of IL-6 by human macrophages in vitro as a single agent but also the production of TNF-α, one of the other major pro-inflammatory cytokines mediating inflammatory-related pathologies such as Crohn's disease (CD) [36] and rheumatoid arthritis (RA) [37].

Finally, the aromatic isothiocyanate SNB, which has been shown before to participate in the anti-proliferative and antimicrobial effects of mustard (*Sinapis*) extracts [38], showed the lowest anti-inflammatory bioactivity in our in vitro model of human macrophages. Although it was able to block the production of TNF-α and IL-6 when cells were treated with the highest dose tested, the level of inhibition reached by SNB was the lowest amongst the tested compounds. Nevertheless, this is the first time that SNB has been proven to present anti-inflammatory bioactivity upon human macrophages as a single agent, which adds to its capacity of acting as an antioxidant agent, which could make SNB an interesting phytochemical to consider for nutritional interventions [39].

Altogether, our results show how the studied phytochemicals may function individually as single agents with remarkable anti-inflammatory activities at low doses ($\leq$50 μM), which would be like those expected to be found in physiological conditions after the intake of food matrices rich in these bioactive ingredients. Thus, our data would be relevant to optimizing the incorporation of interesting "phytochemical cocktails" from vegetables and fruits, natural sources of the studied compounds, to nutritional interventions addressing the reduction in inflammation in related chronic pathologies.

4. Materials and Methods

4.1. Compounds

Phenolic di-caffeoylquinic acid (Di-CQA), flavonoid Cy3,5DiG, and aromatic isothiocyanate SNB were obtained from Phytoplan Diehm & Neuberger GmbH (Heidelberg, Germany), while the aliphatic isothiocyanate SFN was obtained from LKT Laboratories, Inc. (St. Paul, MN, USA) (Table 2).

Table 2. Compounds tested as anti-inflammatory agents.

Compound	Type	Abbreviation	Formula	Structure	Natural/Food Source
Sulforaphane	Aliphatic Isothiocyanate	SFN	$C_6H_{11}NOS_2$		Glucoraphanin from broccoli, broccolini, cabbage, etc.
Sinalbin	Aromatic Isothiocyanate	SNB	$C_{14}H_{18}NO_{10}S_2K$	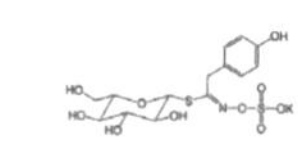	Glucosinalbin from mustards, horseradish, watercress, etc.
Di-caffeoylquinic acid	Hydroxycinnamic acid (phenolic acid)	Di-CQA	$C_{25}H_{24}O_{12}$	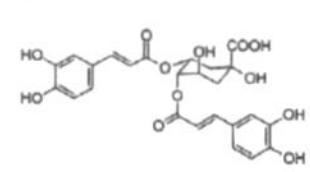	Chlorogenic acids in mustards, artichoke, eggplant, bay leaf, etc.
Cyanidin-3,5-diglucoside	Anthocyanin (flavonoid)	Cy3,5DiG	$C_{27}H_{31}O_{16}Cl$	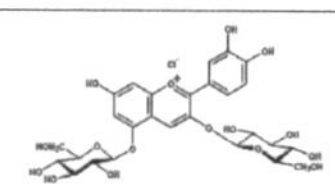	Natural pigments in berries, beans, sprouts, etc.

4.2. Compound Preparation for In Vitro Assays

A 1 mg/mL stock solution of the different compounds was freshly prepared directly in sterile phosphate-buffered saline (PBS) pH = 7.4 (Biowest, Nuaillé, France), with the only exception of SFN, which was first diluted in dimethyl sulfoxide (DMSO) (Merck, Whitehouse Station, NJ, USA), in order to get a final stock solution of 1 mg/mL SFN in PBS containing 10% DMSO. The subsequent dilutions were made at the appropriate concentration in a complete cell-culture medium (CCM) prior to being used in the in vitro assays.

4.3. Cell Culture

Phytochemical bioactivity was tested using macrophage cells obtained from the differentiation of the human acute myeloid leukemia cell line HL-60 (ATCC® CCL-240™, American Type Culture Collection, Rockville, MD, USA) as previously described [40]. In brief, undifferentiated HL-60 cells were first kept growing at an exponential ratio in CCM, RPMI-1640 (Biowest, Nuaillé, France), supplemented with 10% fetal bovine serum (GIBCO Invitrogen, Paisley, UK) and 1% penicillin/streptomycin (GIBCO) prior to their differentiation. Growing conditions were set at 37 °C and under an atmosphere of 5% CO_2 in the incubator. All the in vitro assays were performed on cells always differentiated between passages 5 and 20.

4.4. Cell Viability Assays

The viability of human macrophage-like cells obtained by the differentiation of HL-60 and treated with different doses of the tested phytochemicals was measured by MTT assay [37]. Briefly, after cell differentiation with 10 ng/mL phorbol myristate acetate (PMA; Sigma Chemical Co., St. Louis, MO, USA) in 96-well plates, the cells were pre-incubated with 10, 50, and 100 μg/mL of either of the tested phytochemicals, and after 30 min, the cells were treated with LPS (0.1 μg/mL) and incubated for 24 h. MTT (Alfa Aesar, Thermo Fisher, Karlsruhe, Germany) was then added to the final concentration of 483 μM and incubated for 1 h at 37 °C and under a 5% CO_2 atmosphere. Then, an acidified isopropanol solubilization solution of 0.04 M hydrochloric acid and 0.1% NP-40 detergent was added to each well to lysate the cells and dissolve the purple formazan crystal formed inside the cells, obtaining a purple-colored solution. The absorbance was measured at 550 nm in a plate-reading spectrophotometer (Spectrostar Nano; BMG Labtech, Ortenberg, Germany). The cell viability was determined by comparison with control conditions, setting the latter as 100% viability and thus 0% cytotoxicity. Control cells were maintained under basal conditions and treated only with CCM. Three replicates were performed for each experimental condition.

4.5. In Vitro Anti-Inflammatory Assays

The anti-inflammatory bioactivity of the phytochemicals was analyzed as previously described [17]. Briefly, the potential to modulate the secretion of pro-inflammatory cytokines in inflammatory-like conditions induced by bacterial lipopolysaccharide (LPS) in vitro was studied in a model of human macrophage-like cells obtained by the differentiation of HL-60 cells (human myeloid leukemia) with a treatment of 10 ng/mL of PMA (Sigma Chemical Co., St. Louis, MO, USA) for 24 h upon a ratio of $2 \cdot 10^5$ cells/well in 96-wells plates in CCM. Cells rested then in CCM without PMA for another 24 h. After differentiation and resting periods, tested compounds (or vehicles) were added to cells at doses of 10 μM, 25 μM, and 50 μM, and plates were pre-incubated for 30 min at 37 °C. Finally, LPS (Escherichia coli O111:B44; Sigma Chemical Co.) was added at a final concentration of 0.1 μg/mL to induce a low degree of pro-inflammatory conditions like those found in chronic inflammatory diseases. Plates were then incubated at 37 °C with 5% CO_2 for 24 h, and supernatants were collected and kept at −20 °C until further analysis. Three replicates were performed for each experiment.

4.6. Enzyme-Linked Immunosorbent Assays

Stored supernatants from in vitro assays performed on human macrophage-like HL-60 cells were used to determine the levels of the pro-inflammatory cytokines TNF-α and IL-6 using colorimetric enzyme-linked immunosorbent assay (ELISA) kits (Invitrogen, Thermo Fisher Scientific, Waltham, MA, USA). The absorbance in each well was determined using a microplate reader (Spectrostar Nano; BMG Labtech, Ortenberg, Germany) at 450 nm and corrected at 570 nm. Each cytokine concentration was calculated using the corresponding standard curve. The modulation of cytokines was attained by calculating the level of release inhibition when compared to control conditions. To compare the effect of the different tested phytochemicals, the concentrations of the cytokines obtained from enzyme-linked immunosorbent assays (ELISA) (within the ranges 40–560 pg/mL for TNF-α and 1.5–110 pg/mL for IL-6) were normalized and represented as the percentage of cytokine production for each individual assay by setting the negative control of the corresponding experiment as the reference, valued as 100%. The effect of the tested compounds upon cytokine levels in inflammatory-like conditions induced by LPS is presented as the percentage fold-change of normalized results compared to the negative control (100%, represented as a dotted line in the result figures).

4.7. Statistical Analysis

Data are reported as normalized percentages compared to the negative control (100%). Results are reported as the mean ± SEM and are graphically represented as histograms. Calculations were performed using GraphPad Prism 9 software. Statistical differences were analyzed by ANOVA and Tukey's HSD tests. Values of p under 0.05 were considered to indicate statistical significance.

5. Conclusions

In conclusion, some of the food phytochemicals individually tested in this study (SFN, Di-CQA, and Cy3,5DiG) have shown to be potentially active anti-inflammatory agents from a natural origin, as they were able to inhibit the production of key pro-inflammatory cytokines responsible for the establishment and maintenance of systemic chronic inflammation. Therefore, these data would be of great interest for the design of nutritional interventions that aim to ameliorate inflammatory diseases.

Author Contributions: Conceptualization, A.J.R.-A., B.R.-M., and D.A.M.; methodology, A.J.R.-A., B.R.-M., and D.A.M.; software, A.J.R.-A. and M.A.N.-S.; validation, A.J.R.-A., M.A.N.-S., B.R.-M., and D.A.M.; formal analysis, A.J.R.-A., L.B., and M.A.N.-S.; investigation, A.J.R.-A., L.B., P.M.P.-M., M.A.N.-S., A.O.-B., M.A.S.-M., B.R.-M., and D.A.M.; resources, A.J.R.-A., B.R.-M., and D.A.M.; data curation, A.J.R.-A. and M.A.N.-S.; writing—original draft preparation, A.J.R.-A., M.A.N.-S., B.R.-M., and D.A.M.; writing—review and editing, A.J.R.-A., L.B., P.M.P.-M., M.A.N.-S., B.R.-M., and D.A.M.; visualization, A.J.R.-A. and M.A.N.-S.; supervision, A.J.R.-A., M.A.N.-S., B.R.-M., and D.A.M.; project administration, A.J.R.-A., B.R.-M., and D.A.M.; funding acquisition, A.J.R.-A., B.R.-M., and D.A.M. All authors have read and agreed to the published version of the manuscript.

Funding: This research was funded by Fundación Séneca (grant numbers 20855/PI/18 (D.A.M.) and 22080/JLI/22 (M.A.N.-S)) and by the Biomedical Research Institute of Murcia (IMIB/CI/09, A.J.R-A). M.A.N.-S. is supported by the "Miguel Servet Type I" program (CP23/00051, ISCIII, Spain; co-funded by the Fondo Europeo de Desarrollo Regional-FEDER). M.A.M-S holds a PFIS predoctoral fellowship from the ISCIII (FI21/0003, ISCIII, Spain; co-funded by the Fondo Europeo de Desarrollo Regional-FEDER). B.R-M is supported by the "Miguel Servet Type I" program (CP19/00098, ISCIII, Spain; co-funded by the Fondo Europeo de Desarrollo Regional-FEDER). The funding organization played no role in the design of the study, review and interpretation of the data, or final approval of the manuscript.

Institutional Review Board Statement: Not applicable.

Informed Consent Statement: Not applicable.

Data Availability Statement: All research data are presented in the article.

Conflicts of Interest: The authors declare no conflicts of interest.

References

1. GBD 2017 Causes of Death Collaborators. Global, regional, and national age-sex-specific mortality for 282 causes of death in 195 countries and territories, 1980–2017: A systematic analysis for the Global Burden of Disease Study 2017. *Lancet* **2018**, *392*, 1736–1788. [CrossRef]
2. Scrivo, R.; Vasile, M.; Bartosiewicz, I.; Valesini, G. Inflammation as "common soil" of the multifactorial diseases. *Autoimmun. Rev.* **2011**, *10*, 369–374. [CrossRef] [PubMed]
3. Germolec, D.R.; Shipkowski, K.A.; Frawley, R.P.; Evans, E. Markers of Inflammation. *Methods Mol. Biol.* **2018**, *1803*, 57–79. [CrossRef] [PubMed]
4. Nathan, C.; Ding, A. Nonresolving inflammation. *Cell* **2010**, *140*, 871–882. [CrossRef]
5. Green, G.A. Understanding NSAIDs: From aspirin to COX-2. *Clin. Cornerstone* **2001**, *3*, 50–60. [CrossRef]
6. Knights, K.M.; Mangoni, A.A.; Miners, J.O. Defining the COX inhibitor selectivity of NSAIDs: Implications for understanding toxicity. *Expert Rev. Clin. Pharmacol.* **2010**, *3*, 769–776. [CrossRef]
7. Fine, M. Quantifying the impact of NSAID-associated adverse events. *Am. J. Manag. Care* **2013**, *19*, s267–s272. [PubMed]

8. Poyik, T.; Rasane, P.; Singh, J.; Kaur, S.; Kaur, J.; Gunjal, M.; Avinashe, H.; Dubey, N.; Mahato, D.S. Bioactive Compounds of Mustard, its Role in Consumer Health and in the Development of Potential Functional Foods. *Curr. Nutr. Food Sci.* **2023**, *19*, 950–960. [CrossRef]
9. García-Ibáñez, P.; Núñez-Sánchez, M.A.; Oliva-Bolarín, A.; Martínez.Sánchez, M.A.; Ramos-Molina, B.; Ruiz-Alcaraz, A.J.; Moreno, D.A. Anti-inflammatory potential of digested Brassica sprout extracts in human macrophage-like HL-60 cells. *Food Funct.* **2023**, *14*, 112–121. [CrossRef]
10. Guijarro-Real, C.; Hernandez-Canovas, L.; Abellan-Victorio, A.; Ben-Romdhane, O.; Moreno, D.A. The Combination of Monochromatic LEDs and Elicitation with Stressors Enhances the Accumulation of Glucosinolates in Mustard Sprouts with Species-Dependency. *Plants* **2022**, *11*, 2961. [CrossRef]
11. Russo, M.; Spagnuolo, C.; Russo, G.L.; Skalicka-Wozniak, K.; Daglia, M.; Sobarzo-Sanchez, E.; Nabavi, S.F.; Nabavi, S.M. Nrf2 targeting by sulforaphane: A potential therapy for cancer treatment. *Crit. Rev. Food Sci. Nutr.* **2018**, *58*, 1391–1405. [CrossRef] [PubMed]
12. Fahey, J.W.; Zhang, Y.; Talalay, P. Broccoli sprouts: An exceptionally rich source of inducers of enzymes that protect against chemical carcinogens. *Proc. Natl. Acad. Sci.USA* **1997**, *94*, 10367–10372. [CrossRef] [PubMed]
13. Gezer, C. Potential health effects if the popular compound of artichoke: Cyanarin. *Prog. Nutr.* **2017**, *19*, 5–9.
14. Tajik, N.; Tajik, M.; Mack, I.; Enck, P. The potential effects of chlorogenic acid, the main phenolic components in coffee, on health: A comprehensive review of the literature. *Eur. J. Nutr.* **2017**, *56*, 2215–2244. [CrossRef] [PubMed]
15. Oh, J.H.; Karadeniz, F.; Lee, J.I.; Seo, Y.; Kong, C.S. Protective effect of 3,5-dicaffeoyl-epi-quinic acid against UVB-induced photoaging in human HaCaT keratinocytes. *Mol. Med. Rep.* **2019**, *20*, 763–770. [CrossRef] [PubMed]
16. Wang, H.; Zhao, Y.; Zhang, Y.; Yang, T.; Zhao, S.; Sun, N.; Tan, H.; Zhang, H.; Wang, C.; Fan, H. Effect of Chlorogenic Acid via Upregulating Resolvin D1 Inhibiting the NF-kappaB Pathway on Chronic Restraint Stress-Induced Liver Inflammation. *J. Agric. Food Chem.* **2022**, *70*, 10532–10542. [CrossRef] [PubMed]
17. Overall, J.; Bonney, S.A.; Wilson, M.; Beermann, A.; Grace, M.H.; Esposito, D.; Lila, M.A.; Komarnytsky, S. Metabolic Effects of Berries with Structurally Diverse Anthocyanins. *Int. J. Mol. Sci.* **2017**, *18*, 422. [CrossRef]
18. Ruiz-Alcaraz, A.J.; Martinez-Sanchez, M.A.; Garcia-Penarrubia, P.; Martinez-Esparza, M.; Ramos-Molina, B.; Moreno, D.A. Analysis of the anti-inflammatory potential of Brassica bioactive compounds in a human macrophage-like cell model derived from HL-60 cells. *Biomed. Pharmacother.* **2022**, *149*, 112804. [CrossRef]
19. Wei, L.Y.; Zhang, J.K.; Zheng, L.; Chen, Y. The functional role of sulforaphane in intestinal inflammation: A review. *Food Funct.* **2022**, *13*, 514–529. [CrossRef]
20. Ruhee, R.T.; Suzuki, K. The Integrative Role of Sulforaphane in Preventing Inflammation, Oxidative Stress and Fatigue: A Review of a Potential Protective Phytochemical. *Antioxidants* **2020**, *9*, 521. [CrossRef]
21. Li, C.; Xu, M.M.; Wang, K.; Adler, A.J.; Vella, A.T.; Zhou, B. Macrophage polarization and meta-inflammation. *Transl. Res.* **2018**, *191*, 29–44. [CrossRef] [PubMed]
22. dos Santos, M.D.; Gobbo-Neto, L.; Albarella, L.; de Souza, G.E.; Lopes, N.P. Analgesic activity of di-caffeoylquinic acids from roots of Lychnophora ericoides (Arnica da serra). *J. Ethnopharmacol.* **2005**, *96*, 545–549. [CrossRef]
23. Siriwatanametanon, N.; Heinrich, M. The Thai medicinal plant Gynura pseudochina var. hispida: Chemical composition and in vitro NF-kappaB inhibitory activity. *Nat. Prod. Commun.* **2011**, *6*, 627–630. [CrossRef] [PubMed]
24. Cao, W.; Wu, T.; Liang, F.; Fang, Y.; Cheng, Y.; Pan, S.; Xu, X. Protective effects of di-caffeoylquinic acids from Artemisia selengensis Turcz leaves against monosodium urate-induced inflammation via the modulation of NLRP3 inflammasome and Nrf2 signaling pathway in THP-1 macrophages. *J. Food Biochem.* **2022**, *46*, e14252. [CrossRef] [PubMed]
25. Fragoso-Medina, J.A.; Lopez Vaquera, S.R.; Dominguez-Uscanga, A.; Luna-Vital, D.; Garcia, N. Single anthocyanins effectiveness modulating inflammation markers in obesity: Dosage and matrix composition analysis. *Front. Nutr.* **2023**, *10*, 1255518. [CrossRef]
26. Peng, Y.; Yan, Y.; Wan, P.; Dong, W.; Huang, K.; Ran, L.; Mi, J.; Lu, L.; Zeng, X.; Cao, Y. Effects of long-term intake of anthocyanins from Lycium ruthenicum Murray on the organism health and gut microbiota in vivo. *Food Res. Int.* **2020**, *130*, 108952. [CrossRef]
27. Deepa, P.; Hong, M.; Sowndhararajan, K.; Kim, S. A Review of the Role of an Anthocyanin, Cyanidin-3-O-beta-glucoside in Obesity-Related Complications. *Plants* **2023**, *12*, 3889. [CrossRef]
28. Li, X.; Sun, M.; Long, Y. Cyanidin-3-O-Glucoside Attenuates Lipopolysaccharide-Induced Inflammation in Human Corneal Epithelial Cells by Inducing Let-7b-5p-Mediated HMGA2/PI3K/Akt Pathway. *Inflammation* **2020**, *43*, 1088–1096. [CrossRef]
29. Molonia, M.S.; Occhiuto, C.; Muscara, C.; Speciale, A.; Bashllari, R.; Villarroya, F.; Saija, A.; Cimino, F.; Cristani, M. Cyanidin-3-O-glucoside restores insulin signaling and reduces inflammation in hypertrophic adipocytes. *Arch. Biochem. Biophys.* **2020**, *691*, 108488. [CrossRef]
30. Molonia, M.S.; Quesada-Lopez, T.; Speciale, A.; Muscara, C.; Saija, A.; Villarroya, F.; Cimino, F. In Vitro Effects of Cyanidin-3-O-Glucoside on Inflammatory and Insulin-Sensitizing Genes in Human Adipocytes Exposed to Palmitic Acid. *Chem. Biodivers.* **2021**, *18*, e2100607. [CrossRef]
31. Wu, C.F.; Wu, C.Y.; Lin, C.F.; Liu, Y.W.; Lin, T.C.; Liao, H.J.; Chang, G.R. The anticancer effects of cyanidin 3-O-glucoside combined with 5-fluorouracil on lung large-cell carcinoma in nude mice. *Biomed. Pharmacother.* **2022**, *151*, 113128. [CrossRef] [PubMed]
32. Yao, Y.; Zhang, X.; Xu, Y.; Zhao, Y.; Song, F.; Tian, Z.; Zhao, M.; Liang, Y.; Ling, W.; Mao, Y.H.; et al. Cyanidin-3-O-beta-Glucoside Attenuates Platelet Chemokines and Their Receptors in Atherosclerotic Inflammation of ApoE(-/-) Mice. *J. Agric. Food Chem.* **2022**, *70*, 8254–8263. [CrossRef] [PubMed]

33. Trinei, M.; Carpi, A.; Menabo, R.; Storto, M.; Fornari, M.; Marinelli, A.; Minardi, S.; Riboni, M.; Casciaro, F.; DiLisa, F.; et al. Dietary intake of cyanidin-3-glucoside induces a long-lasting cardioprotection from ischemia/reperfusion injury by altering the microbiota. *J. Nutr. Biochem.* **2022**, *101*, 108921. [CrossRef] [PubMed]
34. Zhou, Y.; Wang, S.; Wan, T.; Huang, Y.; Pang, N.; Jiang, X.; Gu, Y.; Zhang, Z.; Luo, J.; Yang, L. Cyanidin-3-O-beta-glucoside inactivates NLRP3 inflammasome and alleviates alcoholic steatohepatitis via SirT1/NF-kappaB signaling pathway. *Free. Radic. Biol. Med.* **2020**, *160*, 334–341. [CrossRef] [PubMed]
35. Mansoor, H.U.H.; Ahmed, A.; Rasool, F. The inhibitory potential of chemical constituents of Ficus carica targeting interleukin-6 (IL-6) mediated inflammation. *Cell Biochem. Funct.* **2023**, *41*, 573–589. [CrossRef]
36. Bamias, G.; Corridoni, D.; Pizarro, T.T.; Cominelli, F. New insights into the dichotomous role of innate cytokines in gut homeostasis and inflammation. *Cytokine* **2012**, *59*, 451–459. [CrossRef]
37. Jang, D.I.; Lee, A.H.; Shin, H.Y.; Song, H.R.; Park, J.H.; Kang, T.B.; Lee, S.R.; Yang, S.H. The Role of Tumor Necrosis Factor Alpha (TNF-alpha) in Autoimmune Disease and Current TNF-alpha Inhibitors in Therapeutics. *Int. J. Mol. Sci.* **2021**, *22*, 2719. [CrossRef]
38. Boscaro, V.; Boffa, L.; Binello, A.; Amisano, G.; Fornasero, S.; Cravotto, G.; Gallicchio, M. Antiproliferative, Proapoptotic, Antioxidant and Antimicrobial Effects of Sinapis nigra L. and Sinapis alba L. Extracts. *Molecules* **2018**, *23*, 3004. [CrossRef]
39. Natella, F.; Maldini, M.; Leoni, G.; Scaccini, C. Glucosinolates redox activities: Can they act as antioxidants? *Food Chem.* **2014**, *149*, 226–232. [CrossRef]
40. Guirado, A.; Lopez Sanchez, J.I.; Ruiz-Alcaraz, A.J.; Bautista, D.; Galvez, J. Synthesis and biological evaluation of 4-alkoxy-6,9-dichloro[1,2,4]triazolo[4,3-a]quinoxalines as inhibitors of TNF-alpha and IL-6. *Eur. J. Med. Chem.* **2012**, *54*, 87–94. [CrossRef]

International Journal of
Molecular Sciences

MDPI

Article

Exploring the Effects and Potential Mechanisms of Hesperidin for the Treatment of CPT-11-Induced Diarrhea: Network Pharmacology, Molecular Docking, and Experimental Validation

Xinyao Shu †, Ruitong Xu †, Peiyu Xiong, Junyu Liu, Zubing Zhou, Tao Shen * and Xiaobo Zhang *

School of Basic Medicine, Chengdu University of Traditional Chinese Medicine, Chengdu 611137, China; shuxinyao0341@163.com (X.S.); xuruitong@stu.cdutcm.edu.cn (R.X.); xiongpeiyu@stu.cdutcm.edu.cn (P.X.); junyuliu1007@stu.cdutcm.edu.cn (J.L.); zhouzubing@stu.cdutcm.edu.cn (Z.Z.)

* Correspondence: shentaotcm@aliyun.com (T.S.); zhangxiaobo@stu.cdutcm.edu.cn (X.Z.)

† These authors contributed equally to this work.

Abstract: Chemotherapy-induced diarrhea (CID) is a potentially serious side effect that often occurs during anticancer therapy and is caused by the toxic effects of chemotherapeutic drugs on the gastrointestinal tract, resulting in increased frequency of bowel movements and fluid contents. Among these agents, irinotecan (CPT-11) is most commonly associated with CID. Hesperidin (HPD), a flavonoid glycoside found predominantly in citrus fruits, has anti-oxidation properties and anti-inflammation properties that may benefit CID management. Nevertheless, its potential mechanism is still uncertain. In this study, we firstly evaluated the pharmacodynamics of HPD for the treatment of CID in a mouse model, then used network pharmacology and molecular docking methods to excavate the mechanism of HPD in relieving CID, and finally further proved the predicted mechanism through molecular biology experiments. The results demonstrate that HPD significantly alleviated diarrhea, weight loss, colonic pathological damage, oxidative stress, and inflammation in CID mice. In addition, 74 potential targets for HPD intervention in CID were verified by network pharmacology, with the top 10 key targets being AKT1, CASP3, ALB, EGFR, HSP90AA1, MMP9, ESR1, ANXA5, PPARG, and IGF1. The Kyoto Encyclopedia of Genes and Genomes (KEGG) enrichment analysis revealed that the PI3K–Akt pathway, FoxO pathway, MAPK pathway, TNF pathway, and Ras pathway were most relevant to the HPD potential treatment of CID genes. The molecular docking results showed that HPD had good binding to seven apoptosis-related targets, including AKT1, ANXA5, CASP3, HSP90AA1, IGF1, MMP9, and PPARG. Moreover, we verified apoptosis by TdT-mediated dUTP nick-end labeling (TUNEL) staining and immunohistochemistry, and the hypothesis about the proteins above was further verified by Western blotting in vivo experiments. Overall, this study elucidates the potential and underlying mechanisms of HPD in alleviating CID.

Keywords: chemotherapy-induced diarrhea; hesperidin; network pharmacology; molecular docking; experimental validation

1. Introduction

According to the International Agency for Research on Cancer's "Global Cancer Statistics 2018: GLOBOCAN Estimates of Incidence and Mortality Worldwide for 36 Cancers in 185 Countries" in 2018, there were 9.65 million cancer-related deaths and 18.08 million new instances of the disease globally. Asian patients accounted for over 50% of these fatalities and newly diagnosed cases [1]. This significant cancer burden in China and across Asia imposes substantial physical and financial strain on patients and their families. Chemotherapy remains a primary treatment modality for cancer, but it is associated with considerable adverse effects [2].

Chemotherapy-induced diarrhea (CID) is a common and potentially severe side effect, resulting from the toxic effects of chemotherapeutic agents on the gastrointestinal (GI) tract, leading to increased stool frequency and liquidity [3]. CID often occurs during cancer treatment with CPT-11 as the first-line agent [4]. The incidence of CID varies depending on the chemotherapy regimen. For instance, among patients undergoing treatment regimens that include 5-fluorouracil (5-FU) and CPT-11, the incidence of chemotherapy-induced diarrhea (CID) ranges from 30% to 80%. The incidence is particularly high in patients receiving combination therapies or high-dose treatments [5]. CID ranges from mild to severe, characterized by increased stool frequency and liquidity. Severe diarrhea can result in dehydration, electrolyte disturbances, and kidney failure. Patients often experience abdominal cramping, urgency, and in severe cases, fever and bloody stools. The underlying cause of CID is damage to the intestinal mucosa caused by chemotherapeutic agents. This damage disrupts the epithelial barrier, leading to inflammation and altered absorption and secretion in the intestines. Chemotherapy can also modify the gut microbiome, exacerbating the condition [6]. The etiology of CID includes direct toxic effects of chemotherapy on the GI mucosa, resulting in mucositis and subsequent diarrhea. Specific drugs, such as CPT-11, 5-FU, and capecitabine, are known to cause high rates of CID due to their mechanisms of action on rapidly dividing cells in the GI tract [7]. Management of CID involves both preventive and therapeutic approaches. In addition to treatments such as loperamide and phenethylpiperidine, the growth inhibitor analogue octreotide and probiotics, including herbs such as ginger and ginseng, have shown potential benefit in the management of CID [8].

Hesperidin (HPD), a flavonoid glycoside predominantly found in citrus fruits, such as oranges, lemons, and grapefruits [9], has been extensively studied for its health benefits, including antioxidant, anti-inflammatory, and cardiovascular protective effects (Figure 1). It improves vascular function, lowers blood pressure, and supports overall heart health [10]. These effects are primarily attributed to its ability to enhance endothelial function and reduce oxidative stress [11]. HPD also shows promise in treating diarrhea, particularly due to its anti-inflammatory properties. By modulating inflammatory responses in the gut, HPD can reduce the severity and duration of diarrhea [12]. Its ability to improve blood flow and support vascular health also contributes to better gut health, potentially reducing the incidence and severity of diarrhea [13]. However, the mechanism of HPD treatment for CID remains poorly understood and requires further investigation.

Network pharmacology is a multidisciplinary approach that integrates systems biology, bioinformatics, and pharmacology to study the interactions between drugs, targets, and disease networks. This approach enables a more comprehensive understanding of the mechanisms of action of drugs, considering their effects on multiple targets and pathways rather than single-target interactions [14]. In recent years, network pharmacology has gained attention for its potential in drug discovery, particularly in traditional Chinese medicine, where complex herbal formulations involve multiple compounds acting on various targets [15]. For example, studies have demonstrated how network pharmacology can be used to predict potential therapeutic effects, identify bioactive compounds, and explore the mechanisms behind complex diseases like cancer and cardiovascular disorders [16]. By integrating molecular networks, computational tools, and big data, network pharmacology offers a novel approach to precision medicine, moving beyond the "one drug, one target" paradigm [17].

In this study, CPT-11 was utilized to create a CID mouse model to assess the therapeutic efficacy of HPD. In addition, we predicted the key targets and pathways for HPD ameliorating CID by network pharmacology and molecular docking analyses. Further, we confirmed the predicted key proteins and pathways through in vivo experiments. This study informs the discovery of drugs for the treatment of CID and provides novel insights into the therapeutic potential of HPD for CID.

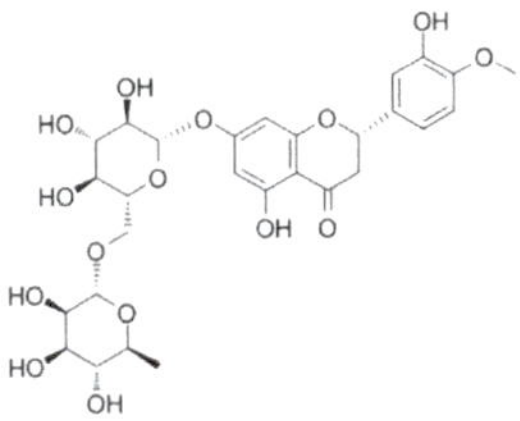

Figure 1. The chemical structure of HPD.

2. Results

2.1. Amelioration of HPD on CID

2.1.1. HPD Ameliorates Diarrhea in CID Mice

In this experiment, the control group exhibited normal conditions, whereas the CPT-11 group exhibited poor mental status, lusterless and disheveled fur, reduced food intake, and severe cases of blood in the stool. Moreover, the CPT-11 group displayed decreased body weight, increased DAI scores, and shortened colon length. However, as shown in Figure 2A–D, the variations were significantly improved by HPD treatment.

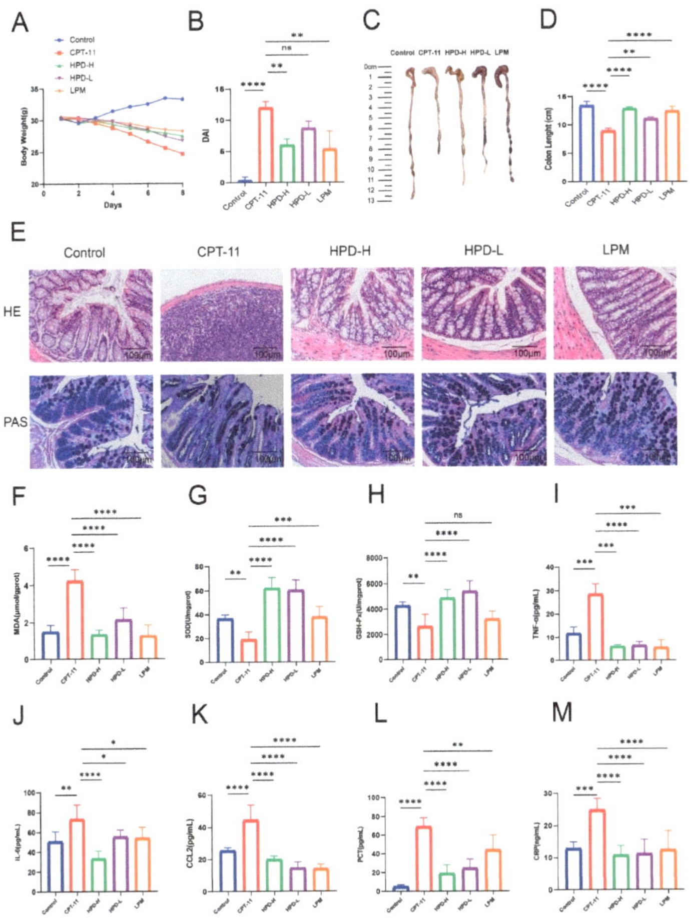

Figure 2. Effects of HPD on mice with CID. (**A**) Body weight; (**B**) disease activity index; (**C**) representative pictures of colon tissues; (**D**) colon length; (**E**) representative H&E staining and PAS staining

of colon tissues (magnification, 100×; scale bar, 100 μm); (**F–H**) levels of MDA, SOD, and GSH-Px; (**I–M**) levels of TNF-α, IL-6, CCL2, PCT, and CRP. Data were presented as the mean ± SD. * $p < 0.05$, ** $p < 0.01$, *** $p < 0.001$, **** $p < 0.0001$ and ns indicates no statistically significant difference ($p \geq 0.05$).

2.1.2. HPD Attenuates Intestinal Pathology in CID Mice

The H&E staining showed that the CPT-11 group had mucosal epithelial disruption and inflammatory cell infiltration. Nevertheless, HPD obviously transformed the pathological changes (Figure 2E). The PAS staining showed that goblet cells in the control group were full and round, whereas the goblet cells in the CPT-11 group exhibited varying degrees of atrophy and reduction (Figure 2E). The HPD-H, HPD-L, and LMP groups showed reversal of goblet cell atrophy and an increase in mucus secretion.

2.1.3. HPD Attenuates Intestinal Oxidative Stress in CID Mice

Due to the potent antioxidant effect of HPD, we aimed to verify its effects on CID in mice. We monitored the oxidative product MDA and the antioxidant enzymes SOD and GSH-Px (Figure 2F–H). In the CPT-11 group, the MDA levels were elevated; however, this elevation was inhibited by HPD administration. The levels of SOD and GSH-Px decreased in the CPT-11 group, but HPD administration increased the levels of SOD and GSH-Px.

2.1.4. HPD Attenuates Intestinal Inflammation in CID Mice

We also assessed the changes in the levels of inflammatory cytokines. The levels of TNF-α and IL-6 in the CPT-11 group were notably increased, but HPD inhibited this elevation (Figure 2I,J). In addition, the CPT-11 group showed increased expression of CCL2, whereas these levels were markedly lower in the HPD groups (Figure 2K).

To monitor intestinal bacterial infections in the CID mice, PCT and CRP concentrations in the colon tissues were measured. The levels of PCT and CRP were notably higher in the CPT-11 group, but HPD inhibited this elevation (Figure 2L,M). These results demonstrate that HPD could alleviate CID in mice.

2.2. The Effect of HPD on CID Analyzed by Network Pharmacology

2.2.1. Network Analysis of Targets

HPD target genes were identified from the PharmMapper database, excluding genes lacking official gene symbols. Among the 290 active targets identified, 2 duplicates were removed, leaving 288 included in our analysis. A search of the GeneCards and OMIM databases for "chemotherapy-induced diarrhea" yielded 974 active targets. After removing 2 duplicates, 972 valid targets were included in the analysis. A total of 74 overlapping genes between the HPD and CID targets were identified (Figure 3A). The regulatory network of HPD and CID was built by Cytoscape v3.10.1 (San Francisco, CA, USA) (Figure 3B). We built the protein–protein interaction (PPI) network of the proteins mentioned above via STRING. We employed Cytoscape v3.10.1 (San Francisco, CA, USA) for visualization (Figure 4A). This network included 73 nodes and 1318 edges. The top ten core targets were AKT1, CASP3, ALB, EGFR, HSP90AA1, MMP9, ESR1, ANXA5, PPARG, and IGF1. The CytoHubba plugin was subsequently applied to screen for central targets. By integrating the results from the four algorithms, the targets that overlapped across these screening methods were identified as the central targets (Figure 4B–E).

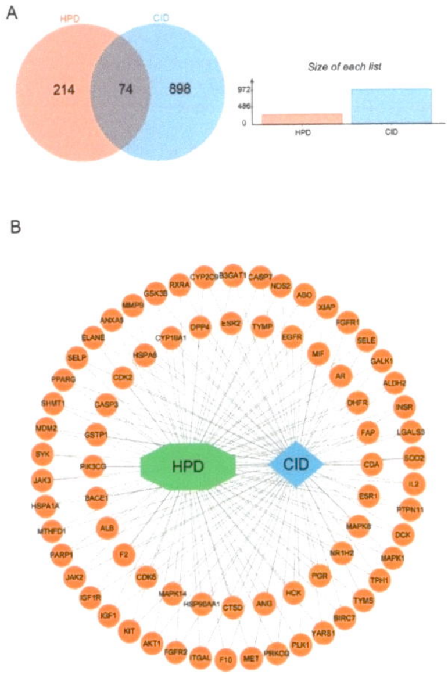

Figure 3. Targets and network analysis of HPD for treating CID. (**A**) Venn diagram of the intersection targets of HPD and CID; (**B**) the regulatory network of HPD acting on CID.

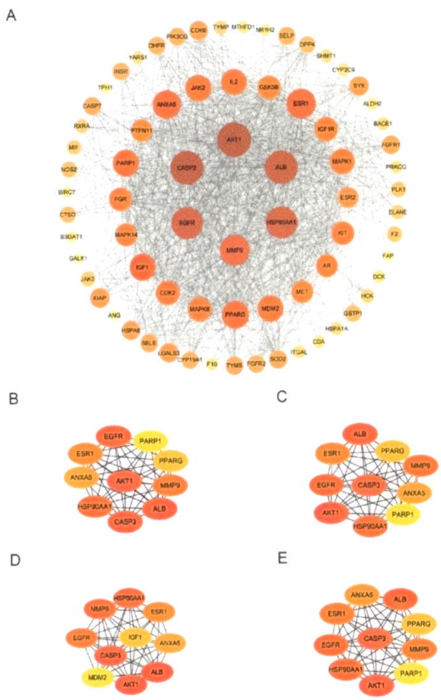

Figure 4. (**A**) PPI network of the potential targets; (**B**) closeness; (**C**) degree; (**D**) MCC; and (**E**) MNC. Deeper colors represent proteins with higher importance in the disease, while lighter colors indicate lower importance.

2.2.2. Enrichment Analysis of Overlapped Target

In the GO enrichment analysis, which included 298 BP terms, 43 CC terms, and 84 MF terms, the top 10 significantly enriched terms were selected for presentation (Figure 5A). The results indicated that HPD's potential therapeutic genes for CID were predominantly enriched in processes such as phosphorylation, negative regulation of the apoptotic process, and functions including enzyme binding and ATP binding.

The KEGG pathway enrichment analysis identified 135 pathways ($p < 0.05$), from which 20 pathways were found to be significant ($p < 0.05$) (Figure 5B). According to the KEGG analysis, the PI3K–Akt pathway, FoxO pathway, MAPK pathway, TNF pathway, and Ras pathway were most associated with HPD's potential treatment of CID.

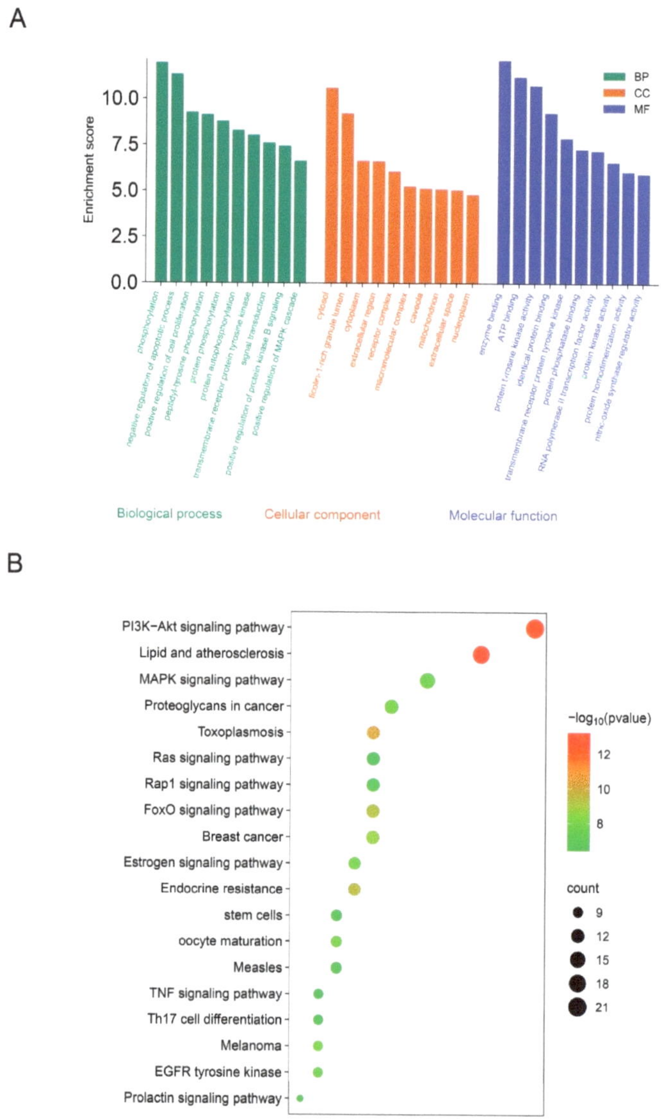

Figure 5. (**A**) Histogram of the GO enrichment analysis involving BP, CC, and MF; (**B**) bubble chart of KEGG pathway.

2.3. Molecular Docking Analysis

AutoDockTools 1.5.6 software was used for molecular docking, and we explored the interaction and binding modes of HPD with target proteins at the molecular level. HPD was docked into the active pockets of the top ten proteins (AKT1, CASP3, ALB, EGFR, HSP90AA1, MMP9, ESR1, ANXA5, PPARG, and IGF1), and the theoretical binding modes were visualized using PyMOL v1.7.2.1 (New York City, NY, USA). The results demonstrated strong binding of HPD to seven core targets: AKT1, ANXA5, CASP3, HSP90AA1, IGF1, MMP9, and PPARG, with binding free energies of −7.3 kcal/mol, −6.9 kcal/mol, −6.1 kcal/mol, −5.2 kcal/mol, −8.1 kcal/mol, −9.3 kcal/mol, and −8.6 kcal/mol, respectively. The molecules formed a conventional hydrogen bond, carbon hydrogen bond, unfavorable donor–donor, Pi–cation, Pi–anion, Pi–alkyl, Pi–sigma, and Pi–Pi stacked, which highlighted the role of HPD in the regulation of the key targets. We performed a visual analysis of the first five targets (Figure 6). Based on these findings, we predicted that these target proteins may contribute to HPD in relieving CID, prompting further experimental validation.

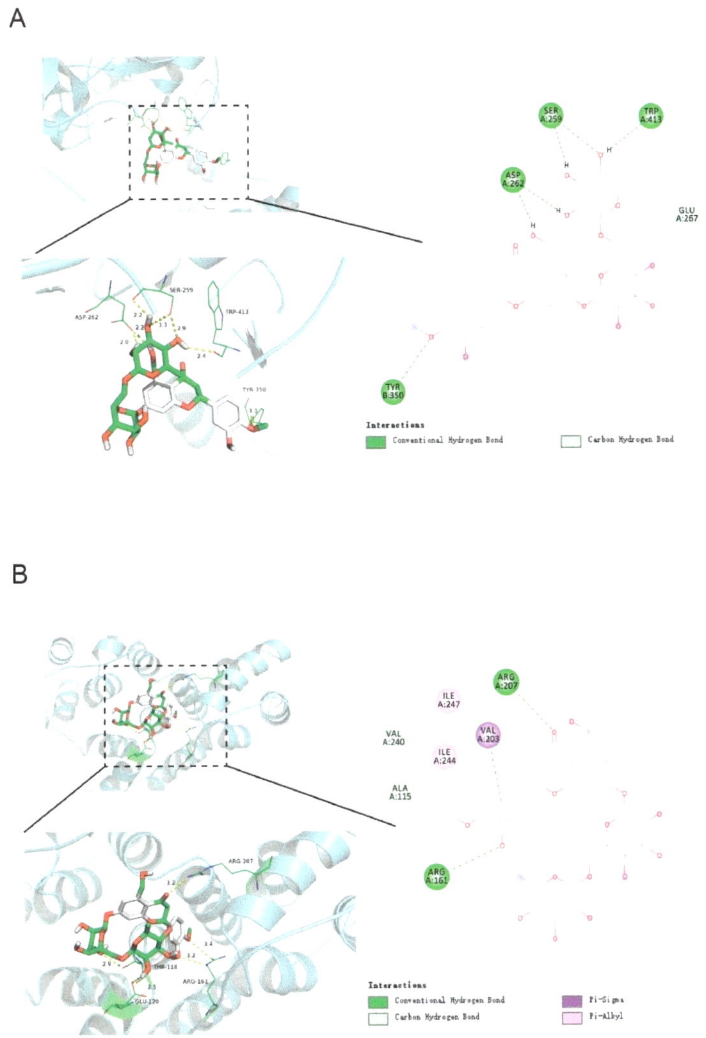

Figure 6. *Cont.*

C

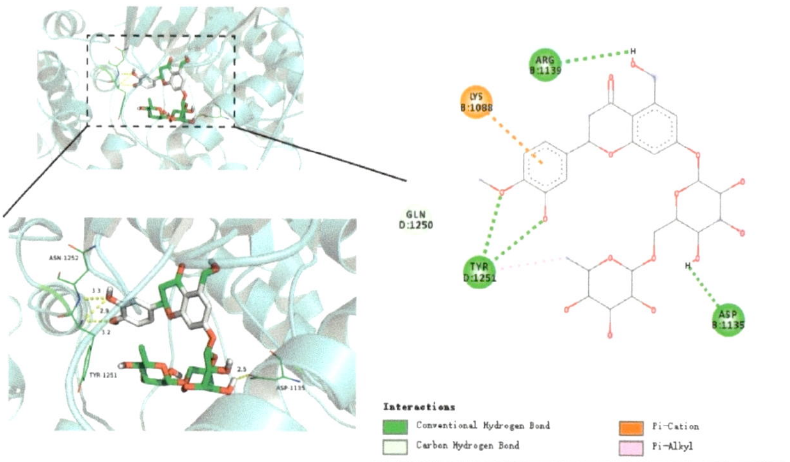

D

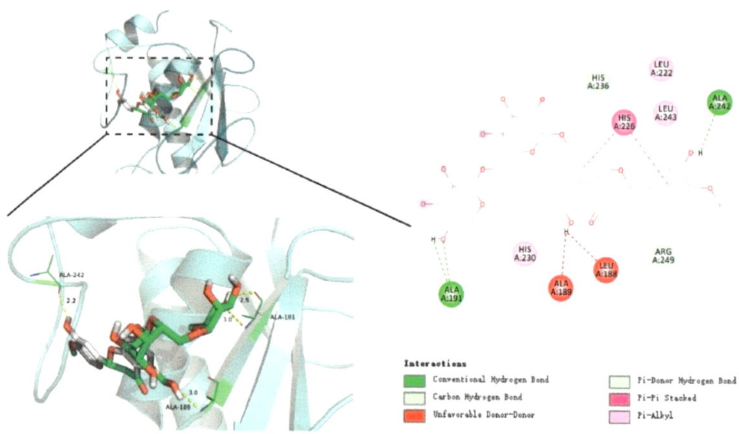

Figure 6. *Cont.*

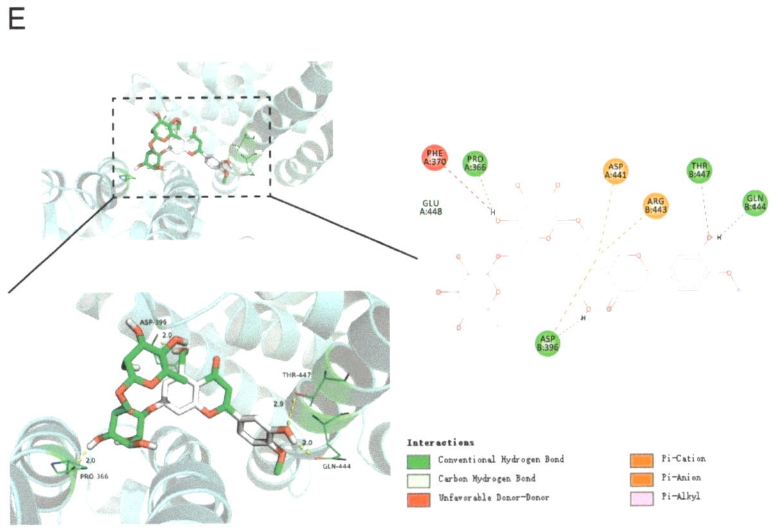

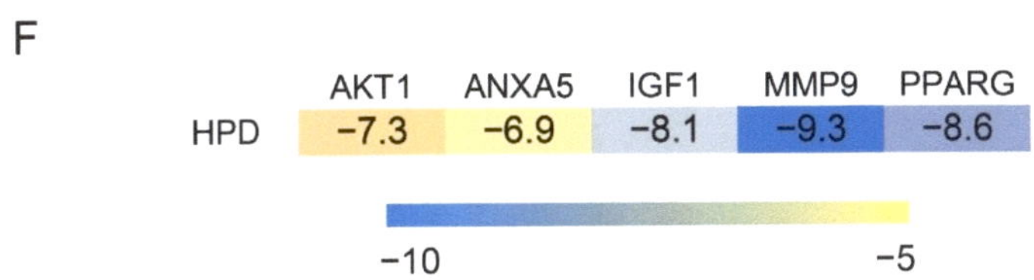

Figure 6. Schematic diagram of the interaction between HPD and target proteins. (**A**) AKT1; (**B**) ANXA5; (**C**) IGF1; (**D**) MMP9; (**E**) PPARG; (**F**) heat map of molecular docking binding energy.

2.4. *Verification of HPD on Alleviating CID*

2.4.1. HPD Attenuates Enterocyte Apoptosis in CID Mice

The seven proteins above are closely linked to apoptosis, and apoptosis of intestinal epithelial cells is a critical pathological feature leading to damage of the intestinal mucosal barrier [18]. Therefore, we conducted a TUNEL staining assay (Figure 7A,B). The results revealed a significantly higher rate of apoptosis in the CPT-11 group; however, HPD treatment effectively reversed this change.

Furthermore, Bcl-2 and Bax are intricately involved in apoptosis, with Bcl-2 being anti-apoptotic and Bax pro-apoptotic. The reciprocal interactions between these proteins affect cellular survival or death [19]. An immunohistochemistry analysis of Bcl-2 and Bax proteins is presented in Figure 7C,D. In the CPT-11 group, Bax expression was markedly increased, and Bcl-2 expression was decreased, indicating enhanced apoptosis. Treatment with HPD significantly reversed these situations, suggesting its potential to mitigate apoptosis in colonic mucosal cells by modulating Bcl-2 family proteins.

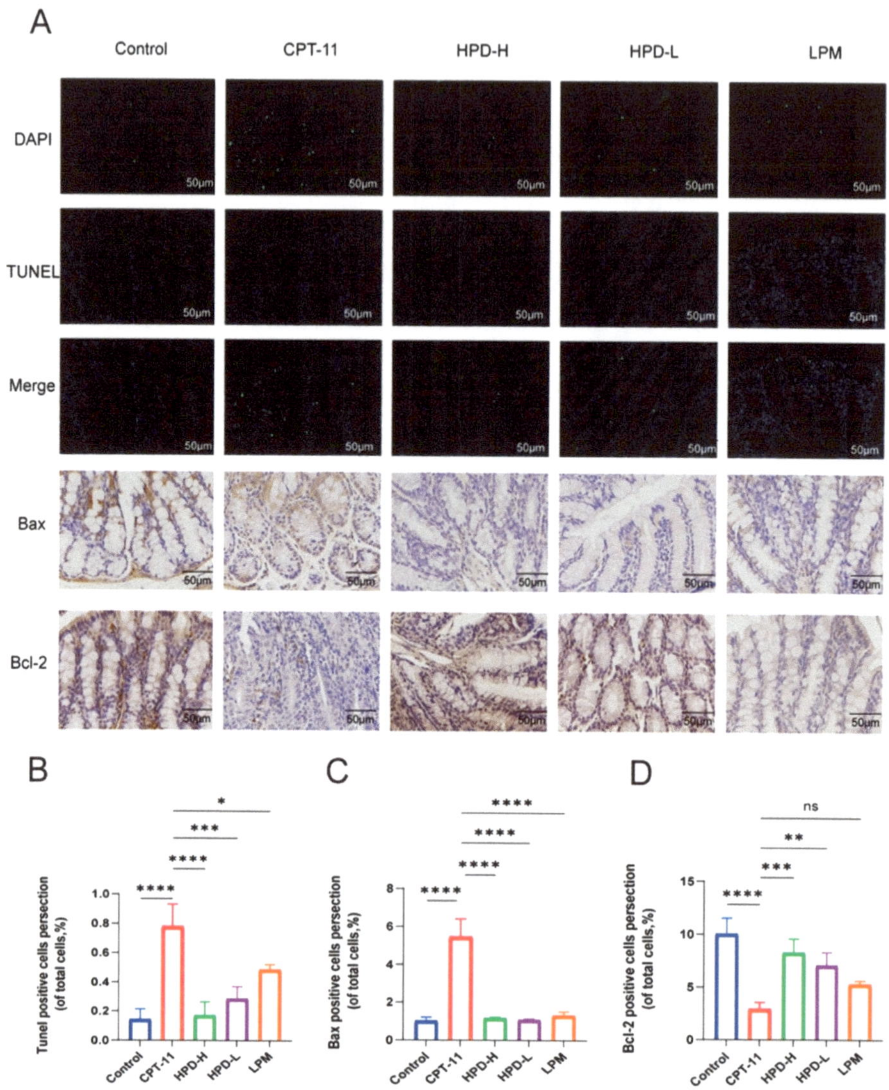

Figure 7. HPD attenuates enterocyte apoptosis in CID mice. (**A**) Representative TUNEL staining of colon tissues, and immunohistochemistry staining of Bax and Bcl-2 protein (magnification, 400×; scale bar, 50 μm). Representative images of the rate of apoptosis by TUNEL staining (**B**), Bax (**C**), and Bcl-2 (**D**) protein level in colon determined by immunohistochemistry. The data were presented as the mean ± SD. * $p < 0.05$, ** $p < 0.01$, *** $p < 0.001$, **** $p < 0.0001$ and ns indicates no statistically significant difference ($p \geq 0.05$).

2.4.2. WB Results for 7 Key Proteins

According to network pharmacology and molecular docking studies, HPD intervention in CID may be strongly associated with seven core target proteins: AKT1, ANXA5, CASP3, HSP90AA1, IGF1, MMP9, and PPARG. To further validate these findings, we conducted Western blotting experiments (Figure 8). The results demonstrated that in the model group, the expression levels of AKT1, p-AKT1, ANXA5, CASP3, HSP90AA1, and IGF1 were increased, while MMP9 was decreased compared to the control group. Administration of HPD-H markedly reversed these effects in the CPT-11 group, indicating its regulatory effect on these proteins.

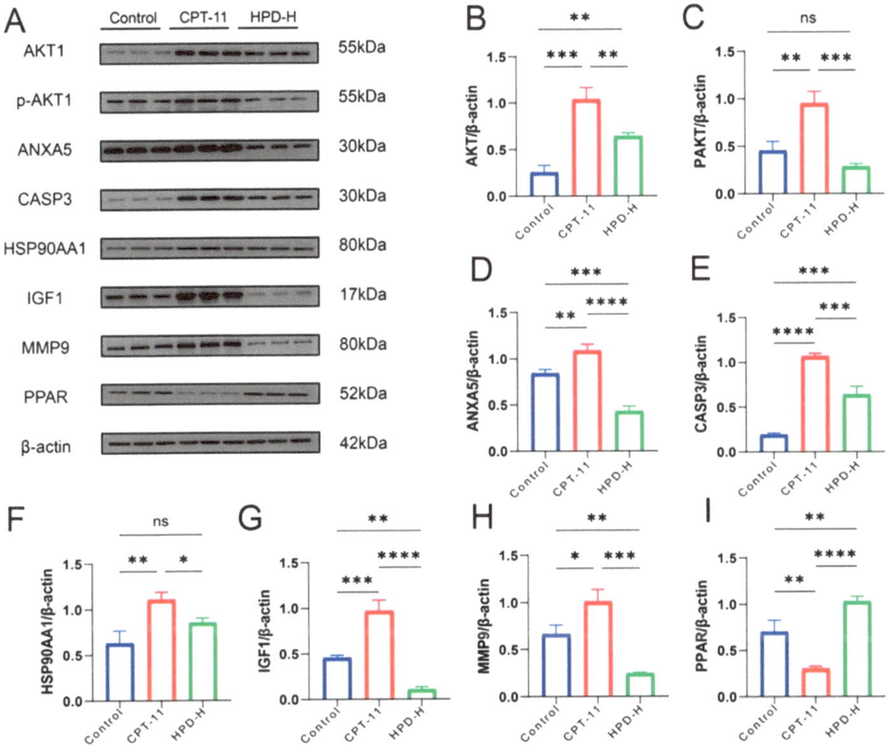

Figure 8. WB results of 7 key proteins. (**A–I**) The expression levels of AKT1, PAKT, ANXA5, CASP3, HSP90AA1, IGF1, MMP9, and PPARG in the colon tissue from mice as identified through Western blotting with β-actin as the internal reference. The data were presented as the mean ± SD. * $p < 0.05$, ** $p < 0.01$, *** $p < 0.001$, **** $p < 0.0001$ and ns indicates no statistically significant difference ($p \geq 0.05$).

3. Discussion

CID is a common complication in cancer patients receiving chemotherapy. A study found that 45 ± 2% of 100 patients receiving chemotherapy experienced diarrhea. Among them, 66 adjustments were made during the chemotherapy treatment between 56 patients, including dose reductions in 34%, dose delays in 12%, therapy discontinuations in 23%, or multiple changes in 17% [20]. CID interferes with patients' daily activities and has a bad impact on patients' quality of life [21]. Notably, CID often occurs during cancer treatment with CPT-11 as the first-line agent. The mechanism of CPT-11-induced diarrhea is primarily associated with its active metabolite, SN-38. CPT-11 is converted into SN-38, which is highly toxic to the intestinal mucosa, leading to apoptosis and mucosal damage, causing severe diarrhea [22]. The toxic form of SN-38 is stabilized in an acidic environment, increasing its absorption and toxicity in the intestines. Studies suggest that alkalizing the intestinal environment, such as with sodium bicarbonate supplementation, can reduce SN-38 toxicity and alleviate irinotecan-induced diarrhea [23]. HPD comes from a wide range of sources and possesses anti-inflammatory and antioxidant activities, and some studies in the literature have reported that HPD can alleviate CID, so we further investigated the effect and mechanism of HPD in the treatment of CID. The effect of HPD on CID was examined in the study, and we explored the mechanism of HPD intervention in CID using network pharmacology and molecular docking. Additionally, we verified the therapeutic role of HPD on CID. We found that CID causes severe colonic damage, oxidative stress, and inflammatory attacks, all of which were significantly reversed by the administration of HPD.

Oxidative stress is a biological phenomenon resulting from disruptions in the redox state caused by environmental conditions both inside and outside the cell, which lead to excessive production of oxides, such as ROS and MDA, as well as decreased SOD activity, ultimately disrupting oxidative protection mechanisms [24]. GSH-Px is a critical antioxidant that inhibits oxygen free radicals and blocks lipid peroxidation reactions, thereby protecting the body from peroxide-induced damage [25]. Chemotherapy can damage the colonic mucosa, and oxidative stress is pivotal in the pathogenesis and treatment of intestinal damage [26]. It has been reported that intestinal arginine and NO metabolism triggers intestinal barrier dysfunction, low-grade endotoxemia, and aging [27]. Excessive oxidative stress activation in the intestinal mucosa damages the mucosal epithelium and impairs mucosal function, thereby exacerbating the inflammatory response. A study found that cystine (Cys2), a substrate of glutathione (GSH), can locally inhibit oxidative stress-induced pro-inflammatory responses and thus improve intestinal barrier function [28]. Thus, regulating antioxidant defenses is crucial in treating CID. HPD is a potent antioxidant with excellent reducing power and chelating activity. Moreover, its antioxidant activity increases with higher concentrations [29]. The study's findings demonstrated that HPD intervention significantly reduced MDA levels while increasing SOD and GSH-Px levels. These results indicate that HPD enhances the antioxidant capacity of colon tissues.

CPT-11 is converted to SN-38 by hepatic metabolism, and SN-38 is converted to glucuronidated SN-38 by uridine diphosphate glucuronosyltransferase-1A1 and 1A9 in the liver, and the free SN-38 in the intestinal lumen can induce intestinal mucosal injury in mice [30]. In addition, CID can cause intestinal structural damage to the mucosal barrier by altering the intestinal flora, causing bacterial entry into the bloodstream and inducing inflammation [31]. Among the inflammatory responses induced by CPT-11, production of the pro-inflammatory factors TNF-α and IL-6 causes connective tissue damage and epithelial basal cell death [32]. A study indicated that increased levels of CCL2 in macrophages could promote colitis [33]. HPD exhibits anti-inflammatory effects in both animal models and human trials [34]. Our study shows that the levels of TNF-α, IL-6, and CCL2 were markedly higher in the CPT-11 group, but HPD reversed these aberrant changes. Meanwhile, we measured the concentrations of PCT and CPR in colon tissue to monitor intestinal bacterial infections in CID mice [35]. The results showed that the levels of PCT and CRP were notably higher in the CPT-11 group, but HPD inhibited this elevation.

Based on the effectiveness of HPD for the treatment of CID, we predicted the potential mechanism through network pharmacology. Findings from network pharmacology suggested that the mechanism by which HPD treated CID involved multiple protein targets and pathways: AKT1, CASP3, ALB, EGFR, HSP90AA1, MMP9, ESR1, ANXA5, PPARG, and IGF1 were the top ten core potential targets. GO and KEGG analyses displayed that the relevant pathways were the PI3K–Akt pathway, FoxO pathway, MAPK pathway, TNF pathway, and Ras pathway. Of these, the PI3K–AKT, Ras, and TNF pathways might have influence on HPD intervention in CID via inflammation suppression and anti-carcinogenesis [36]. A study indicated that the molecular signature of the MAPK pathway may be a potential biomarker for evaluating the risk of intestinal damage [37]. Another study found that a FoxO1 inhibitor could significantly reduce intestinal symptoms in an in vivo experiment [38]. To further evaluate the binding ability of HPD to core targets, we performed molecular docking. Seven key targets were screened by molecular docking, including AKT1, ANXA5, CASP3, HSP90AA1, IGF1, MMP9, and PPARG.

Apoptosis of intestinal epithelial cells is an important pathological feature that leads to damage of the intestinal mucosal barrier and triggers diarrhea. It has been reported that CPT-11-induced apoptosis could eliminate peritoneal resident macrophages, which may impair the function of peritoneal B-1 cells to maintain intestinal homeostasis, which is one of the reasons why CPT-11-treated cancer patients can experience diarrhea [39]. Interestingly, the seven proteins above are closely related to apoptosis. Molecular docking screened seven proteins closely related to apoptosis. AKT1 enhances cell proliferation through cell cycle proteins, such as p21, p27, and cyclin D1, and inhibits apoptosis via proteins like p53 [40]. ANXA5 is a cell membrane-associated protein with important roles in apoptosis, coagulation, inflammation, and biomedical fields [41]. MMP9 increases the expression of pro-apoptotic proteins and DNA damage [42]. Caspase-3 [43], HSP90AA1 [44], IGF-1 [45], and PPARG [46] can regulate a variety of apoptotic signaling molecules, such as Bcl-2, to inhibit apoptosis, and these functions are important for maintaining cell survival and removing damaged cells. A recent study showed that HPD modulates NF-κB signaling, PI3/Akt/PTE signaling, and microRNA to improve apoptosis in vivo and in vitro [47]. Meanwhile, our study shows that the protein expression of AKT1, p-AK1, ANXA5, CASP3, HSP90AA1, IGF1, and MMP9 were increased, and PPARG was decreased in the CPT-11 group. However, administration of HPD-H markedly reversed these effects in the CPT-11 group. Therefore, we hypothesize that HPD may alleviate CID by inhibiting the expression of AKT1, ANXA5, CASP3, HSP90AA1, IGF-1, and MMP9, while enhancing the expression of PPARG.

In addition, we observed apoptosis by TUNEL staining and immunohistochemical analysis. The results of the TUNEL staining showed that the rate of apoptosis was notably increased in the CPT-11 group; however, HPD was able to reverse this change. The endogenous apoptosis pathway is mainly executed by the Bcl-2 apoptosis family proteins, of which the pro-apoptotic protein Bax and the anti-apoptotic protein Bcl-2 are the best indicators of apoptosis [48]. The immunohistochemical results showed that the expression of Bax was significantly higher, while Bcl-2 was significantly lower in the CPT-11 group. This result was indicative of apoptosis. HPD treatment reduced the expression of Bax and enhanced the expression of Bcl-2, suggesting that it can attenuate apoptosis in colonic mucosal cells by regulating Bcl-2 family proteins. Despite the prospective findings, this study had several limitations. As a next step, we will further validate the deeper mechanisms through in vitro experiments using inhibitors and agonists on key proteins of the pathway.

4. Materials and Methods

4.1. Chemicals and Reagents

CPT-11 was supplied by Tianjin JinYao Pharmaceutical Co., Ltd. (Tianjin, China). HPD, with a purity greater than 98%, was sourced from citrus peels and procured from Xi'an Yiyang Bio-TECH (Xi'an, China). The malondialdehyde colorimetric assay kit (MDA, WU148LD87626), total superoxide dismutase activity assay kit (SOD, WU134H066375), glutathione peroxidase activity assay kit (GSH-Px, WU120H4D5123), mouse TNF-α ELISA kit (TNF, WU12P8808590), mouse interleukin 6 ELISA kit (IL-6, WU10042L5917), mouse monocyte chemotactic protein 1 ELISA kit (CCL2, WU11V2R07171), mouse procalcitonin ELISA kit (PCT, WU078V441694), and mouse C-reactive protein ELISA kit (CRP, WU06Z4040286) were from Elabscience Biotechnology Co., Ltd. (Wuhan, China). The AKT1 pAb (AKT, M29N003), IGF1 pAb (IGF1, N25MA23), MMP9 pAb (MMP9, M12DE14), and PPAR gamma pAb (PPAR, M04AP01) were from Chengdu Zen-Bioscience Co., Ltd. (Chengdu, China). The Annexin V pAb (ANXA5, 00107879), Caspase 3/p17/p19 pAb (Caspase-3, 00136098), and HSP90 pAb (HSP90, 00138548) were from Proteintech Group, Inc. (Wuhan, China). The Phospho-Akt mAb (PAKT, 25) was from Cell Signaling Technology, Inc. (Danvers, MA, USA). Other chemicals and reagents were sourced from commercial suppliers.

4.2. Experimental Animals

Healthy male ICR mice were obtained from a licensed supplier, SPF (Beijing) Biotechnology Co., Ltd. (Beijing, China). These animals were maintained in plastic cages under controlled environmental conditions: a temperature of 23 ± 2 °C, relative humidity of 50 ± 10%, and a 12 h light/dark cycle. The mice were provided with unrestricted access to food and water. Efforts were made to minimize animal usage and alleviate their distress.

4.3. Experimental Design

A total of 60 mice were divided into six groups (n = 10 in each group) at random: Control group, CPT-11 group, high-dose HPD (HPD-H) group (80 mg/kg), low-dose HPD (HPD-L) group (20 mg/kg), and loperamide (LPM) group (50 mg/kg). Based on the literature [49], from days 1 to 4, the Control group received intraperitoneal injections of saline, while the other groups received intraperitoneal injections of CPT-11 (40 mg/kg/day). From days 1 to 8, the Control and CPT-11 groups were administered saline orally, while the LPM, HPD-L, and HPD-H groups received the corresponding doses of their respective solutions orally.

4.4. Assessment of Severity of CID

The severity of CID was evaluated using the disease activity index (DAI), which is measured by hematochezia, weight loss, and stool consistency. Weight loss is scored from zero to four, with zero points for no change or weight gain, one point for weight loss between 1% and 5%, two points for a loss between 5% and 10%, three points for a loss between 10% and 15%, and four points for a loss exceeding 15%. Stool consistency is rated as zero points for normal stools, two points for loose stools, and four points for diarrhea. Hematochezia is also scored, with zero points indicating no bleeding, two points for positive occult blood, and four points for visible bleeding. The overall DAI value is obtained by summing the scores of these three categories [50].

4.5. Histopathological Examination

Colon tissue samples were fixed in 4% paraformaldehyde, embedded in paraffin, and sectioned into 5 μm slices. These sections were then subjected to hematoxylin and eosin (H&E) staining, terminal deoxynucleotidyl transferase dUTP nick-end labeling (TUNEL) staining, and periodic acid-Schiff (PAS) staining for histological analysis.

4.6. ELISA Determination

The colon tissue samples were homogenized in phosphate-buffered saline (PBS) and centrifuged at 3000 rpm for 20 min. The supernatant was subsequently collected, and the optical density (OD) values were measured using an ELISA reader according to the provided instructions. This procedure was employed to quantify the levels of MDA, SOD, GSH-Px, TNF, IL-6, CCL2, PCT, and CRP in the tissue samples.

4.7. Network Pharmacology Analysis

4.7.1. HPD Target Prediction and Disease-Related Target Screening

Target prediction for HPD was conducted using the PharmMapper Database [51] (https://www.lilab-ecust.cn/pharmmapper/ Accessed on 5 December 2023), with the species restricted to "*Homo sapiens*". Additionally, relevant disease targets were identified by inputting the keywords "chemotherapy-induced diarrhea" into the GeneCards Database [52] (http://www.genecards.org/ Accessed on 5 December 2023) and OMIM Database [53] (http://omim.org/ Accessed on 5 December 2023). Subsequently, false-positive and duplicate targets were removed and integrated. The intersection of HPD targets and CID targets was determined using R v4.3.2 (Vienna, Austria) packages to identify potential HPD targets for treating the disease. The common gene targets were visualized in a Venn diagram [54] (https://www.bioinformatics.com.cn/ Accessed on 5 December 2023). Finally, Cytoscape v3.10.1 (San Francisco, CA, USA) [55] was utilized to construct the regulatory network of HPD's action on CID.

4.7.2. Protein–Protein Interaction Network Analysis

The protein–protein interaction (PPI) network of the target genes was retrieved from the Search Tool for the Retrieval of Interacting Genes/Proteins (STRING) database [56] (https://cn.string-db.org/ Accessed on 5 December 2023.) with a minimum required interaction score of $\geq$0.4, restricted to the species "*Homo sapiens*". The network visualization was conducted using Cytoscape v3.10.1 (San Francisco, CA, USA). In this PPI network, nodes denote the target proteins, and edges represent the predicted or experimentally validated interactions between these proteins. To identify core targets related to CID, the CytoHubba plugin in Cytoscape v3.10.1 (San Francisco, CA, USA) was employed using the closeness, degree, maximal clique centrality (MCC), and maximum neighborhood component (MNC) algorithms. The top 10 targets identified by each method were analyzed, and the overlapping targets among these methods were considered the central targets [57].

4.7.3. GO and KEGG Pathway Enrichment Analyses

Gene ontology (GO) enrichment analysis encompasses biological process (BP), molecular function (MF), and cellular component (CC) categories. The Kyoto Encyclopedia of Genes and Genomes (KEGG) serves as a bioinformatics resource for identifying significantly altered metabolic pathways enriched in the gene list. The clusterProfiler package in R v4.3.2 (Vienna, Austria) was utilized to conduct a GO enrichment analysis ($p < 0.05$) and a KEGG pathway analysis ($p < 0.05$) on the target genes of HPD acting on CID. The results were visualized using the bioinformatics platform [58] (http://www.bioinformatics.com.cn/).

4.8. Molecular Docking

Based on the results from the previous network pharmacology screening, the 3D structures of the top ten target proteins were retrieved from the protein database (PDB) [59] (http://www.rcsb.org/ Accessed on 10 December 2023) and the corresponding PDB format files were downloaded. PyMOL v1.7.2.1 (New York City, NY, USA) [59] was employed to remove ligands, ions, and solvent molecules from these structures. The chemical structure of HPD (PubChem CID 10621) was obtained from the PubChem database [60] (https://pubchem.ncbi.nlm.nih.gov/ Accessed on 10 December 2023). Additionally, AutoDockTools 1.5.6 software was used to convert the formats of HPD and the target genes to pdbqt format for molecular docking. In molecular docking, the selection of protein crystals and binding sites is critical for ensuring docking accuracy. High-resolution structures with a resolution below 2.5 Å should be prioritized for structural precision [61]. It is essential to consider the integrity of the protein, prioritizing full-length proteins or those containing key functional domains while avoiding excessive mutations or modifications. Additionally, structures bound with known ligands help capture active conformations [23]. For selecting binding sites, known active sites should be prioritized; in their absence, homology modeling, pocket detection tools, and molecular dynamics simulations can be used to predict and validate potential binding pockets [62].

4.9. Immunohistochemical Analysis

The colon paraffin sections were routinely deparaffinized and washed three times with distilled water for 3 min each time; then, the endogenous peroxidase was sealed with 3% hydrogen peroxide, the sections were incubated with microwave heating antigenic repair of 10% normal goat serum, drops of p53 antibody diluted 1:100 with PBS were added, and the samples were stored overnight in a refrigerator at 4 °C. Goat anti-rabbit lgG secondary antibody was added dropwise by oscillation and washed with PBS, the diamine benzylamine was developed for color development, the hematoxylin was restained, and the sections were routinely mounted. The film was blocked routinely. After observation, photos were taken, and the images were analyzed by Image J v1.54f (Bethesda, MD, USA).

4.10. Western Blotting

The colon tissue was homogenized in lysis buffer containing protease inhibitors thoroughly, then centrifuged at 4 °C (12,000 rpm, 15 min). We collected the supernatant and measured the protein concentration using a BCA protein assay kit. The samples were incubated in a metal bath at 95 °C for 15 min. We then separated the protein samples by electrophoresis and transferred them to a PVDF membrane. After blocking for 1 h, we then applied the primary antibodies: Beta-actin (1:10,000), GAPDH (1:10,000), AKT1 (1:3000),

ANXA5 (1:3000), Caspase-3 (1:1000), HSP90 (1:3000), IGF1 (1:1000), MMP9 (1:1000), PPAR (1:1000), and PAKT (1:1000). The samples were incubated at 4 °C overnight. We incubated the PVDF membrane with the corresponding secondary antibodies for 1 h, then developed and exposed them using the ChemiScope 6100 chemiluminescence imaging analysis system. We then analyzed the grayscale using Image J v1.54f (Bethesda, MD, USA).

4.11. Data Analysis

All the data were expressed as mean ± SD. Comparisons among multiple groups were conducted using one-way analysis of variance (ANOVA). A p-value of <0.05 was considered statistically significant. Statistical analyses and graphing were performed using GraphPad Prism 9.0 software (San Diego, CA, USA).

5. Conclusions

During the study, we found that HPD notably alleviated CID in mice. Then, a network pharmacology analysis displayed that HPD modulates five pathways associated with CID. Furthermore, a molecular docking analysis indicated that HPD interacts with seven key targets (AKT1, ANXA5, CASP3, HSP90AA1, IGF1, MMP9, PPARG) involved in the regulation of apoptosis. It was supported that HPD ameliorated apoptosis and modulated these seven critical targets by in vivo experiments. In summary, these findings provide preliminary scientific evidence supporting the use of HPD in treating CID. Moreover, we plan to follow up with further research to explore the mechanism of HPD for CPT-11-induced diarrhea.

Author Contributions: Investigation, X.S. and R.X.; methodology, X.S., J.L. and Z.Z.; formal analysis, X.S.; visualization, X.S.; validation; writing—original draft, X.S.; writing—review and editing, X.S., R.X., T.S. and X.Z.; formal analysis, P.X.; conceptualization, T.S. and X.Z.; project administration, T.S. and X.Z.; funding acquisition, X.Z. All authors have read and agreed to the published version of the manuscript.

Funding: The present research was supported by the school-level project "Xinglin Scholars" Discipline Talent Research Promotion Plan of Chengdu University of Traditional Chinese Medicine (No. BSH2021016).

Institutional Review Board Statement: All the animal experiments were conducted with the approval of the Committee of Laboratory Animal Welfare Ethics at Chengdu University of Traditional Chinese Medicine, under the review number 2023041 (The date of approval is 28 December 2023).

Informed Consent Statement: All the authors agree to be accountable for all aspects of the work.

Data Availability Statement: Raw data relevant to the conclusions of this study will be provided by the corresponding authors upon reasonable request.

Acknowledgments: We would like to express out gratitude to Chengdu University of Traditional Chinese Medicine College of Fundamental Medicine for providing site assistance.

Conflicts of Interest: The authors declare no conflicts of interest.

Abbreviations

5-FU, 5-fluorouracil; BP, biological process; CC, cellular component; CCL2, C-C motif chemokine ligand 2; CID, chemotherapy-induced diarrhea; CRP, C-reactive; HPD, hesperidin; DAI, disease activity index; GO, gene ontology; GSH-Px, glutathione peroxidase; H&E, hematoxylin and eosin; IL-6, interleukin-6; KEGG, Kyoto Encyclopedia of Genes and Genomes; LPM, loperamide; MDA, malondialdehyde; MF, molecular functions; NO, nitric oxide; OD, optical density; PAS, periodic acid-Schiff; PBS, phosphate-buffered saline; PCT, procalcitonin; PPI, protein–protein interaction; ROS, reactive oxygen species; SD, standard deviation; SOD, superoxide dismutase; TNF-α, tumor necrosis factor-α; TUNEL, TdT-mediated dUTP nick-end labeling; GI, gastrointestinal.

References

1. Bray, F.; Ferlay, J.; Soerjomataram, I.; Siegel, R.L.; Torre, L.A.; Jemal, A. Global cancer statistics 2018: GLOBOCAN estimates of incidence and mortality worldwide for 36 cancers in 185 countries. *CA Cancer J. Clin.* **2018**, *68*, 394–424. [CrossRef] [PubMed]
2. Zraik, I.M.; Heß-Busch, Y. [Management of chemotherapy side effects and their long-term sequelae]. *Urologe* **2021**, *60*, 862–871. [CrossRef]
3. Maroun, J.A.; Anthony, L.B.; Blais, N.; Burkes, R.; Dowden, S.D.; Dranitsaris, G.; Samson, B.; Shah, A.; Thirlwell, M.P.; Vincent, M.D.; et al. Prevention and management of chemotherapy-induced diarrhea in patients with colorectal cancer: A consensus statement by the Canadian Working Group on Chemotherapy-Induced Diarrhea. *Curr. Oncol.* **2007**, *14*, 13–20. [CrossRef] [PubMed]
4. Lv, J.; Jia, Y.; Li, J.; Kuai, W.; Li, Y.; Guo, F.; Xu, X.; Zhao, Z.; Lv, J.; Li, Z. Gegen Qinlian decoction enhances the effect of PD-1 blockade in colorectal cancer with microsatellite stability by remodelling the gut microbiota and the tumour microenvironment. *Cell Death Dis.* **2019**, *10*, 415. [CrossRef]
5. Jansman, F.G.; Sleijfer, D.T.; de Graaf, J.C.; Coenen, J.L.; Brouwers, J.R. Management of chemotherapy-induced adverse effects in the treatment of colorectal cancer. *Drug Saf.* **2001**, *24*, 353–367. [PubMed]
6. Kawasaki, Y.; Kakimoto, K.; Tanaka, Y.; Shimizu, H.; Nishida, K.; Numa, K.; Kinoshita, N.; Tatsumi, Y.; Nakazawa, K.; Koshiba, R.; et al. Relationship between Chemotherapy-Induced Diarrhea and Intestinal Microbiome Composition. *Digestion* **2023**, *104*, 357–369. [CrossRef]
7. Wadler, S.; Benson, A.B., III; Engelking, C.; Catalano, R.; Field, M.; Kornblau, S.M.; Mitchell, E.; Rubin, J.; Trotta, P.; Vokes, E. Recommended guidelines for the treatment of chemotherapy-induced diarrhea. *J. Clin. Oncol.* **1998**, *16*, 3169–3178. [CrossRef]
8. Yu, Y.; Kong, R.; Cao, H.; Yin, Z.; Liu, J.; Nan, X.; Phan, A.T.; Ding, T.; Zhao, H.; Wong, S.T.C. Two birds, one stone: Hesperetin alleviates chemotherapy-induced diarrhea and potentiates tumor inhibition. *Oncotarget* **2018**, *9*, 27958. [CrossRef]
9. Pyrzynska, K.A.-O. Hesperidin: A Review on Extraction Methods, Stability and Biological Activities. *Nutrients* **2022**, *14*, 2387. [CrossRef]
10. Abuelsaad, A.S.; Mohamed, I.; Allam, G.; Al-Solumani, A.A. Antimicrobial and immunomodulating activities of hesperidin and ellagic acid against diarrheic Aeromonas hydrophila in a murine model. *Life Sci.* **2013**, *93*, 714–722. [CrossRef]
11. Habauzit, V.; Morand, C. Evidence for a protective effect of polyphenols-containing foods on cardiovascular health: An update for clinicians. *Ther. Adv. Chronic Dis.* **2012**, *3*, 87–106. [CrossRef]
12. Xu, L.; Yang, Z.L.; Li, P.; Zhou, Y.Q. Modulating effect of Hesperidin on experimental murine colitis induced by dextran sulfate sodium. *Phytomedicine* **2009**, *16*, 989–995. [CrossRef]
13. Shafik, N.A.-O.; Gaber, R.A.; Mohamed, D.A.; Ebeid, A.M. Hesperidin modulates dextran sulfate sodium-induced ulcerative colitis in rats: Targeting sphingosine kinase-1- sphingosine 1 phosphate signaling pathway, mitochondrial biogenesis, inflammation, and apoptosis. *J. Biochem. Mol. Toxicol.* **2019**, *33*, e22312. [CrossRef] [PubMed]
14. Hopkins, A.L. Network pharmacology: The next paradigm in drug discovery. *Nat. Chem. Biol.* **2008**, *4*, 682–690. [CrossRef] [PubMed]
15. Zhang, X.; Zhang, J.; Zhou, Z.; Xiong, P.; Cheng, L.; Ma, J.; Wen, Y.; Shen, T.; He, X.; Wang, L.; et al. Integrated network pharmacology, metabolomics, and transcriptomics of Huanglian-Hongqu herb pair in non-alcoholic fatty liver disease. *J. Ethnopharmacol.* **2024**, *10*, 117828. [CrossRef]
16. Zhou, Z.; Chen, B.; Chen, S.; Lin, M.; Chen, Y.; Jin, S.; Chen, W.; Zhang, Y. Applications of Network Pharmacology in Traditional Chinese Medicine Research. *Evid. Based Complement. Altern. Med.* **2020**, *2020*, 1646905. [CrossRef]
17. Zhao, L.; Zhang, H.; Li, N.; Chen, J.; Xu, H.; Wang, Y.; Liang, Q. Network pharmacology, a promising approach to reveal the pharmacology mechanism of Chinese medicine formula. *J. Ethnopharmacol.* **2023**, *309*, 116306. [CrossRef] [PubMed]
18. Wang, Y.; Wang, Y.; Shen, W.; Wang, Y.; Cao, Y.; Nuerbulati, N.; Chen, W.; Lu, G.; Xiao, W.; Qi, R. Grape Seed Polyphenols Ameliorated Dextran Sulfate Sodium-Induced Colitis via Suppression of Inflammation and Apoptosis. *Pharmacology* **2020**, *105*, 9–18. [CrossRef]
19. Hu, Q.; Zhang, W.; Wu, Z.; Tian, X.; Xiang, J.; Li, L.; Li, Z.; Peng, X.; Wei, S.; Ma, X.; et al. Baicalin and the liver-gut system: Pharmacological bases explaining its therapeutic effects. *Pharmacol. Res.* **2021**, *165*, 105444. [CrossRef]
20. Arbuckle, R.B.; Huber, S.L.; Zacker, C. The consequences of diarrhea occurring during chemotherapy for colorectal cancer: A retrospective study. *Oncologist* **2000**, *5*, 250–259. [CrossRef]
21. Richardson, G.; Dobish, R. Chemotherapy induced diarrhea. *J. Oncol. Pharm. Pract.* **2007**, *13*, 181–198. [CrossRef] [PubMed]
22. Ikegami, T.; Ha, L.; Arimori, K.; Latham, P.; Kobayashi, K.; Ceryak, S.; Matsuzaki, Y.; Bouscarel, B. Intestinal alkalization as a possible preventive mechanism in irinotecan (CPT-11)-induced diarrhea. *Cancer Res.* **2002**, *62*, 179–187. [PubMed]
23. Huang, S.Y.; Zou, X. Advances and challenges in protein-ligand docking. *Int. J. Mol. Sci.* **2010**, *11*, 3016–3034. [CrossRef] [PubMed]
24. Cheng, Y.; Zan, J.; Song, Y.; Yang, G.; Shang, H.; Zhao, W. Evaluation of intestinal injury, inflammatory response and oxidative stress following intracerebral hemorrhage in mice. *Int. J. Mol. Med.* **2018**, *42*, 2120–2128. [CrossRef] [PubMed]
25. Izadparast, F.; Riahi-Zajani, B.; Yarmohammadi, F.; Hayes, A.W.; Karimi, G. Protective effect of berberine against LPS-induced injury in the intestine: A review. *Cell Cycle.* **2022**, *21*, 2365–2378. [CrossRef]

26. Zhang, M.; Xia, F.; Xia, S.; Zhou, W.; Zhang, Y.; Han, X.; Zhao, K.; Feng, L.; Dong, R.; Tian, D.; et al. NSAID-Associated Small Intestinal Injury: An Overview from Animal Model Development to Pathogenesis, Treatment, and Prevention. *Front. Pharmacol.* **2022**, *13*, 818877. [CrossRef]
27. Brandt, A.; Baumann, A.; Hernández-Arriaga, A.; Jung, F.; Nier, A.; Staltner, R.; Rajcic, D.; Schmeer, C.; Witte, O.W.; Wessner, B.; et al. Impairments of intestinal arginine and NO metabolisms trigger aging-associated intestinal barrier dysfunction and 'inflammaging'. *Redox Biol.* **2022**, *58*, 102528. [CrossRef]
28. Hasegawa, T.; Mizugaki, A.; Inoue, Y.; Kato, H.A.-O.X.; Murakami, H. Cystine reduces tight junction permeability and intestinal inflammation induced by oxidative stress in Caco-2 cells. *Amino Acids* **2021**, *53*, 1021–1032. [CrossRef]
29. Ahmadi, A.; Shadboorestan, A. Oxidative stress and cancer; the role of hesperidin, a citrus natural bioflavonoid, as a cancer chemoprotective agent. *Nutr. Cancer* **2016**, *68*, 29–39. [CrossRef]
30. Bailly, C. Irinotecan: 25 years of cancer treatment. *Pharmacol. Res.* **2019**, *148*, 104398. [CrossRef]
31. Di Tommaso, N.; Gasbarrini, A.; Ponziani, F.A.-O. Intestinal Barrier in Human Health and Disease. *Int. J. Environ. Res. Public Health* **2021**, *18*, 12836. [CrossRef] [PubMed]
32. Chen, J.; Li, M.; Chen, R.; Xu, Z.; Yang, X.; Gu, H.; Zhang, L.; Fu, C.; Zhang, J.; Wu, Y. Gegen Qinlian standard decoction alleviated irinotecan-induced diarrhea via PI3K/AKT/NF-κB axis by network pharmacology prediction and experimental validation combination. *Chin. Med.* **2023**, *18*, 46. [CrossRef] [PubMed]
33. He, J.; Song, Y.; Li, G.; Xiao, P.; Liu, Y.; Xue, Y.; Cao, Q.; Tu, X.; Pan, T.; Jiang, Z.; et al. Fbxw7 increases CCL2/7 in CX3CR1hi macrophages to promote intestinal inflammation. *J. Clin. Invest.* **2019**, *129*, 3877–3893. [CrossRef] [PubMed]
34. Miles, E.A.; Calder, P.C. Effects of Citrus Fruit Juices and Their Bioactive Components on Inflammation and Immunity: A Narrative Review. *Front. Immunol.* **2021**, *12*, 712608. [CrossRef] [PubMed]
35. Pierrakos, C.; Velissaris, D.; Bisdorff, M.; Marshall, J.C.; Vincent, J.A.-O. Biomarkers of sepsis: Time for a reappraisal. *Crit. Care* **2020**, *24*, 287. [CrossRef]
36. Liu, X.; Fan, Y.; Du, L.; Mei, Z.; Fu, Y. In Silico and In Vivo Studies on the Mechanisms of Chinese Medicine Formula (Gegen Qinlian Decoction) in the Treatment of Ulcerative Colitis. *Front. Pharmacol.* **2021**, *12*, 665102. [CrossRef]
37. Zhao, X.; Kang, B.; Lu, C.; Liu, S.; Wang, H.; Yang, X.; Chen, Y.; Jiang, B.; Zhang, J.; Lu, Y.; et al. Evaluation of p38 MAPK pathway as a molecular signature in ulcerative colitis. *J. Proteome Res.* **2011**, *10*, 2216–2225. [CrossRef]
38. Wei, Y.; Fan, Y.; Huang, S.; Lv, J.; Zhang, Y.; Hao, Z. Baizhu shaoyao decoction restores the intestinal barrier and brain-gut axis balance to alleviate diarrhea-predominant irritable bowel syndrome via FoxO1/FoxO3a. *Phytomedicine* **2024**, *122*, 155163. [CrossRef]
39. Huang, M.Y.; Pan, H.; Liang, Y.D.; Wei, H.X.; Xu, L.H.; Zha, Q.B.; He, X.H.; Ouyang, D.Y. Chemotherapeutic agent CPT-11 eliminates peritoneal resident macrophages by inducing apoptosis. *Apoptosis* **2016**, *21*, 130–142. [CrossRef]
40. Hinz, N.; Jücker, M.A.-O. Distinct functions of AKT isoforms in breast cancer: A comprehensive review. Cell Commun. *Signal* **2019**, *17*, 154. [CrossRef]
41. Li, Y.Z.; Wang, Y.Y.; Huang, L.; Zhao, Y.Y.; Chen, L.H.; Zhang, C. Annexin A protein family in atherosclerosis. *Clin. Chim. Acta* **2022**, *531*, 406–417. [CrossRef]
42. Lei, K.; Wei, Q.; Cheng, Y.; Wang, Z.; Wu, H.; Zhao, F.; Ding, W.; Shi, F. OONO-/MMP2/MMP9 pathway-mediated apoptosis of porcine granulosa cells is associated with DNA damage. *Reproduction* **2023**, *165*, 431–443. [CrossRef] [PubMed]
43. Porter, A.G.; Jänicke, R.U. Emerging roles of caspase-3 in apoptosis. *Cell Death Differ.* **1999**, *6*, 99–104. [CrossRef] [PubMed]
44. Yan, H.; Li, Y.; Yang, B.; Long, F.; Yang, Z.; Tang, D. Exploring the mechanism of action of Yiyi Fuzi Baijiang powder in colorectal cancer based on network pharmacology and molecular docking studies. *Biotechnol. Genet. Eng. Rev.* **2023**, 1–21. [CrossRef]
45. Gobé, G.; Zhang, X.J.; Cuttle, L.; Pat, B.; Willgoss, D.; Hancock, J.; Barnard, R.; Endre, R.B. Bcl-2 genes and growth factors in the pathology of ischaemic acute renal failure. *Immunol. Cell Biol.* **1999**, *77*, 279–286. [CrossRef] [PubMed]
46. Yang, J. PPAR-γ silencing inhibits the apoptosis of A549 cells by upregulating Bcl-2. Chin. *J. Lung Cancer* **2013**, *16*, 125–130.
47. Muhammad, T.A.-O.; Ikram, M.; Ullah, R.; Rehman, S.U.; Kim, M.O. Hesperetin, a Citrus Flavonoid, Attenuates LPS-Induced Neuroinflammation, Apoptosis and Memory Impairments by Modulating TLR4/NF-κB Signaling. *Nutrients* **2019**, *11*, 648. [CrossRef]
48. Ashkenazi, A.; Fairbrother, W.J.; Leverson, J.D.; Souers, A.J. From basic apoptosis discoveries to advanced selective BCL-2 family inhibitors. *Nat. Rev. Drug Discov.* **2017**, *16*, 273–284. [CrossRef]
49. Wang, J.; Feng, W.; Zhang, S.; Chen, L.; Tang, F.; Sheng, Y.; Ao, H.; Peng, C. Gut microbial modulation in the treatment of chemotherapy-induced diarrhea with Shenzhu Capsule. *BMC Complement. Altern. Med.* **2019**, *19*, 126. [CrossRef]
50. Li, W.; Zhang, Y.; Wang, Q.; Wang, Y.; Fan, Y.; Shang, E.; Jiang, S.; Duan, J. 6-Gingerol ameliorates ulcerative colitis by inhibiting ferroptosis based on the integrative analysis of plasma metabolomics and network pharmacology. *Food Funct.* **2024**, *15*, 6054–6067. [CrossRef]
51. Wang, X.; Shen, Y.; Wang, S.; Li, S.; Zhang, W.; Liu, X.; Lai, L.; Pei, J.; Li, H. *PharmMapper* 2017 update: A web server for potential drug target identification with a comprehensive target pharmacophore database. *Nucleic Acids Res.* **2017**, *45*, W356–W360. [CrossRef] [PubMed]
52. Rebhan, M.; Chalifa-Caspi, V.; Prilusky, J.; Lancet, D. GeneCards: Integrating information about genes, proteins and diseases. *Trends Genet.* **1997**, *13*, 163. [CrossRef] [PubMed]

53. Amberger, J.S.; Bocchini, C.A.; Scott, A.F.; Hamosh, A. OMIM.org: Leveraging knowledge across phenotype-gene relationships. *Nucleic Acids Res.* **2019**, *47*, D1038–D1043. [CrossRef] [PubMed]
54. Bardou, P.; Mariette, J.; Escudié, F.; Djemiel, C.; Klopp, C. jvenn: An interactive Venn diagram viewer. *BMC Bioinform.* **2014**, *15*, 293. [CrossRef]
55. Shannon, P.; Markiel, A.; Ozier, O.; Baliga, N.S.; Wang, J.T.; Ramage, D.; Amin, N.; Schwikowski, B.; Ideker, T. Cytoscape: A software environment for integrated models of biomolecular interaction networks. *Genome Res.* **2003**, *13*, 2498–2504. [CrossRef]
56. Szklarczyk, D.; Gable, A.L.; Lyon, D.; Junge, A.; Wyder, S.; Huerta-Cepas, J.; Simonovic, M.; Doncheva, N.T.; Morris, J.H.; Bork, P.; et al. STRING v11: Protein-protein association networks with increased coverage, supporting functional discovery in genome-wide experimental datasets. *Nucleic Acids Res.* **2019**, *47*, D607–D613. [CrossRef]
57. Cao, Y.; Wang, C.; Dong, L. Exploring the Mechanism of White Peony in the Treatment of Lupus Nephritis Based on Network Pharmacology and Molecular Docking. *Arch. Esp. Urol.* **2023**, *76*, 123–131. [CrossRef]
58. Tang, D.; Chen, M.; Huang, X.; Zhang, G.; Zeng, L.; Zhang, G.; Wu, S.; Wang, Y. SRplot: A free online platform for data visualization and graphing. *PLoS ONE* **2023**, *18*, e0294236. [CrossRef]
59. Burley, S.K.; Bhikadiya, C.; Bi, C.; Bittrich, S.; Chen, L.; Crichlow, G.V.; Christie, C.H.; Dalenberg, K.; Di Costanzo, L.; Duarte, J.M.; et al. RCSB Protein Data Bank: Powerful new tools for exploring 3D structures of biological macromolecules for basic and applied research and education in fundamental biology, biomedicine, biotechnology, bioengineering and energy sciences. *Nucleic Acids Res.* **2021**, *49*, D437–D451. [CrossRef]
60. Seeliger, D.; de Groot, B.L. Ligand docking and binding site analysis with PyMOL and Autodock/Vina. *J. Comput. Aided Mol. Des.* **2010**, *24*, 417–422. [CrossRef]
61. Wang, R.; Fang, X.; Lu, Y.; Wang, S. The PDBbind database: Collection of binding affinities for protein-ligand complexes with known three-dimensional structures. *J. Med. Chem.* **2017**, *48*, 3126–3139. [CrossRef] [PubMed]
62. Kozakov, D.; Grove, L.E.; Hall, D.R.; Bohnuud, T.; Mottarella, S.E.; Luo, L.; Xia, B.; Beglov, D.; Vajda, S. The FTMap family of web servers for determining and characterizing ligand-binding hot spots of proteins. *Nat. Protoc.* **2015**, *10*, 733–755. [CrossRef] [PubMed]

International Journal of
Molecular Sciences

MDPI

Article

Network Pharmacology and Bioinformatics Study of Geniposide Regulating Oxidative Stress in Colorectal Cancer

Yingzi Wu, Jinhai Luo and Baojun Xu *

Guangdong Provincial Key Laboratory IRADS, Department of Life Sciences, BNU-HKBU United International College, Zhuhai 519087, China
* Correspondence: baojunxu@uic.edu.cn; Tel.: +86-756-2620636

Abstract: This study aims to identify the mechanism of geniposide regulating oxidative stress in colorectal cancer (CRC) through network pharmacology and bioinformatics analysis. Targets of geniposide, oxidative stress-related targets and targets related to CRC were applied from databases. The hub genes for geniposide regulating oxidative stress in CRC were identified with the protein–protein interaction (PPI) network. Furthermore, we applied Gene ontology (GO) and Kyoto Encyclopedia of Genes and Genomes (KEGG) enrichment to analyze the hub genes from a macro perspective. We verified the hub genes by molecular docking, GEPIA, HPA and starBase database. We identified five hub genes: *IL1B, GSK3B, NOS3, RELA and CDK4*. GO analysis results suggested that the anti-colorectal cancer effect of geniposide by regulating oxidative stress is possibly related to the influence of multiple biological processes, including response to temperature stimulus, response to alkaloid, nitric oxide biosynthetic process, nitric oxide metabolic process, reactive nitrogen species metabolic process, cellular response to peptide, etc. KEGG enrichment analysis results indicated that the PI3K–Akt signaling pathway, IL-17 signaling pathway, p53 signaling pathway, NF-κB signaling pathway and NOD-like receptor signaling pathway are likely to be the significant pathways. Molecular docking results showed that the geniposide had a good binding activity with the hub genes. This study demonstrates that geniposide can regulate oxidative stress in CRC, and induction of oxidative stress is one of the possible mechanisms of anti-recurrence and metastasis effects of geniposide against CRC.

Keywords: oxidative stress; CRC; network pharmacology; molecular docking; bioinformatics; geniposide

1. Introduction

Colorectal cancer is one of the top three malignancies with the highest mortality rate today [1]. Clinical treatments for colorectal cancer mainly include endoscopy, surgical resection, radiotherapy, immunotherapy and palliative chemotherapy targeted therapy, extensive surgery, and local ablation of metastases, which can extend the survival of patients to a certain extent [1]. However, due to the occurrence of tumor drug resistance, recurrence and metastasis, colorectal cancer is still a difficult problem in medical treatment [2]. It is urgent to explore effective and low-toxicity therapeutic methods to combat colorectal cancer [3].

Fructus Gardeniae (FG) is the dried fruit of *Gardenia jasminoides* Ellis (GjE), belonging to the family Rubiaceae [4]. Geniposide ($C_{17}H_{24}O_{10}$), a well-known iridoid glycoside compound, is one of the main bioactive components of FG [5]. Previous studies have shown that geniposide has many positive effects, such as anti-inflammatory, anti-tumor, blood glucose and lipid regulation, antibacterial, anti-depression, anti-oxidation, immune regulation, and so on [6]. Geniposide has also been shown to greatly reduce the growth of various kinds of cancer cell lines, including diffuse large B-cell lymphoma cells [7], medulloblastoma cells [8] and gastric MKN45 cells [9]. Though many emerging benefits of geniposide have been found, the studies on geniposide anti-colorectal cancer are still rare.

Oxidative stress is an imbalance between the production of reactive oxygen species (ROS) and the body's antioxidant defenses. This imbalance contributes to chronic inflammation, which can lead to precancerous conditions. ROS can influence cellular functions, including gene regulation and apoptosis, making oxidative stress a significant factor in cancer development [10–12]. Network pharmacology is an emerging, systematic approach to exploring the complex mechanisms of traditional Chinese medicine (TCM). It deviates from the traditional 'one drug, one target' model, highlighting that multiple substances can interact with various genes or proteins. This approach allows for a more holistic understanding of medicine and disease interactions, providing insights into multiple medications and targets [13,14]. In order to determine the utility of geniposide as a medicine, we decided to explore its impact on CRC using a network pharmacological method.

In this study, network pharmacology was used to identify the pharmacologic target network of geniposide, and bioinformatics was used to compile information on the relevant targets of colorectal cancer. The intersection targets were subjected to GO and KEGG enrichment analysis. To clarify the pathophysiology of colorectal cancer and offer a fresh perspective on clinical intervention therapy, we verified the hub genes by using molecular docking, GEPIA, HPA, and the starBase database. The study flowchart is shown in Figure 1.

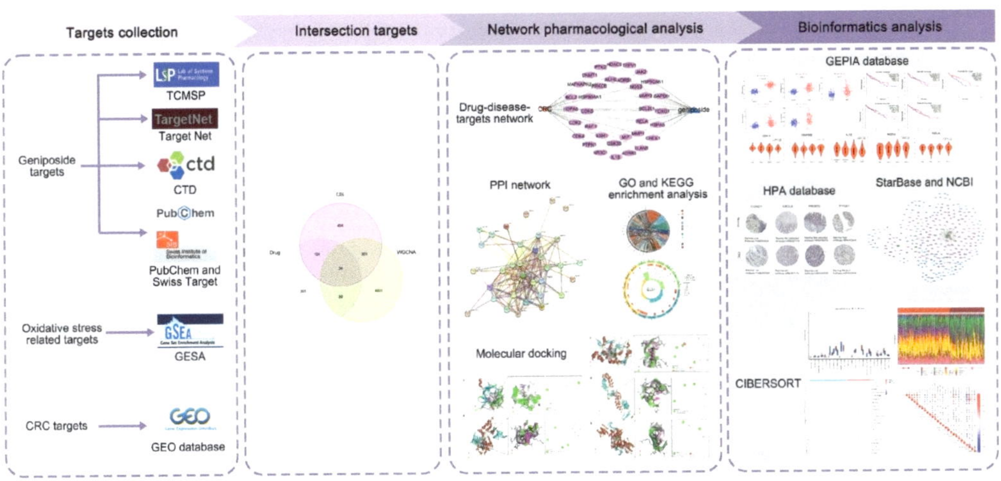

Figure 1. A flowchart of this study on geniposide regulating oxidative stress in colorectal cancer.

2. Results

2.1. Construction of Target Network Map of Geniposide

We retrieved 102, 473, 24 and 7 geniposide-related targets, respectively, through the Swiss Target Prediction, TargetNet, CTD (Comparative Toxicogenomics Database) and TCMSP databases (Traditional Chinese Medicine Systems Pharmacology Database). After merging and deduplicating the related targets of the four databases, 548 geniposide-related targets in total were obtained. Subsequently, we used the Cytoscape software to construct a network diagram of geniposide and its related targets, as shown in Figure 2.

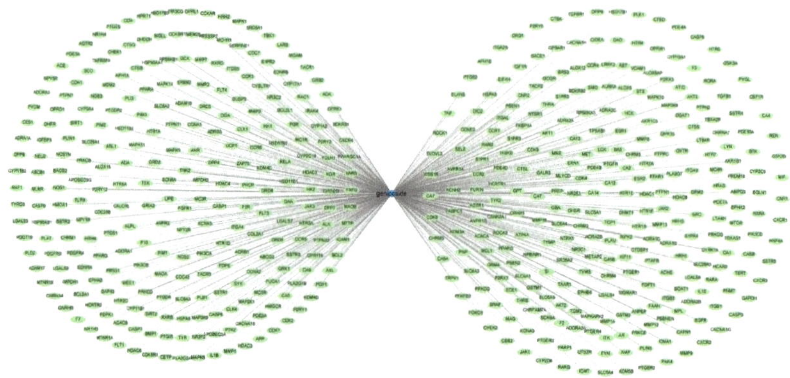

Figure 2. Geniposide-targets network, 548 geniposide-related targets in total were obtained.

2.2. Screening of Gene Modules and Construction of Co-Expression Network

To identify the hub gene modules associated with CRC, we used the WGCNA algorithm to establish co-expression networks and modules in healthy people and CRC patients. We calculated the variance of the expression of each gene in GSE44076 and then selected the top 25% of genes with the most significant variance for further analysis. Co-expressed gene modules were identified when the soft power value was set to 10, and the scale-free R2 equaled 0.9, as illustrated in Figure 3A. Using the dynamic cutting algorithm, a total of nine co-expression modules with different colors were obtained when presenting the heat map of the topological overlap matrix (TOM), as shown in Figure 3B–D. Subsequently, these genes in the nine color modules were sequentially applied to analyze the similarity and adjacency of the co-expression of the module-clinical features (normal group and CRC group). Furthermore, the green module showed the most substantial relationship with CRC, which included 5157 genes, as shown in Figure 3E. Finally, we observed a positive correlation of green module genes in different CRC samples, as indicated in Figure 3F.

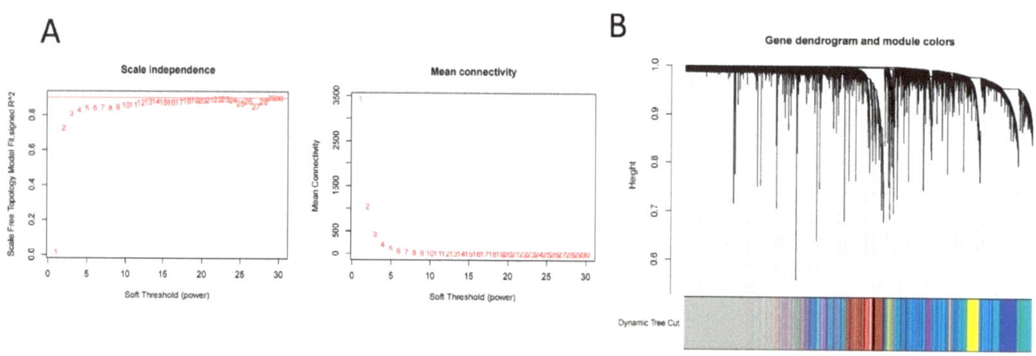

Figure 3. *Cont.*

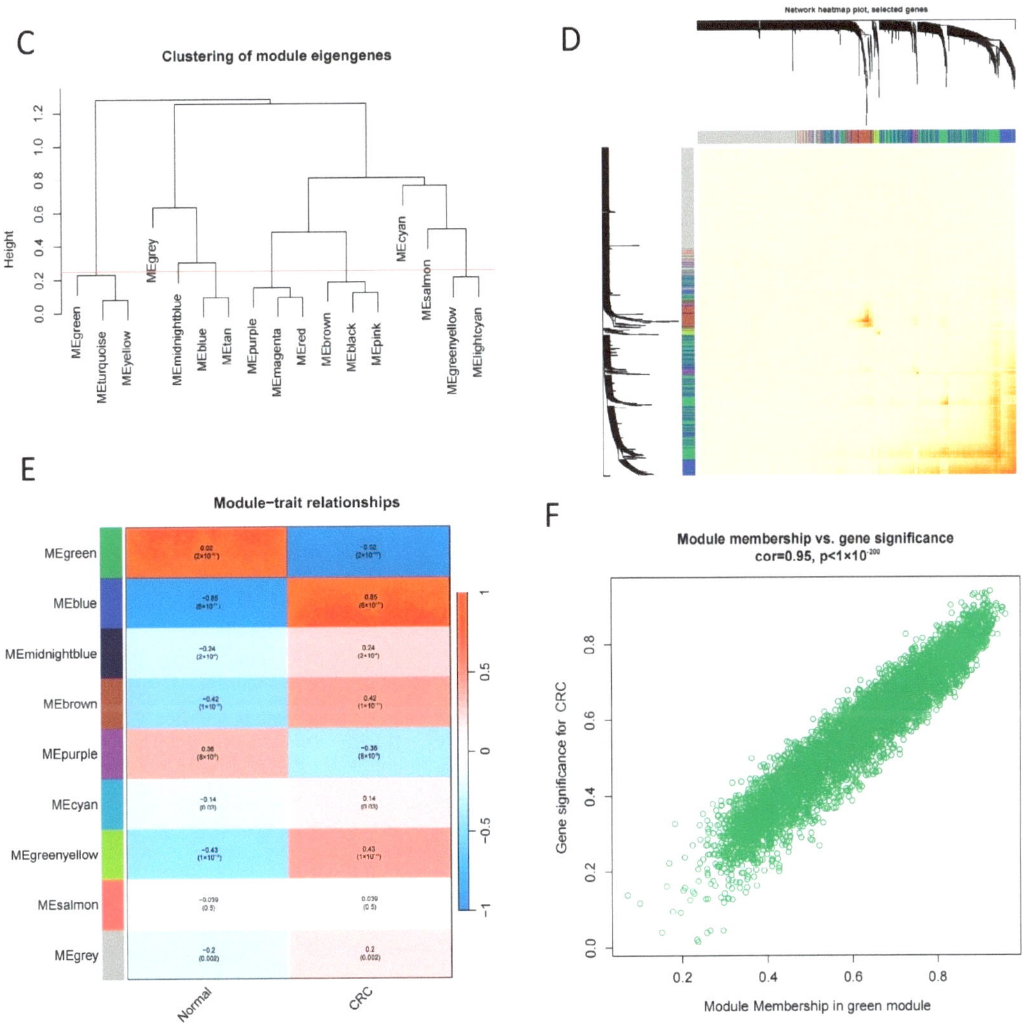

Figure 3. WGCNA analysis chart; (**A**) Network topology for different soft-thresholding powers. Numbers in the plots indicate the corresponding soft thresholding powers. (**B**) Gene dendrogram obtained by clustering the dissimilarity based on consensus Topological Overlap with the corresponding module colors indicated by the color row. (**C**) Dendrogram of consensus module eigengenes obtained by WGCNA on the consensus correlation. The red line is the merging threshold, and groups of eigengenes below the threshold represent modules whose expressions profiles should be merged due to their similarity. (**D**) The heatmap depicts the TOM among all modules included in the analysis. The progressively darker red color represents an increasing overlap. (**E**) Module–trait associations. Red represents positive correlation, while blue represents negative correlation. (**F**) Module membership of green module, which represents a positive correlation of green module genes in different CRC samples.

2.3. Screening of Core Genes

In order to explore the role of geniposide regulating oxidative stress against colorectal cancer, we intersected the targets of geniposide with the most critical module genes and oxidative stress-related genes in CRC, and obtained a total of 34 core genes as shown in

Figure 4, including *DNMT1, HSP90AA1, PTPN1, ESR1, PTK2, RELA, CDK4, CDK5, ADRB1, MAPKAPK2, TRPV1, CDK2, NOS3, JAK2, MIF, GSK3B, ACHE, MMP3, NR3C1, HSP90AB1, ADRB3, MMP1, BCL2, ELANE, HDAC2, RAC1, PRKCB, BCL2L1, GAPDH, HSPA8, HSPA5, CHEK1, FOXO1, IL1B*. Then, we constructed a compound–disease–targets network diagram using the Cytoscape software, as illustrated in Figure 5.

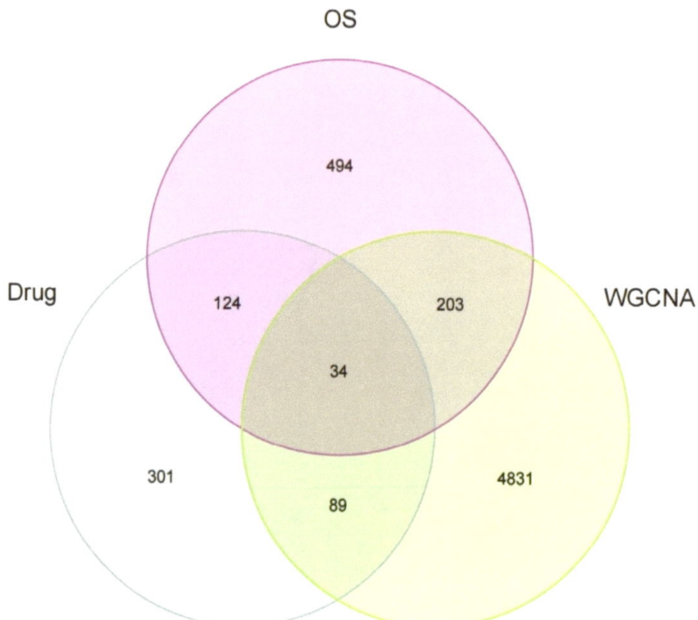

Figure 4. The intersecting targets between potential targets of geniposide, oxidative stress-related genes, and CRC-related genes.

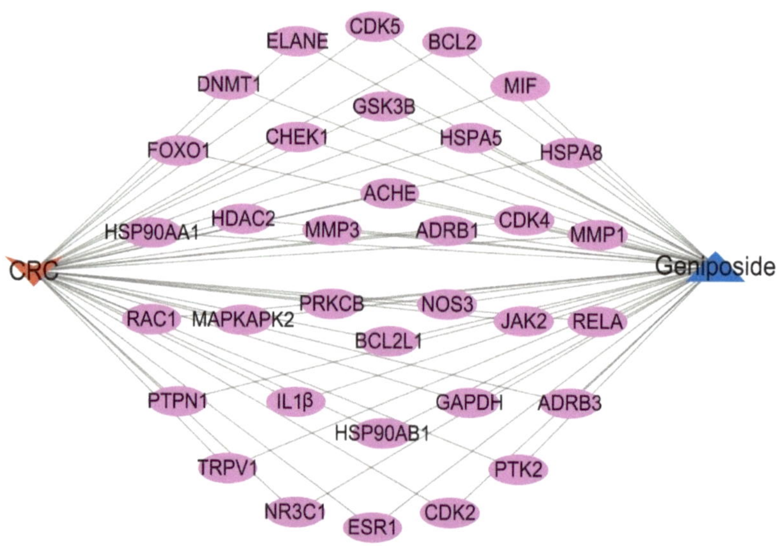

Figure 5. Compound–disease–targets network, purple represents the targets related to compound as well as disease.

2.4. Core Genes PPI Network Construction and Enrichment Analysis

We imported the above 34 core genes into the String database to obtain the PPI network between the core genes, as shown in Figure 6.

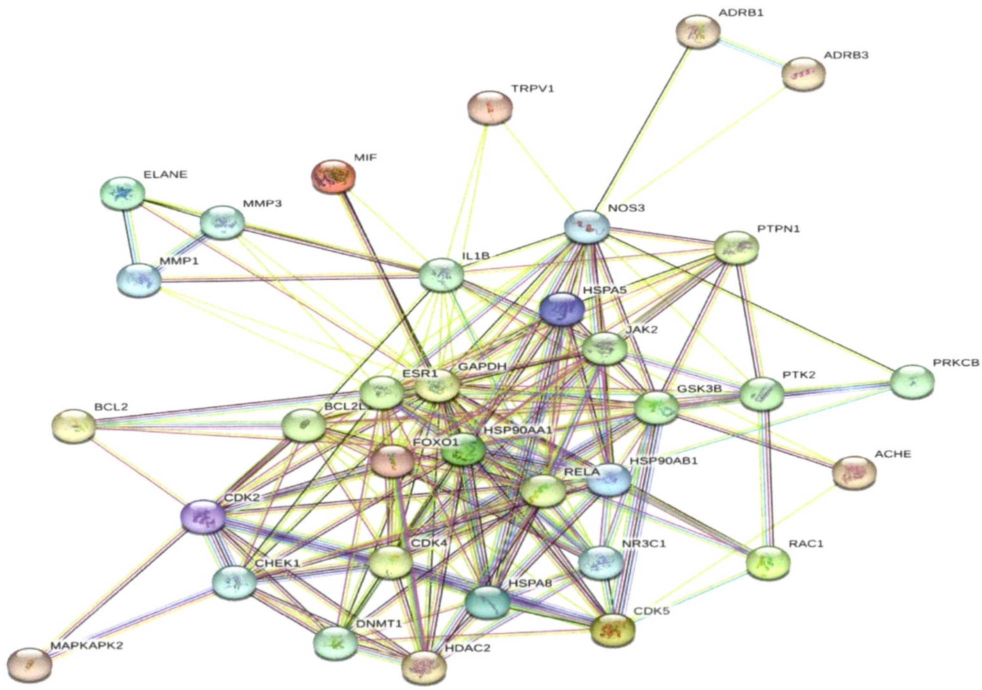

Figure 6. Protein–protein interaction Network.

Through the "cluster profiler" package of the R language software, the GO function enrichment analysis of the 34 core genes of geniposide anti-colorectal cancer by regulating oxidative stress was carried out. A total of 1582 GO entries were obtained (p-value < 0.05), and the main category was the biological process (BP), cell composition (CC), and molecular function (MF), as illustrated in Table 1. Among them were 1395 BP entries, mainly involving response to temperature stimulus, response to alkaloid and nitric oxide biosynthetic process, and so on. There were 61 CC entries, mainly related to melanosome, pigment granule, ficolin-1-rich granule, and so on. Also,126 MF entries were mainly related to histone kinase activity, ubiquitin protein ligase binding, ubiquitin-like protein ligase binding, and so on. The above analysis showed that geniposide is likely to act on these cell structures, affect these biological processes, and exert these molecular functions to achieve the purpose of anti-colorectal cancer by regulating oxidative stress, as illustrated in Figure 7.

By utilizing the "cluster profiler" package of the R language software, the KEGG pathway enrichment analysis of the 34 core genes of geniposide on CRC was carried out. The results showed 87 signal pathways (p-value < 0.05) in which geniposide interfered with CRC, mainly involving pathways in the cancer signaling pathway, PI3K–Akt signaling pathway, IL-17 signaling pathway, p53 signaling pathway, NF-κB signaling pathway, and NOD-like receptor signaling pathway, as shown in Figure 8.

Table 1. GO enrichment results.

Classification	Serial Number	Biological Function
BP	GO:0009266	response to temperature stimulus
BP	GO:0043279	response to alkaloid
BP	GO:0006809	nitric oxide biosynthetic process
BP	GO:0046209	nitric oxide metabolic process
BP	GO:2001057	reactive nitrogen species metabolic process
BP	GO:1901653	cellular response to peptide
CC	GO:0042470	melanosome
CC	GO:0048770	pigment granule
CC	GO:0101002	ficolin-1-rich granule
CC	GO:0098978	glutamatergic synapse
CC	GO:1904813	ficolin-1-rich granule lumen
CC	GO:0000307	cyclin-dependent protein kinase holoenzyme complex
MF	GO:0035173	histone kinase activity
MF	GO:0031625	ubiquitin protein ligase binding
MF	GO:0044389	ubiquitin-like protein ligase binding
MF	GO:0031072	heat shock protein binding
MF	GO:0140662	ATP-dependent protein folding chaperone
MF	GO:0048156	tau protein binding

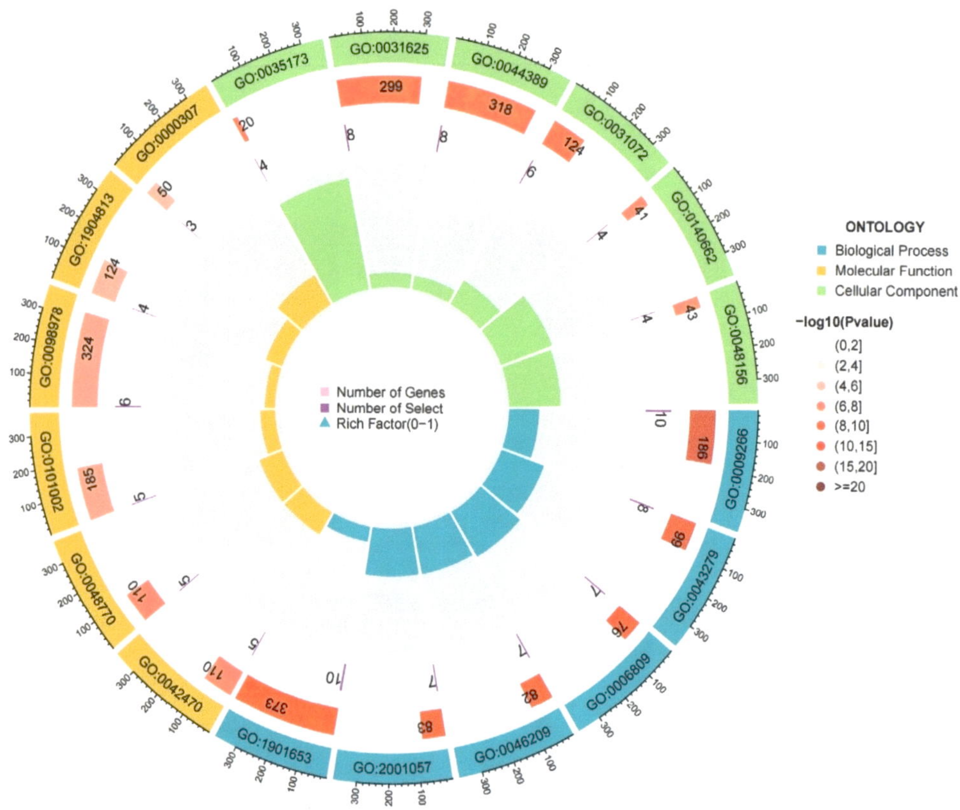

Figure 7. GO enrichment analysis; blue represents Biological Process (BP), yellow represents Molecular Function (MF), and green represents Cellular Component (CC).

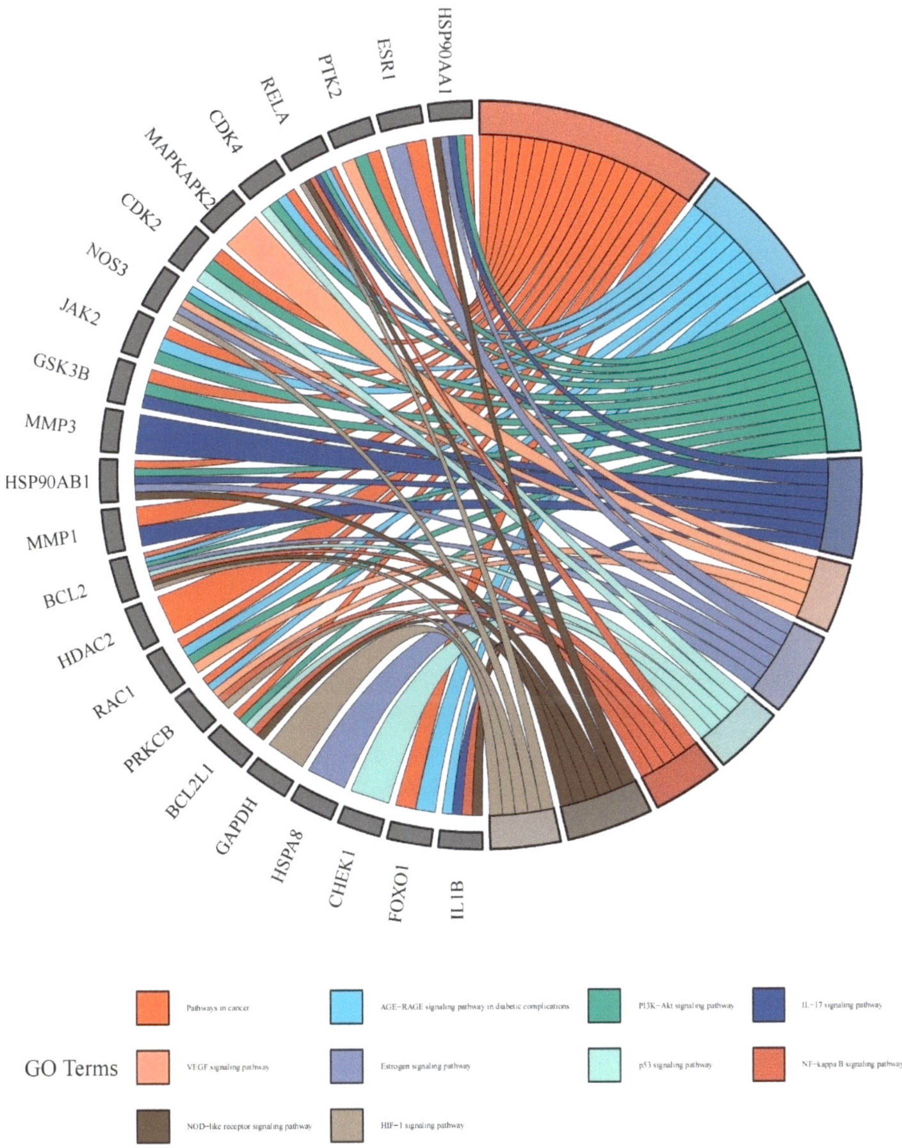

Figure 8. KEGG enrichment analysis; red represents signaling pathway in cancer, dark blue represents IL-17 signaling pathway, light green represents p53 signaling pathway, red represents NF-κB signaling pathway, and brown represents NOD-like receptor signaling pathway.

Finally, we imported the compound–disease–therapeutic targets-core pathways network file into Cytoscape software to draw the pathway network diagram, which displayed more intuitively the multi-component-multi-target action characteristics of active ingredients of traditional Chinese medicine in the treatment of CRC as illustrated in Figure 9.

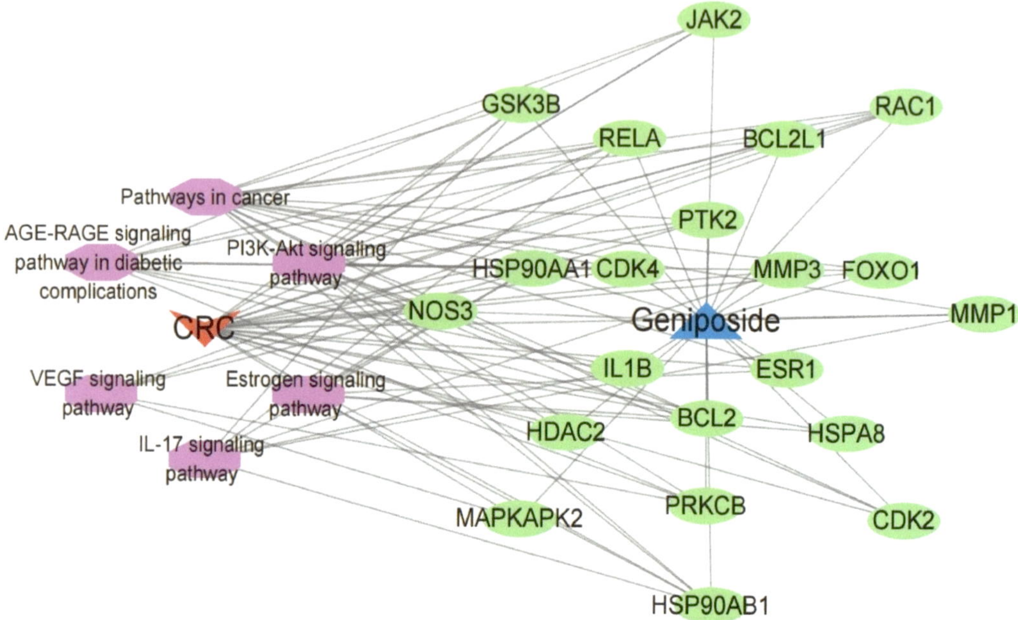

Figure 9. Compound–disease–therapeutic targets-core pathways network; red triangles represent diseases, blue triangles represent compounds, purple ovals represent pathways, and green ovals represent targets.

2.5. Hub Genes Screening and Correlation Analysis of Pyroptosis-Related Genes

As shown in Figure 10, we screened 34 core genes and found that when the degree value was greater than the average, 18 genes were selected (*IL1B, GSK3B, NOS3, RELA, CDK4, HSP90AB1, NR3C1, FOXO1, JAK2, CDK2, HSPA8, CDK5, HSPA5, CHEK1, HDAC2, PTK2, DNMT1, PTPN1*). As shown in Figure 10B,C, according to degree, all of the top five genes (*IL1B, GSK3B, NOS3, RELA, and CDK4*) are above 15, which were selected as hub genes for subsequent analysis. The functional enrichment of GO and KEGG suggested that the mechanism of geniposide's anti-colorectal cancer regulating oxidative stress may be related to cell pyroptosis, and oxidative stress is one of the vital induction methods of cell pyroptosis. In order to explore the correlation between hub genes and pyroptotic genes, we conducted a correlation analysis between hub genes and pyroptotic hotspot genes (*IL1A, IL1B, IL18, GZMB, GZMA, GSDMB, CYCS, CHMP2B, CASP5, BAK1*). The results showed that *1L1B, CDK4* and *RELA* are closely related to pyroptosis hotspot genes, as shown in Figure 11.

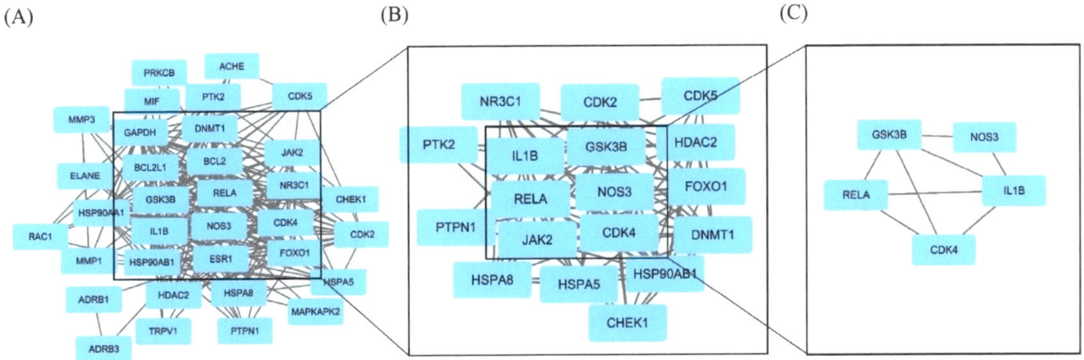

Figure 10. Hub genes screening map; (**A**) represents interaction relationship of 34 core genes; (**B**) represents interaction relationship of genes whose degrees are higher than the average degree; (**C**) represents interaction relationship of hub genes.

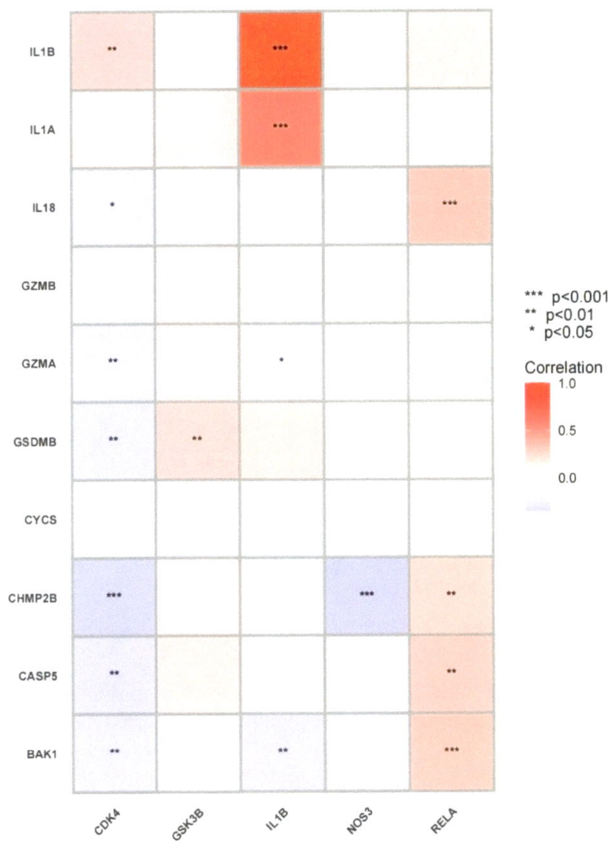

Figure 11. Correlation analysis between hub genes and pyroptosis hotspot genes; the horizontal coordinate represents hub genes, and the vertical coordinate represents hotspot genes of pyroptosis.

2.6. Hub Genes ceRNA Network Construction

We constructed a ceRNA network based on hub genes with starBase, the miRDB database and the miRnada database. The network includes 319 nodes (5 hub genes, 157 miRNAs and 157 lncRNAs) (Figure 12). In detail, a total of 119 lncRNAs could competitively bind miRNA hsa-miR-4318, hsa-miR-3147, and hsa-miR-29a-5p to regulate *GSK3B*. In the ceRNA network, 11 lncRNAs can combine with miRNA hsa-miR-495-3p, hsa-miR-326, and hsa-miR-4303 to regulate the hub gene *IL1B*. miRNA hsa-miR-4318, hsa-miR-4280, hsa-miR-103b, hsa-miR-506-3p, hsa-miR-198, hsa-miR-568, hsa-miR-149-3p, and hsa-miR-486-5p bind to regulate the hub gene *CDK4*. miRNA hsa-miR-4318, hsa-miR-4254, hsa-miR-761, hsa-miR-4297, hsa-miR-3202, hsa-miR-1303, hsa-miR-4324, and hsa-miR-214-3p bind to regulate the hub gene *NOS3*. In addition, 18 lncRNAs could combine with has-miR-156-5p, hsa-miR-858-3p and hsa-miR-7-5p to regulate the hub gene *RELA*.

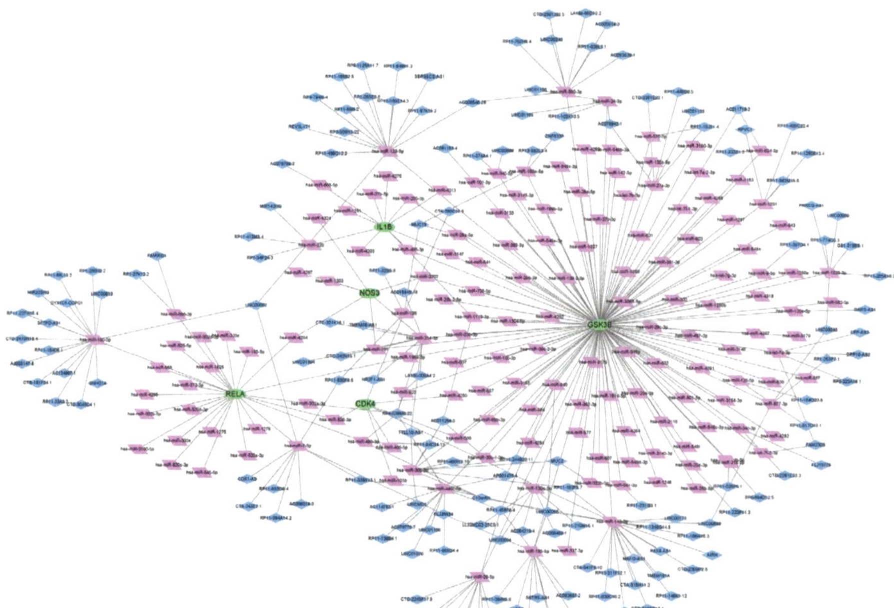

Figure 12. ceRNA network diagram of hub genes; purple represents miRNA, blue represents lncRNA, and green represents mRNA.

2.7. PPI Network Analysis, Differential Expression Lever, Pathological Staging, and OS Analysis of Hub Genes

To understand the interrelationships among the hub genes, we conducted a PPI network analysis on them, as shown in Figure 13. Subsequently, we conducted differential expression levels, pathological stage, and OS in the GEPIA database. The results showed that the expression of hub genes in the CRC group was more remarkable than that in the normal group with p-value < 0.05, as shown in Figure 14A. Moreover, the prognosis of the *IL1B* high expression group was statistically significant, as illustrated in Figure 14B. Pathological staging showed that as the disease progressed, the mRNA levels of *IL1B* gradually decreased while the mRNA levels of *NOS3* gradually increased, as shown in Figure 14C. The results of the HPA database showed that the other hub genes were expressed to varying degrees in colorectal cancer tissue, except for *IL1B*, which had no corresponding data in the database (Figure 15).

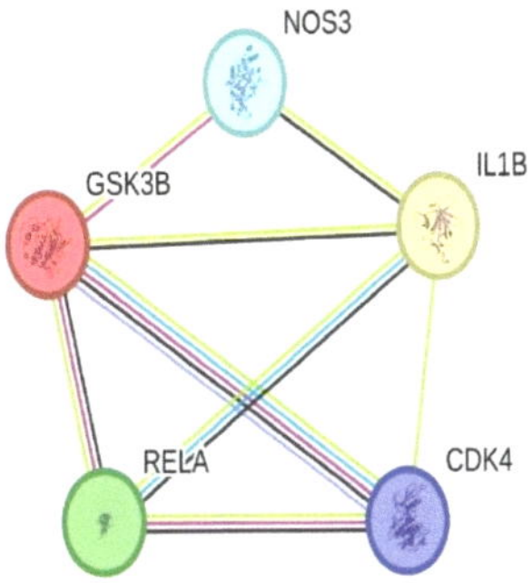

Figure 13. Hub genes PPI network.

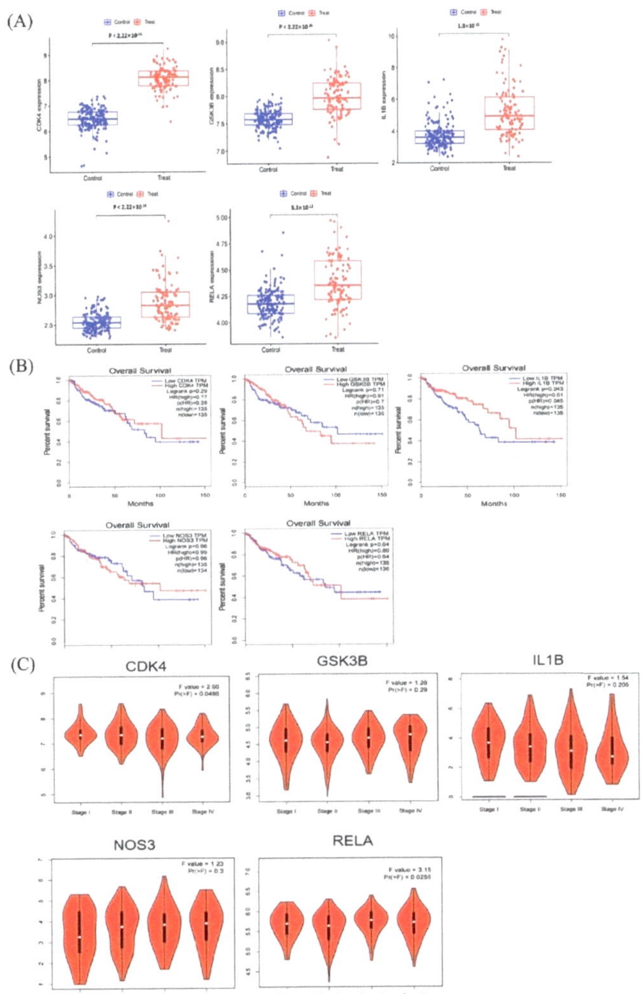

Figure 14. mRNA expression level, pathological stage, and OS in the GEPIA database. (**A**) Box plots showing the mRNA expression levels of *CDK4, GSK3B, IL1B, NOS3*, and *RELA*. Red represents tumor

group, while blue represents normal group. (**B**) The line charts show the overall survival (OS) of hub genes in GEPIA. The survival curve compares the patients with high (red) and low (blue) expression in CRC. (**C**) The violin diagram indicates the stage plot of mRNA expression level and pathological stage in the GEPIA database.

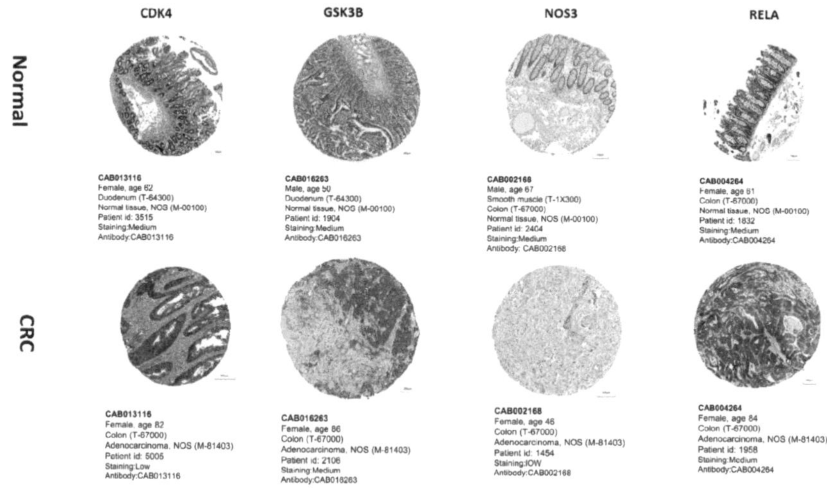

Figure 15. The protein expression levels of hub genes with microscope's magnification of 200 μm in the HPA database.

2.8. Immune Infiltration Analysis of Hub Genes

The difference in immune microenvironment between the normal group and the CRC group was analyzed by CYBERSORT. The difference in immune cell content between the normal group and the CRC group can be visually seen in Figure 16A,B. Statistical "wilcox.test" method was used to compare the difference between the two groups (Figure 16C). Eosinophil, Macrophages M0, Macrophages M1, Macrophages M2, Mast cells activated, Mast cells resting, Neutrophils, Plasma cells, T cells CD4 memory activated, T cells CD4 memory resting, T cells CD4 naive, T cells CD8, T cells gamma delta, T cells regulatory (Tregs) and so on were significantly different (p-value < 0.05) between the two groups. Among them, Eosinophil, Macrophages M2, Plasma cells, T cells CD4 memory resting, T cells CD8, T cells gamma delta, T cells regulatory (Tregs) were the most invasive immune cells in the CRC group. The M1/M2 ratio imbalance had significance in the formation of malignancies, immune escape, and subsequent metastatic treatment resistance. The relationship between the hub genes and immune cells is shown in Figure 16D, which suggested that there was a strong positive correlation between the hub genes that were all strongly positively correlated with Macrophages M0, Macrophages M1, and Mast cells activated, while a strong negative correlation with Plasma cells, T cells CD8, T cells CD4 memory resting, T cells gamma delta, Macrophages M2, Mast cells resting, and Eosinophils.

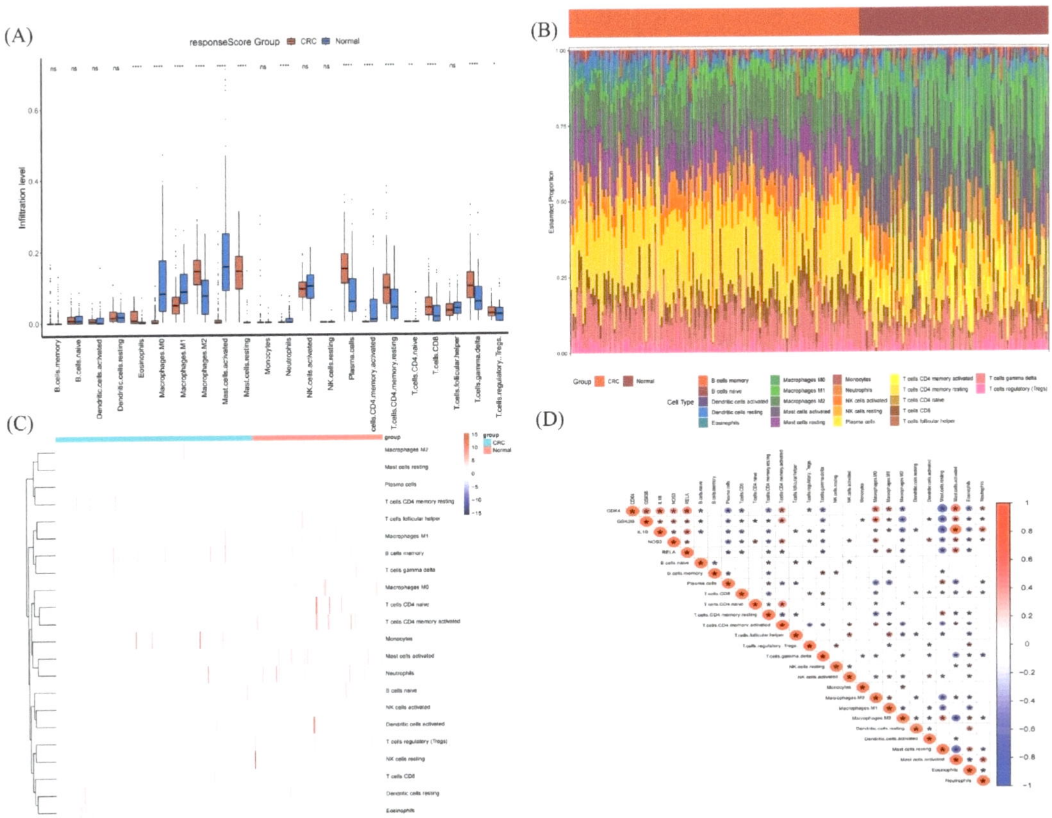

Figure 16. Immune infiltration analysis of hub genes; (**A**) reflects the difference in immune cell infiltration between normal group and cancer group; (**B**) shows the difference in immune cell infiltration between each sample; (**C**) represents the difference in immune cell infiltration between the normal group and the tumor group, it was statistically analyzed, and "ns" indicated that the difference was statistically insignificant; (**D**) indicates that five genes were strongly correlated with each other, with red representing positive correlation, blue representing negative correlation, and "*" indicating statistically significant correlation, and "*" means p-value < 0.05, "**" means p-value < 0.01, "****" means p-value < 0.001.

2.9. Molecular Docking Results

Results of *IL1B*_1I1B: The binding energy between the small molecule and the receptor was −6.2kcal/mol, which demonstrated that it had a good interaction force and formed hydrogen bond associations mainly with GLY A49 and LYS A103, as shown in Figure 17A. Results of *GSK3B*_1J1B: The binding energy between the small molecule and the receptor was −7.5 kcal/mol, which demonstrated that it had a good binding effect. Small molecules interacted with receptor proteins, mainly forming hydrogen bonds with ASN A152, GLNA151, ARGA107, ILEA28, and LYS A103, as shown in Figure 17B. Results of *NOS3*_3E7S: The binding energy between the small molecule and the receptor was −7.0 kcal/mol, which indicated that it has a good binding effect. Small molecules interacted with receptor proteins, mainly with GLUA284, ARGA106, TRPA370, and TRPA279, and obvious hydrogen bond interactions were found, as shown in Figure 17C. Results of *RELA*_1NFI: The binding energy of the small molecule to the receptor was −6.3 kcal/mol, which indicated that it had a good binding effect. Small molecules interacted with receptor proteins mainly through the formation of hydrogen bonds and hydrophobic bond forces, including hydrogen bonds THRA164, ASNA131, and GLUE92, as shown in Figure 17D. Results of *CDK4*_3G33: The binding energy of the small molecule to the receptor was −6.4 kcal/mol, which indicated that it had a good binding effect. Small molecules interacted with receptor proteins, including hydrogen bonds formed by GLUB74 and GLUD175, as shown in Figure 17E. The docking scores of geniposide and hub genes can be seen in Table 2.

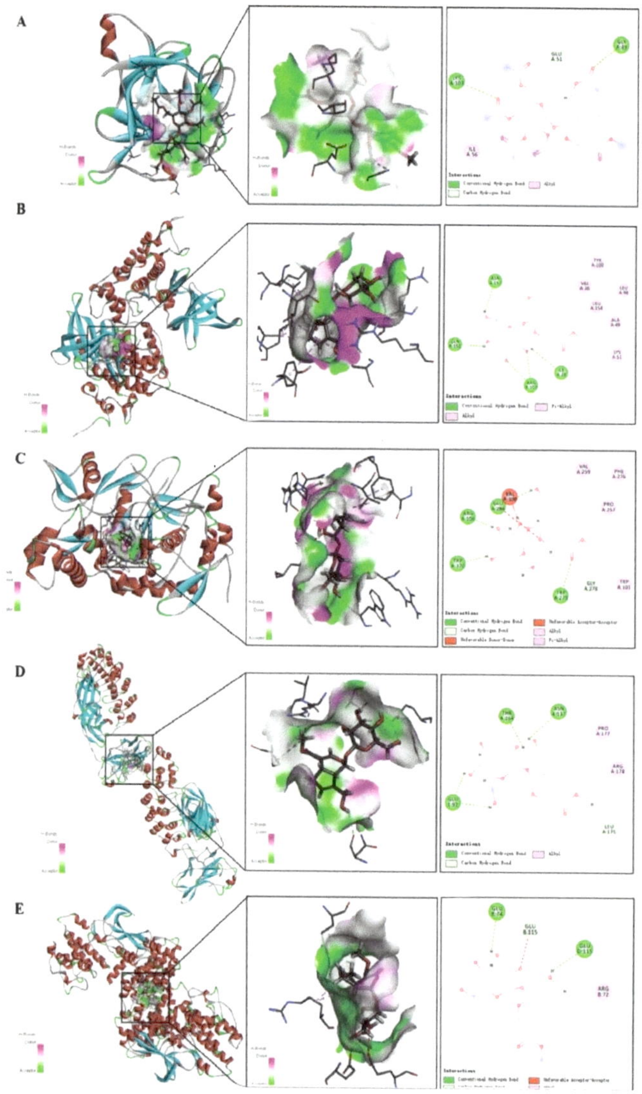

Figure 17. Schematic diagram of docking between geniposide and *IL1B, GSK3B, NOS3, RELA* and *CDK4* (**A**–**E**). (**A**) Represents the molecular binding of geniposide with *IL1B*; (**B**) represents the molecular binding of geniposide with *GSK3B*. (**C**) Represents the molecular binding of geniposide with *NOS3*; (**D**) represents the molecular binding of geniposide with *RELA*; (**E**) represents the molecular binding of geniposide with *CDK4*.

Table 2. Docking scores of geniposide and hub genes.

Compound	PDB ID	Gene	Binding Energy
geniposide	1I1B	*IL1B*	−6.2
geniposide	1J1B	*GSK3B*	−7.5
geniposide	3E7S	*NOS3*	−7.0
geniposide	1NFI	*RELA*	−6.3
geniposide	3G33	*CDK4*	−6.4

3. Discussion

The prevalence of colorectal cancer has been rising annually in recent years because of changes in human living habits, and the age at which it occurs is getting younger. In the past 10 years, more and more natural products have been confirmed to have anti-tumor effects. Geniposide is the primary active component in the herbal remedy Gardenia jasminoides used in traditional Chinese medicine. It has activities in anti-inflammatory, antimicrobial, anti-tumor, anti-oxidation, and immunological regulatory, according to preliminary research. Oxidative stress is produced when the ratio of reactive oxygen species (ROS) production to antioxidant effectiveness is out of balance. Excessive ROS synthesis can cause cellular damage and is one of the factors resulting in many disorders, including cancer. This study aims to explore the molecular mechanism of geniposide in the treatment of colorectal cancer by regulating oxidative stress and the bioinformatics analysis of its hub genes to provide a reference for the development of new treatment strategies. Through the Swiss Target Prediction, TargetNet, CTD and TCMSP databases, we collected 548 targets of geniposide. Then, we searched the CRC dataset in the GEO database and used the WGCNA algorithm to establish the co-expression network and modules of healthy people and CRC patients and a collection of 5157 genes with positive correlations among different CRC samples was identified. Since the occurrence of oxidative stress is closely related to the occurrence and development of malignant tumors, we need to explore the potential mechanism of whether geniposide can treat CRC by regulating oxidative stress. We employed the GSEA website to collect 855 oxidative stress-related genes and took the intersection of the three to obtain the hub genes of geniposide treatment on CRC.

In addition, GO enrichment analysis showed that the biological functions mainly involved positive regulation of response to temperature stimulus, response to alkaloid, nitric oxide biosynthetic process, nitric oxide metabolic process, reactive nitrogen species metabolic process, cellular response to peptide, and so on. From the enrichment results of GO, the synthesis and metabolism of nitric oxide and the metabolic process of reactive nitrogen were significantly enriched. Nitric oxide (NO) is synthesized by many cell types involved in immunity and inflammation. It also controls the functional activity, growth, and death of a wide range of immunological and inflammatory cell types, including Macrophages, T lymphocytes, antigen-presenting cells, Mast cells, Neutrophils, and natural killer cells [15]. Inflammation and cancer are related in numerous ways, including proliferation, invasion, angiogenesis, and metastasis [16]. Furthermore, inflammation-mediated cytokine release regulated by diverse cells within the tumor microenvironment plays a significant role in these processes [17], which reveals that the progression of inflammation can, to some extent, trigger tumor growth. Cell components mainly included melanosome, pigment granule, ficolin-1-rich granule, glutamatergic synapse, ficolin-1-rich granule lumen, cyclin-dependent protein kinase holoenzyme complex, and so on. Molecular functions mainly included histone kinase activity, ubiquitin protein ligase binding, ubiquitin-like protein ligase binding, heat shock protein binding, ATP-dependent protein folding chaperone, tau protein binding and other related gene targets.

KEGG pathway enrichment analysis showed that there were 87 signaling pathways (p-value < 0.05) in which geniposide-regulated oxidative stress intervened in CRC, mainly involving pathways in the cancer signaling pathway, PI3K–Akt signaling pathway, IL-17 signaling pathway, p53 signaling pathway, NF-κB signaling pathway and NOD-like receptor signaling pathway. Despite the fact that the PI3K–AKT signaling pathway governs various cellular processes such as differentiation, metabolism, survival, and apoptosis, its abnormal activation contributes to the development of colorectal cancer malignancy [18]. IL-17 is a very flexible pro-inflammatory cytokine that is required for a range of functions, such as host defense, tissue repair, inflammatory disease pathogenesis, and cancer progression [19]. Previous research has revealed that IL-17 activates a variety of intracellular signaling pathways, including NF-κB and MAPKs such as p38, JNK, and ERK1/2, all of which are involved in the regulation of tumor incidence and progression. The tumor suppressor gene TP53 is one of the most frequently altered genes in numerous malignancies, including colorectal cancer [20]. According to some studies, the p53 signaling pathway can decrease cancers via modulating cell cycle arrest [21], angiogenesis inhibition [22], DNA repair and angiogenesis [23], apoptosis [22], etc. As a result, altering the p53 signaling pathway may be a therapeutic approach for colorectal cancer treatment. The NF-κB signaling pathway controls cell proliferation, apoptosis, angiogenesis, inflammation, metastasis, and treatment resistance in CRC. When this route is activated, it causes the production of proliferation-related genes, including cyclin D1, cyclin E, and cyclin-dependent kinase (CDK)-2, as well as IL-6 and Myc. Because abnormal NF-κB regulation is frequently observed in tumor cells, inhibiting this cascade may decrease cell proliferation [24]. One of the main routes involved in the regulation of pyroptosis is the NOD-like receptor signaling pathway, and NLRP3 is the most studied NLR in recent years. NLRP3 recruits ASC and caspase-1 during activation, which is required for the cleavage and maturation of the inflammatory cytokines IL-1B and IL-18, as well as the subsequent inflammatory cell death known as pyroptosis [25]. Li Liang et al. reported that oxymatrine has an anti-cancer effect via LRPPRC inhibition, mitophagy induction, and NLRP3 inflammasome suppression [26]. Qin et al. found that Atractylide I has anti-colorectal cancer properties via suppressing Drp1-mediated mitochondrial fission and NLRP3 inflammasome activation in colitis-associated colorectal cancer [27]. Therefore, modulating the NOD-like receptor signaling pathway could be one possibility for anti-colorectal cancer therapy. These signaling pathways may collaborate in the molecular mechanism of geniposide-regulated oxidative stress response against colorectal cancer.

Pyroptosis is intimately related to oxidative stress and has recently become a research hotspot. The primary signaling pathways enriched by KEGG were involved in regulating the occurrence and progression of pyroptosis. Therefore, we also conducted a study on the correlation between hub genes and hotspot pyrogenic genes, the result of which suggested that hub genes were closely related to pyroptosis hotspot genes, especially *IL1B, CDK4* and *RELA*. The activation of pyroptosis promotes the cleavage of the pro-inflammatory cytokines IL-1 and IL-18 into active and secretory forms [25]. The ceRNA network of hub genes further suggested that there was a strong positive correlation between the five hub genes, which were all strongly positively correlated with Macrophages M0, Macrophages M0, Macrophages M1, and Mast cells activated, while a negative correlation with Plasma cells, T cells CD8, T cells CD4 memory resting, T cells gamma delta, Macrophages M2, Mast cells resting, and Eosinophils. As an important cell component in the tumor microenvironment, Macrophages play an important role in the composition and functional composition of the tumor microenvironment. M2 Macrophages in the microenvironment surrounding the tumor express a high level of IL10, which limits the tumor microenvironment's immune response. Moreover, Plasma cells and T CD8 are crucial immune cells that assist the immune system in removing antigens, viruses, and tumors, as well as suppress the incidence and proliferation of cancer cells [28].

Our PPI network analysis results showed that *IL1B, GSK3B, NOS3, RELA* and *CDK4* were related to oxidative stress induced by geniposide against colorectal cancer. We also performed a bioinformatics analysis of these hub genes. Analysis of *GSK3B* shows the dual activity of inhibiting or promoting tumors [29]. Considering that *GSK3B* suppresses the Wnt signaling pathway, it is presumed to be an inhibitor of tumors in general [30]. In contrast, current research suggests that *GSK3B* can activate the NF-κB signaling cascade by increasing NF-κB transcriptional activity in the nucleus, hence promoting malignancy [31]. *RELA*(P65) is a transcription factor in the nuclear factor κB (NF-κB) family [32,33]. NF-κB activation promotes CRC by speeding up cell proliferation and angiogenesis, inhibiting apoptosis, and promoting cell invasion and metastasis, and NF-κB is involved in all stages of CRC development [33]. *CDK4* is one of the cell cycle regulators that controls the G1 to S phase of the cell cycle. Because multiple cancer treatments may reduce *CDK4* expression, this protein has been regarded as a critical target protein in a variety of malignancies [34]. Several investigations have shown that *CDK4* is overexpressed in cancer [35,36] and inhibiting *CDK4* overexpression improves clinical treatment efficacy for breast cancer, melanoma, liposarcoma, and mantle cell lymphoma [37], which suggests that CDK4 could be a therapeutic target for cancer [38]. *NOS3* is known to be involved in NO synthesis, primarily in endothelial cells, and is linked to cardiovascular disorders such as hypertension, atherosclerosis, and diabetes mellitus [39]. It has been observed that *NOS3* inhibits apoptosis and promotes angiogenesis, proliferation, invasiveness, and immunosuppression in malignant tumors [40]. Furthermore, *NOS3* levels were found to be inversely associated with progression-free survival and overall survival (OS) in patients with metastatic colorectal cancer [41].

Hub gene *IL-1B* is produced and secreted by a range of tumor cell types, including immune cells, fibroblasts, and cancer cells. It has also been connected to a wide range of physiological activities. It has the capacity to control cytokine production and gene expression as well as cellular adhesion and migration, angiogenesis, and immune response [42]. Additionally, *IL-1B* often encourages the growth and invasion of cancer cells as well as neo-angiogenesis and immune cells that infiltrate tumors [43]. Furthermore, a mouse model of adenomatous polyposis coli (APC) colon cancer showed significant levels of *IL-1B* and *IL-1A* [44]. Furthermore, it has recently been demonstrated that IL-1R1 deficit in Neutrophils promotes bacterial invasion and tumor aggressiveness, while IL-1R1 deficiency in epithelial cells decreases tumorigenesis in an APC model [45]. Inhibiting IL-1B, however, may or may not be helpful for patients based on its polymorphism, depending on the cancer kind or stage, the primary type of immune cells present in the tumor microenvironment, and the anti-cancer treatment utilized [45]. Therefore, before considering *IL1B* as a potential method to treat colorectal cancer, it is important to further examine the correlations between *IL1B* polymorphism and prognosis. In summary, in this study, we employed four databases—Swiss Target Prediction, Target Net, CTD, and TCMSP—to identify genes and targets associated with geniposide. Each of these databases has its unique strengths and coverage, and their combined use ensures a more comprehensive and accurate target identification. GO and KEGG enrichment analyses provided useful insights into the biological processes and signaling pathways these genes may be involved in. Notably, enrichment results in cancer signaling pathways and the PI3K–Akt signaling pathway may suggest potential mechanisms of geniposide's anti-CRC effects. Molecular docking results indicated possible interaction mechanisms between geniposide and its primary targets, providing direction for subsequent in vitro and in vivo experiments. However, these docking results are predictive and will require experimental validation through techniques such as X-ray crystallography or NMR in the future. Our data showed a significant increase in certain immune cells, such as M2 Macrophages and Tregs, in the CRC group. These cells are generally associated with immune suppression and cancer progression. These results suggest that geniposide may exert its anti-cancer effects by modulating the immune response in the tumor microenvironment. Despite the promising findings, our study has several limitations that should be acknowledged. First, the target identification relies heavily on in silico databases, which might not encompass all possible interactions or biological activities of geniposide. Second, our work is predominantly computational and lacks experimental validation. Molecular docking results are purely predictive and should be confirmed through in vitro and in vivo experiments. Third, while our study points to alterations in immune cell populations within the CRC tumor microenvironment, the mechanistic relationship between geniposide and immune modulation remains unclear and warrants further investigation. Lastly, our analyses were based on publicly available datasets, the quality and comprehensiveness of which we could not control. Future work should aim to address these limitations, perhaps by employing a wider variety of databases, conducting laboratory experiments for validation, and exploring the impact of geniposide on immune cells through more targeted assays.

Geniposide displayed high binding activity with the hub genes, as demonstrated in molecular docking results. We discovered that the mRNA levels of five hub genes were significantly expressed in CRC tissues, and the prognostic value of ILIB was significantly different (p-value < 0.05). With the evolution of pathological stages (Stage I to Stage IV), the mRNA expression of *IL1B* gradually decreased, while the mRNA expression of *NOS3* gradually increased. The above analysis results were basically consistent with literature reports.

4. Materials and Methods

4.1. Prediction of Related Targets of Geniposide

We utilized PubChem (https://pubchem.ncbi.nlm.nih.gov/, accessed on 2 July 2023) to search for geniposide-related SMILES numbers and used Swiss Target Prediction databases (http://www.swisstargetprediction.ch/, accessed on 2 July 2023) and TargetNet database (http://targetnet.scbdd.com/home/index/, accessed on 2 July 2023) [46] with the default page setting to predict their corresponding targets. Then, we used the CTD database (https://ctdbase.org/, accessed on 2 July 2023) [47] and the TCMSP database (https://old.tcmsp-e.com/tcmsp.php, accessed on 2 July 2023) [48] with "geniposide" as a keyword to obtain geniposide-related targets. Finally, the UniProt database (http://www.uniprot.org/, accessed on 2 July 2023) [49] was used to correct the entered protein target name to its Official Symbol and set the species as "human".

4.2. Colorectal Cancer Data Collection and Weighted Gene Co-Expression Network Analysis (WGCNA)

We took "Colorectal cancer" as keywords in Gene Expression Omnibus (GEO, http://www.ncbi.nlms.nih.gov/geo, accessed on 3 July 2023) [50] to retrieve the CRC dataset. As a training dataset for discovery, the GSE44076 dataset is constructed from the GPL13667 [HG-U219] Affymetrix Human Genome U219 Array platform and contains colon tissues from 149 CRC patients and 99 healthy people. All data used in the study were obtained from GEO. Therefore, neither ethical approval nor informed consent was required. We applied the R package of "WGCNA" (version 1.72-1, accessed on 3 July 2023), which was performed to determine the co-expression module. We applied only the top 25% of genes with the largest difference in subsequent WGCNA analyses to ensure the accuracy of the quality results. We chose an optimal soft power to construct a weighted adjacency matrix and further transformed it into a topological overlap matrix (TOM). The module was obtained using the TOM similarity and difference metric (1-TOM) based on the hierarchical clustering tree algorithm when the minimum module size was set to 100. Each module was assigned to a random color. Module eigengene represented the global gene expression profile for each module. Relations between modules and disease states were represented by modular significance (MS). Correlations between a gene and its clinical phenotype were described with genetic significance (GS).

4.3. Collection of Oxidative Stress Genes and Acquisition of Core Genes

To explore whether the potential mechanism of geniposide in the treatment of CRC is related to oxidative stress, we collected 855 oxidative stress- related genes from previous literature [51]. Then, we took the three intersections to get the core genes of geniposide in the treatment of CRC by using the "VennDiagram" package (version 1.7.3, accessed on 4 July 2023) in R software (version 4.32.12, accessed on 4 July 2023). The core genes were visually networked with the help of Cytoscape software (version 3.10.0, accessed on 4 July 2023).

4.4. Construction and Enrichment Analysis of Protein-Protein Interaction (PPI) Networks in Core Genes

The above-screened core genes were imported into the String database (https://cn.string-db.org/, version 11.5, accessed on 5 July 2023). The protein–protein interaction network (PPI) was constructed with a minimum required interaction score $\geq$ 0.4 as the screening condition. The PPI network diagram and tsv file were downloaded. To further explore the biological functions of core genes, GO gene enrichment analysis and KEGG signaling pathway analysis were performed and visualized on core genes by using the "clusterProfiler" (version3.17, accessed on 5 July 2023), "org.Hs.eg.db" (version 2.10, accessed on 5 July 2023), and "ggplot2" (version 3.4.3, accessed on 5 July 2023) packages in R software to explore the main molecular biological processes and signaling pathways of potential targets of geniposide. Finally, the compound–disease–therapeutic targets-core pathways network diagram was constructed using Cytoscape software (version 3.10.0, accessed on 5 July 2023) with nodes in the network representing compound, disease, targets, and pathways. Nodes were connected by edges that represented different meanings depending on the ways the network was built.

4.5. Screening of Hub Genes and Correlation Analysis of Genes Related to Pyroptosis

After 34 core genes were imported into Cytoscape software, the degree value was calculated, and hub genes were selected according to their degree value. To further understand the potential role of geniposide in the treatment of hub genes against CRC, we analyzed the correlations between hub genes and hotspot genes of pyroptosis. First, we collected 44 pyroptosis-related genes from previous study [52] Subsequently, the pyroptosis-related genes in CRC samples were screened by using the "limma" package (version 3.56.2, accessed on 5 July 2023) in R software with $|\text{logFC}| > 0.585$, and p-value < 0.05 as the screening conditions, and 10 related pyroptosis hotspot genes in CRC were screened (CASP5, CHMP2B, CYCS, GZMB, IL18, GZMA, BAK1, GSDMB, IL1A, IL1B). Finally, a correlation analysis was performed on hub genes and hotspot pyroptosis-related genes.

4.6. Construction of Hub Genes ceRNA Network

StarBase database (http://starbase.sysu.edu.cn, accessed on 5 July 2023), miRDB database (http://www.mirdb.org/, accessed on 5 July 2023) and miRanda database (http://www.microrna.org/, accessed on 5 July 2023 [53]) were used to predict mRNA–miRNA interaction pairs of hub genes, respectively. The predicted miRNAs were finally searched in the spongeScan database (http://spongescan.rc.ufl.edu, accessed on 5 July 2023) and miRNA–lncRNAs were screened to obtain the ceRNA network of mRNA–miRNA–lncRNAs. Among them, the Cytoscape software was used for visual analysis.

4.7. PPI Network Analysis, mRNA Expression Lever Analysis, Overall Survival (OS) Analysis, Pathological Staging and Immunohistochemical Analysis of Hub Genes

We imported the hub genes screened above into the String database (https://string-db.org/, accessed on 7 July 2023) and used "minimum required interaction score $\geq$ 0.4" as the screening condition to construct a PPI network, download and save the PPI network diagram and tsv format file. We used the "limma" package (version 3.56.2, accessed on 5 July 2023) in R software to analyze the differential expression of hub genes in CRC. Then, the hub genes were input into the online tool GEPIA (http://gepia.cancer-pku.cn/index.html, accessed on 7 July 2023) [54] to verify their pathology Staging and overall survival (OS) in TCGA-COAD. The protein expression of core genes was studied in the HPA database (https://www.proteinatlas.org/, accessed on 7 July 2023) [55].

4.8. Immune Cell Infiltration Analysis

We utilized the CIBERSORT (http://cibersortx.stanford.edu, accessed on 8 July 2023) [56] algorithm to determine the relative proportions of 22 invading immune cell types in each tissue. The immunological score of each sample was calculated using the "ESTIMATE" algorithm (version 2.0.0, accessed on 8 July 2023). In addition, the associations of signature genes with the number of infiltrating immune cells were determined by using the "Spearman" (version 1.11.3, accessed on 8 July 2023) rank correlation analysis in R software. The graphing method of the "ggplot2" package (version 3.4.0, accessed on 8 July 2023) was used to visualize the resulting correlations.

4.9. Molecular Docking

Geniposide was imported into Chem3D software (version 14.0.0.17, accessed 10 July 2023) to construct its chemical structure formula and minimize its energy, then saved as a PDB format file. The hub genes were brought into the PDB database (https://www.rcsb.org/, accessed on 10 July 2023), corresponding proteins were downloaded as PDB format files and imported into PyMol software (version 2.5.5, accessed on 10 July 2023) to remove water molecules and small molecules. The protein was imported into the autodocktool software (version 4.2.6, accessed on 10 July 2023) for hydroprocessing and saved as a PDBQT format file. Finally, PyMol and MOE software (version 2.1, accessed on 10 July 2023) were used for molecular docking and visualization analysis [57].

5. Conclusions

In this research, we successfully determined the molecular pathways of geniposide-regulated oxidative stress anti-colorectal cancer therapy. This study also demonstrated that hub genes (*IL1B, GSK3B, NOS3, RELA, and CDK4*) were most likely to be involved in the effects of oxidative stress brought on by geniposide in the prevention of colorectal cancer. The underlying mechanisms through which geniposide curbs colorectal cancer effects were the regulation of the multiple BP, including response to temperature stimulus, response to alkaloid, nitric oxide biosynthetic process, nitric oxide metabolic process, reactive nitrogen species metabolic process, cellular response to peptide, and so on. We identified six critical pathways (pathways in cancer signaling pathway, PI3K–Akt signaling pathway, IL-17 signaling pathway, p53 signaling pathway, NF-κB signaling pathway, and NOD-like receptor signaling pathway) involved in the treatment of colorectal cancer with geniposide by regulating oxidative stress.

Hence, our findings showed a synergistic effect between various anti-colorectal cancer core targets, numerous molecular pathways, and geniposide treatment for colorectal cancer. The findings of molecular docking revealed that geniposide had a regulatory effect on the anti-colorectal cancer core targets. Further bioinformatic analysis showed the impact of hub genes on disease prognosis and other aspects. Furthermore, the results from molecular docking, bioinformatics, and network pharmacology screening were all consistent, proving the validity of network pharmacology in this study. These results, therefore, provide a basis for the future development of geniposide-based anti-colorectal cancer medications.

Author Contributions: Conceptualization, Y.W. and B.X.; methodology, Y.W., B.X. and J.L.; software, Y.W., B.X. and J.L.; validation Y.W.; formal analysis, Y.W. and B.X.; investigation, B.X., resources, B.X.; data curation, Y.W., B.X. and J.L.; writing—original draft preparation, Y.W.; writing—review and editing, Y.W., B.X. and J.L.; visualization, Y.W. and B.X.; supervision, B.X.; project administration, B.X.; revision, Y.W., B.X. and J.L.; funding acquisition, B.X. All authors have read and agreed to the published version of the manuscript.

Funding: This project was jointly supported by the Guangdong Provincial Key Laboratory IRADS (2022B1212010006, R0400001-22) and a grant (project code: UICR0200007-23) from BNU-HKBU United International College, China.

Institutional Review Board Statement: Not applicable.

Informed Consent Statement: Not applicable.

Data Availability Statement: The data, analytic methods, and study materials that support the findings of this study are available from the corresponding author upon reasonable request.

Conflicts of Interest: The authors declare no conflict of interest.

Abbreviations

BP	Biological process
CC	Cellular component
CDK4	Cyclin-dependent kinase 4
CRC	Colorectal cancer
DC	Degree centrality
DL	Drug-likeness
GO	Gene Ontology
GSK3B	Glycogen Synthase Kinase 3 Beta
IL1B	Interleukin IB
KEGG	Kyoto Encyclopedia of Genes and Genomes
MF	Molecular function
NOS3	Nitric oxide synthase
OB	Oral bioavailability
OS	Overall survival
PDB	Protein Data Bank
PPI	Protein–protein Interaction
ROS	Active oxygen species
RNS	Reactive nitrogen species
TCM	Traditional Chinese Medicine
TCMSP	Traditional Chinese Medicine Systems Pharmacology

References

1. Marmol, I.; Sanchez-de-Diego, C.; Pradilla Dieste, A.; Cerrada, E.; Rodriguez Yoldi, M.J. Colorectal carcinoma: A general overview and future perspectives in colorectal cancer. *Int. J. Mol. Sci.* **2017**, *18*, 197. [CrossRef]
2. Dekker, E.; Tanis, P.J.; Vleugels, J.L.A.; Kasi, P.M.; Wallace, M.B. Colorectal cancer. *Lancet* **2019**, *394*, 1467–1480. [CrossRef] [PubMed]
3. Islam, M.R.; Akash, S.; Rahman, M.M.; Nowrin, F.T.; Akter, T.; Shohag, S.; Rauf, A.; Aljohani, A.S.M.; Simal-Gandara, J. Colon cancer and colorectal cancer: Prevention and treatment by potential natural products. *Chem. Biol. Interact.* **2022**, *368*, 110170. [CrossRef] [PubMed]
4. Tian, J.; Qin, S.; Han, J.; Meng, J.; Liang, A. A review of the ethnopharmacology, phytochemistry, pharmacology and toxicology of Fructus Gardeniae (Zhi-zi). *J. Ethnopharmacol.* **2022**, *289*, 114984. [CrossRef] [PubMed]
5. Habtemariam, S.; Lentini, G. Plant-derived anticancer agents: Lessons from the pharmacology of geniposide and its aglycone, genipin. *Biomedicines* **2018**, *6*, 39. [CrossRef]
6. Zhou, Y.X.; Zhang, R.Q.; Rahman, K.; Cao, Z.X.; Zhang, H.; Peng, C. Diverse pharmacological activities and potential medicinal benefits of geniposide. *Evid. Based Complement. Altern. Med.* **2019**, *2019*, 4925682. [CrossRef]
7. Hu, L.; Zhao, J.; Liu, Y.; Liu, X.; Lu, Q.; Zeng, Z.; Zhu, L.; Tong, X.; Xu, Q. Geniposide inhibits proliferation and induces apoptosis of diffuse large B-cell lymphoma cells by inactivating the HCP5/miR-27b-3p/MET axis. *Int. J. Med. Sci.* **2020**, *17*, 2735–2743. [CrossRef]
8. Chen, Z.; Liu, W.; Qin, Z.; Liang, X.; Tian, G. Geniposide exhibits anticancer activity to medulloblastoma cells by downregulating microRNA-373. *J. Biochem. Mol. Toxicol.* **2020**, *34*, e22471.
9. Ma, J.; Ding, Y. Geniposide suppresses growth, migration and invasion of MKN45 cells by down-regulation of lncRNA HULC. *Exp. Mol. Pathol.* **2018**, *105*, 252–259. [CrossRef]
10. Guina, T.; Biasi, F.; Calfapietra, S.; Nano, M.; Poli, G. Inflammatory and redox reactions in colorectal carcinogenesis. *Ann. N. Y. Acad. Sci.* **2015**, *1340*, 95–103. [CrossRef]
11. Gao, M.; Yi, J.; Zhu, J.; Minikes, A.M.; Monian, P.; Thompson, C.B.; Jiang, X. Role of mitochondria in Ferroptosis. *Mol. Cell* **2019**, *73*, 354–363. [CrossRef]
12. Klaunig, J.E.; Kamendulis, L.M.; Hocevar, B.A. Oxidative stress and oxidative damage in carcinogenesis. *Toxicol. Pathol.* **2010**, *38*, 96–109. [CrossRef] [PubMed]
13. Gupta, R.K.; Patel, A.K.; Shah, N.; Chaudhary, A.K.; Jha, U.K.; Yadav, U.C.; Gupta, P.K.; Pakuwal, U. Oxidative stress and antioxidants in disease and cancer: A review. *Asian Pac. J. Cancer Prev.* **2014**, *15*, 4405–4409. [CrossRef] [PubMed]
14. Barrera, G. Oxidative stress and lipid peroxidation products in cancer progression and therapy. *ISRN Oncol.* **2012**, *2012*, 137289. [CrossRef] [PubMed]
15. Coleman, J.W. Nitric oxide in immunity and inflammation. *Int. Immunopharmacol.* **2001**, *1*, 1397–1406. [CrossRef] [PubMed]
16. Diakos, C.I.; Charles, K.A.; McMillan, D.C.; Clarke, S.J. Cancer-related inflammation and treatment effectiveness. *Lancet Oncol.* **2014**, *15*, e493–e503. [CrossRef]

17. Sharma, B.R.; Kanneganti, T.D. NLRP3 inflammasome in cancer and metabolic diseases. *Nat. Immunol.* **2021**, *22*, 550–559. [CrossRef]
18. Stefani, C.; Miricescu, D.; Stanescu, S., II; Nica, R.I.; Greabu, M.; Totan, A.R.; Jinga, M. Growth factors, PI3K/AKT/mTOR and MAPK signaling pathways in colorectal cancer pathogenesis: Where are we now? *Int. J. Mol. Sci.* **2021**, *22*, 10260. [CrossRef] [PubMed]
19. Li, X.; Bechara, R.; Zhao, J.; McGeachy, M.J.; Gaffen, S.L. IL-17 receptor-based signaling and implications for disease. *Nat. Immunol.* **2019**, *20*, 1594–1602. [CrossRef]
20. Wu, L.; Chen, X.; Zhao, J.; Martin, B.; Zepp, J.A.; Ko, J.S.; Gu, C.; Cai, G.; Ouyang, W.; Sen, G.; et al. A novel IL-17 signaling pathway controlling keratinocyte proliferation and tumorigenesis via the TRAF4-ERK5 axis. *J. Exp. Med.* **2015**, *212*, 1571–1587. [CrossRef]
21. Tarasov, V.; Jung, P.; Verdoodt, B.; Lodygin, D.; Epanchintsev, A.; Menssen, A.; Meister, G.; Hermeking, H. Differential regulation of microRNAs by p53 revealed by massively parallel sequencing: miR-34a is a p53 target that induces apoptosis and G1-arrest. *Cell Cycle* **2007**, *6*, 1586–1593. [CrossRef] [PubMed]
22. Attardi, L.D.; Reczek, E.E.; Cosmas, C.; Demicco, E.G.; McCurrach, M.E.; Lowe, S.W.; Jacks, T. PERP, an apoptosis-associated target of p53, is a novel member of the PMP-22/gas3 family. *Genes Dev.* **2000**, *14*, 704–718. [CrossRef] [PubMed]
23. Zhang, K.; Hu, S.; Wu, J.; Chen, L.; Lu, J.; Wang, X.; Liu, X.; Zhou, B.; Yen, Y. Overexpression of RRM2 decreases thrombspondin-1 and increases VEGF production in human cancer cells in vitro and in vivo: Implication of RRM2 in angiogenesis. *Mol. Cancer* **2009**, *8*, 11. [CrossRef]
24. Soleimani, A.; Rahmani, F.; Ferns, G.A.; Ryzhikov, M.; Avan, A.; Hassanian, S.M. Role of the NF-kappaB signaling pathway in the pathogenesis of colorectal cancer. *Gene* **2020**, *726*, 144132. [CrossRef] [PubMed]
25. Fang, Y.; Tian, S.; Pan, Y.; Li, W.; Wang, Q.; Tang, Y.; Yu, T.; Wu, X.; Shi, Y.; Ma, P.; et al. Pyroptosis: A new frontier in cancer. *Biomed. Pharmacother.* **2020**, *121*, 109595. [CrossRef]
26. Liang, L.; Sun, W.; Wei, X.; Wang, L.; Ruan, H.; Zhang, J.; Li, S.; Zhao, B.; Li, M.; Cai, Z.; et al. Oxymatrine suppresses colorectal cancer progression by inhibiting NLRP3 inflammasome activation through mitophagy induction in vitro and in vivo. *Phytother. Res.* **2023**, *37*, 3342–3362. [CrossRef]
27. Qin, Y.; Yu, Y.; Yang, C.; Wang, Z.; Yang, Y.; Wang, C.; Zheng, Q.; Li, D.; Xu, W. Atractylenolide I inhibits NLRP3 inflammasome activation in colitis-associated colorectal cancer via suppressing Drp1-mediated mitochondrial fission. *Front. Pharmacol.* **2021**, *12*, 674340. [CrossRef]
28. Pan, Y.; Yu, Y.; Wang, X.; Zhang, T. Tumor-associated Macrophages in tumor immunity. *Front. Immunol.* **2020**, *11*, 583084. [CrossRef] [PubMed]
29. Ghosh, J.C.; Altieri, D.C. Activation of P53-dependent apoptosis by acute ablation of glycogen synthase kinase-3beta in colorectal cancer cells. *Clin. Cancer Res.* **2005**, *11*, 4580–4588. [CrossRef]
30. Doble, B.W.; Patel, S.; Wood, G.A.; Kockeritz, L.K.; Woodgett, J.R. Functional redundancy of GSK-3alpha and GSK-3beta in Wnt/beta-catenin signaling shown by using an allelic series of embryonic stem cell lines. *Dev. Cell* **2007**, *12*, 957–971. [CrossRef] [PubMed]
31. Shakoori, A.; Mai, W.; Miyashita, K.; Yasumoto, K.; Takahashi, Y.; Ooi, A.; Kawakami, K.; Minamoto, T. Inhibition of GSK-3 beta activity attenuates proliferation of human colon cancer cells in rodents. *Cancer Sci.* **2007**, *98*, 1388–1393. [CrossRef]
32. Hoesel, B.; Schmid, J.A. The complexity of NF-kappaB signaling in inflammation and cancer. *Mol. Cancer* **2013**, *12*, 86. [CrossRef]
33. Patel, M.; Horgan, P.G.; McMillan, D.C.; Edwards, J. NF-kappaB pathways in the development and progression of colorectal cancer. *Transl. Res.* **2018**, *197*, 43–56. [CrossRef] [PubMed]
34. Park, G.H.; Sung, J.H.; Song, H.M.; Jeong, J.B. Anti-cancer activity of Psoralea fructus through the downregulation of cyclin D1 and CDK4 in human colorectal cancer cells. *BMC Complement. Altern. Med.* **2016**, *16*, 373. [CrossRef] [PubMed]
35. Feng, H.; Ge, F.; Du, L.; Zhang, Z.; Liu, D. MiR-34b-3p represses cell proliferation, cell cycle progression and cell apoptosis in non-small-cell lung cancer (NSCLC) by targeting CDK4. *J. Cell. Mol. Med.* **2019**, *23*, 5282–5291. [CrossRef] [PubMed]
36. Aboushousha, T.; Hammam, O.; Aref, A.; Kamel, A.; Badawy, M.; Abdel Hamid, A. Tissue profile of CDK4 and STAT3 as possible innovative therapeutic targets in urinary bladder cancer. *Asian Pac. J. Cancer Prev.* **2020**, *21*, 547–554. [CrossRef]
37. Dickson, M.A.; Tap, W.D.; Keohan, M.L.; D'Angelo, S.P.; Gounder, M.M.; Antonescu, C.R.; Landa, J.; Qin, L.X.; Rathbone, D.D.; Condy, M.M.; et al. Phase II trial of the CDK4 inhibitor PD0332991 in patients with advanced CDK4-amplified well-differentiated or dedifferentiated liposarcoma. *J. Clin. Oncol.* **2013**, *31*, 2024–2028. [CrossRef]
38. Wang, C.; Li, H.; Wu, L.; Jiao, X.; Jin, Z.; Zhu, Y.; Fang, Z.; Zhang, X.; Huang, H.; Zhao, L. Coiled-Coil Domain-Containing 68 Downregulation Promotes Colorectal Cancer Cell Growth by Inhibiting ITCH-Mediated CDK4 Degradation. *Front. Oncol.* **2021**, *11*, 668743. [CrossRef] [PubMed]
39. Zou, D.; Li, Z.; Lv, F.; Yang, Y.; Yang, C.; Song, J.; Chen, Y.; Jin, Z.; Zhou, J.; Jiang, Y.; et al. Pan-cancer analysis of NOS3 identifies its expression and clinical relevance in gastric cancer. *Front. Oncol.* **2021**, *11*, 592761. [CrossRef] [PubMed]
40. Song, Y.; Zhao, X.P.; Song, K.; Shang, Z.J. Ephrin-A1 is up-regulated by hypoxia in cancer cells and promotes angiogenesis of HUVECs through a coordinated cross-talk with eNOS. *PLoS ONE* **2013**, *8*, e74464. [CrossRef]
41. Marisi, G.; Scarpi, E.; Passardi, A.; Nanni, O.; Ragazzini, A.; Valgiusti, M.; Casadei Gardini, A.; Neri, L.M.; Frassineti, G.L.; Amadori, D.; et al. Circulating VEGF and eNOS variations as predictors of outcome in metastatic colorectal cancer patients receiving bevacizumab. *Sci. Rep.* **2017**, *7*, 1293. [CrossRef]

42. Pantschenko, A.G.; Pushkar, I.; Anderson, K.H.; Wang, Y.; Miller, L.J.; Kurtzman, S.H.; Barrows, G.; Kreutzer, D.L. The interleukin-1 family of cytokines and receptors in human breast cancer: Implications for tumor progression. *Int. J. Oncol.* **2003**, *23*, 269–284. [CrossRef]
43. Qian, H.; Zhang, D.; Bao, C. Two variants of Interleukin-1B gene are associated with the decreased risk, clinical features, and better overall survival of colorectal cancer: A two-center case-control study. *Aging* **2018**, *10*, 4084–4092. [CrossRef] [PubMed]
44. Dmitrieva-Posocco, O.; Dzutsev, A.; Posocco, D.F.; Hou, V.; Yuan, W.; Thovarai, V.; Mufazalov, I.A.; Gunzer, M.; Shilovskiy, I.P.; Khaitov, M.R.; et al. Cell-type-specific responses to interleukin-1 control microbial invasion and tumor-elicited inflammation in colorectal cancer. *Immunity* **2019**, *50*, 166–180.e7. [CrossRef]
45. Rebe, C.; Ghiringhelli, F. Interleukin-1 beta and cancer. *Cancers* **2020**, *12*, 1791. [CrossRef] [PubMed]
46. Yao, Z.J.; Dong, J.; Che, Y.J.; Zhu, M.F.; Wen, M.; Wang, N.N.; Wang, S.; Lu, A.P.; Cao, D.S. TargetNet: A web service for predicting potential drug-target interaction profiling via multi-target SAR models. *J. Comput. Aided Mol. Des.* **2016**, *30*, 413–424. [CrossRef] [PubMed]
47. Davis, A.P.; Grondin, C.J.; Johnson, R.J.; Sciaky, D.; Wiegers, J.; Wiegers, T.C.; Mattingly, C.J. Comparative toxicogenomics database (CTD): Update 2021. *Nucleic Acids Res.* **2021**, *49*, D1138–D1143. [CrossRef] [PubMed]
48. Ru, J.; Li, P.; Wang, J.; Zhou, W.; Li, B.; Huang, C.; Li, P.; Guo, Z.; Tao, W.; Yang, Y.; et al. TCMSP: A database of systems pharmacology for drug discovery from herbal medicines. *J. Cheminform.* **2014**, *6*, 13. [CrossRef]
49. UniProt Consortium. UniProt: A hub for protein information. *Nucleic Acids Res.* **2015**, *43*, D204–D212. [CrossRef] [PubMed]
50. Barrett, T.; Wilhite, S.E.; Ledoux, P.; Evangelista, C.; Kim, I.F.; Tomashevsky, M.; Marshall, K.A.; Phillippy, K.H.; Sherman, P.M.; Holko, M.; et al. NCBI GEO: Archive for functional genomics data sets—Update. *Nucleic Acids Res.* **2013**, *41*, D991–D995. [CrossRef] [PubMed]
51. Xu, M.; Zhou, H.; Hu, P.; Pan, Y.; Wang, S.; Liu, L.; Liu, X. Identification and validation of immune and oxidative stress-related diagnostic markers for diabetic nephropathy by WGCNA and machine learning. *Front Immunol.* **2023**, *14*, 1084531. [CrossRef] [PubMed]
52. Yan, M.; Li, W.; Wei, R.; Li, S.; Liu, Y.; Huang, Y.; Zhang, Y.; Lu, Z.; Lu, Q. Identification of pyroptosis-related genes and potential drugs in diabetic nephropathy. *J. Transl. Med.* **2023**, *21*, 490. [CrossRef] [PubMed]
53. Li, J.H.; Liu, S.; Zhou, H.; Qu, L.H.; Yang, J.H. starBase v2.0: Decoding miRNA-ceRNA, miRNA-ncRNA and protein-RNA interaction networks from large-scale CLIP-Seq data. *Nucleic Acids Res.* **2014**, *42*, D92–D97. [CrossRef] [PubMed]
54. Tang, Z.; Li, C.; Kang, B.; Gao, G.; Li, C.; Zhang, Z. GEPIA: A web server for cancer and normal gene expression profiling and interactive analyses. *Nucleic Acids Res.* **2017**, *45*, W98–W102. [CrossRef] [PubMed]
55. Digre, A.; Lindskog, C. The human protein atlas-spatial localization of the human proteome in health and disease. *Protein Sci.* **2021**, *30*, 218–233. [CrossRef]
56. Chen, B.; Khodadoust, M.S.; Liu, C.L.; Newman, A.M.; Alizadeh, A.A. Profiling tumor infiltrating immune cells with CIBERSORT. *Methods Mol. Biol.* **2018**, *1711*, 243–259.
57. Pinzi, L.; Rastelli, G. Molecular docking: Shifting paradigms in drug discovery. *Int. J. Mol. Sci.* **2019**, *20*, 4331. [CrossRef]

International Journal of *Molecular Sciences*

MDPI

Review

Therapeutic Perspectives of Aminoflavonoids—A Review

Monika Stompor-Gorący [1,*], Agata Bajek-Bil [2], Natalia Potocka [3] and Izabela Zawlik [3,4]

1 Department of Pathophysiology, Faculty of Medicine, Collegium Medicum, University of Rzeszów, Warzywna 1a, 35-310 Rzeszów, Poland

2 Faculty of Chemistry, Rzeszów University of Technology, 35-959 Rzeszów, Poland; abajek@prz.edu.pl

3 Laboratory of Molecular Biology, Centre for Innovative Research in Medical and Natural Sciences, Collegium Medicum, University of Rzeszów, 35-959 Rzeszów, Poland; npotocka@ur.edu.pl (N.P.); izawlik@ur.edu.pl (I.Z.)

4 Department of General Genetics, Faculty of Medicine, Collegium Medicum, University of Rzeszów, 35-959 Rzeszów, Poland

* Correspondence: monika.stompor@gmail.com

Abstract: Natural compounds containing nitrogen are a source of many biologically active molecules used as drugs. Due to their multidirectional effects, they represent effective therapeutic compounds in many medical areas. Flavonoids, as well as their bioprecursors, chalcones, that occur in plants possess a number of medicinal benefits. Their synthetic amino derivatives constitute a large group of compounds that exhibit pharmacological activity. Due to the increasing level of drug resistance among patients, new therapeutic agents and options are urgently needed. Therefore, aminoflavonoids may be a promising source of new drugs. In this review, the biological activities of flavonoids, including chalcones, with complexes containing a nitrogen atom and the aminoflavones Ru and Pt are summarized. The purpose of this review is to provide an overview of the synthesis and pharmacological activity of aminoflavonoids and to show how synthetic modifications of these compounds can influence their biological activities. It covers the most recent reports on obtaining aminoflavones, aminochalcones, and their derivatives, along with information about their anticancer, antimicrobial, antimalarial, antiviral, and anti-inflammatory activities.

Keywords: aminoflavonoids; aminochalcone; aminoflavone; anticancer activity; antioxidants

1. Introduction

Flavonoids are biosynthesized in plants from malonate and the aromatic amino acids phenylalanine and tyrosine [1]. The basic flavonoid structure is the flavan nucleus, which consists of 15 carbon atoms arranged in three rings (C_6-C_3-C_6), labeled A, B, and C (Scheme 1).

Various classes of flavonoids differ in the degree of oxidation and the pattern of substitution of the C-ring, while individual compounds in each class differ in the pattern of substitution of the A and B-rings. The flavonoid classes include flavones, flavanones, isoflavones, flavonols, flavanonols, flavan-3-ols, anthocyanidins, biflavones, chalcones, and aurones. In nature, most flavonoids are found as glycosides. The sugar moiety is a major determinant of the absorption of dietary flavonoid glycosides in humans [2].

Both synthetic and natural derivatives of flavonoids form an interesting and important group of molecules, as they possess a wide range of biological activities, such as anticancer, antioxidant, antibacterial, antiplasmodial, antiviral, antimalarial, anti-inflammatory, and many other actions [3–5]. Flavonoid compounds have also been reported to possess insect antifeedant, antiparasitic, antileishmanial, antifungal, and enzyme-inhibitory activities [6].

Additionally, chalcones and dihydrochalcones, being the precursors of natural flavonoids, can be used as potential cancer chemotherapy agents and inhibitors of enzymes involved in the metabolism of thyroid hormones [7,8].

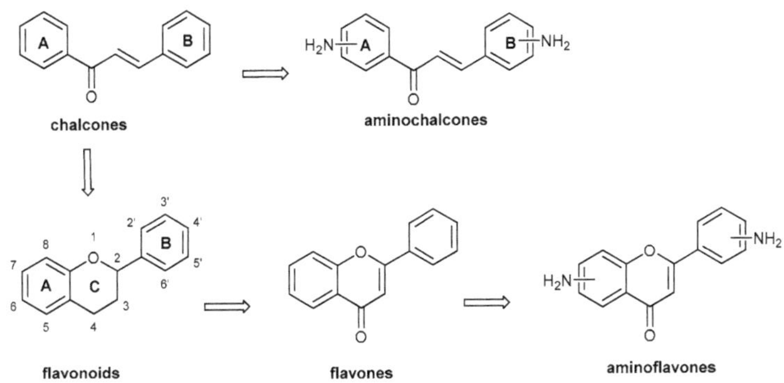

Scheme 1. The general structures of flavonoids.

The pharmacophoric substructures and physicochemical properties of the active compounds based on flavonoids have been defined—among them there are aromatic rings, a basic nitrogen atom, and high lipophilicity. Currently, many alkaloids containing a nitrogen atom are used as drugs. It is predicted that the antitumor effectiveness of a drug increases with the number of hydrogen bonds it forms with the cell membrane components, and hence a large number of known drugs contain nitrogen. There is also research on the functionalization of natural antioxidants, such as xanthohumol, for the synthesis of conjugates with biotin that contain nitrogen and sulfur atoms that are intended for use as anticancer therapeutics [9]. The nitrogen atom may influence the activity of the molecule through its contribution to the hydrogen bond strength [10]. Data show that the nitrogen atom plays an important role in the modulation of the multidrug resistance of cancer cells and in antimicrobial activity [11].

Many investigations [7–10] have shown that the introduction of aliphatic amino groups into the flavonoid moiety resulted in a potent antitumor activity, which is due to the enhancement of the aqueous solubility and drug-like character of candidate compounds. Much attention has been paid to flavonoids, but in the literature, we have not found any comprehensive review on aminoflavonoids. The aim of this review is to summarize the research on the preparation methods and biological activity of aminoflavones, aminochalcones, and their derivatives.

Selected scientific electronic databases, including PubMed, Scopus, and Web of Science, were used for searching the keywords, alone or in combination with other keywords. The flow diagram of the study selection process following the PRISMA guidelines is presented in Scheme 2.

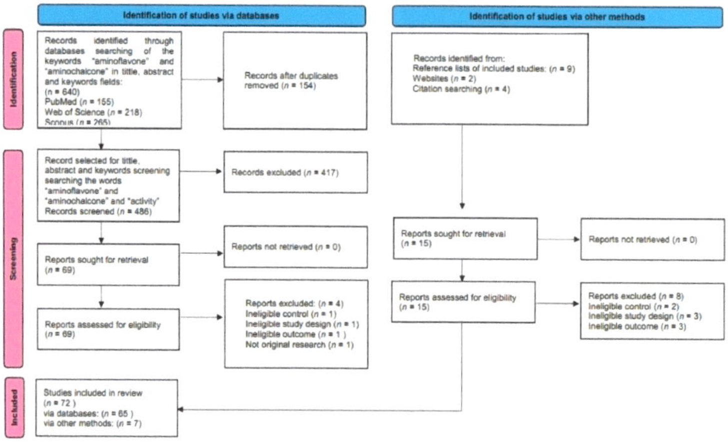

Scheme 2. PRISMA flow diagram of inclusion of the review.

2. Preparation of Aminoflavone Derivatives

Aminoflavones and aminoflavanones are a class of flavonoids that contain amino groups attached to the flavone or flavanone nucleus. They do not occur naturally and therefore must be synthesized via chemical methods [12,13].

A known method for the preparation of biologically active polyphenols with amino groups is the structural modification of natural compounds isolated from plants [14]. Nevertheless, reports on the synthesis and biological properties of aminoflavones are scarce [15–17]. These compounds can be obtained via the reduction of the corresponding nitro derivatives [18], by means of the total synthesis of the aminoflavone derivatives [12,19], by the reduction following the palladium-catalyzed cross-coupling reaction of the corresponding flavone triflate [20], via the reduction of nitroflavones [21], and from hydroxyflavones via tosyloxy- or mesyloxy flavones [22] (Scheme 3). Amino methylated derivatives of flavonoids can also be synthesized via the Mannich reaction upon the electrophilic substitution at C-8 with formaldehyde in the presence of either primary or secondary amines, in 2-propanol as the solvent [23]. Some of the compounds obtained in this way are promising inhibitors of α-glucosidase. An effective copper-mediated amination of bromoflavonoids, leading to a series of corresponding new aminoflavonoids, was also reported [24].

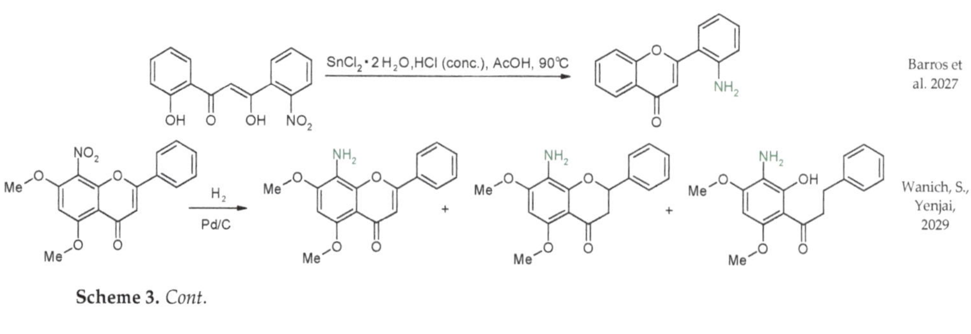

Scheme 3. *Cont.*

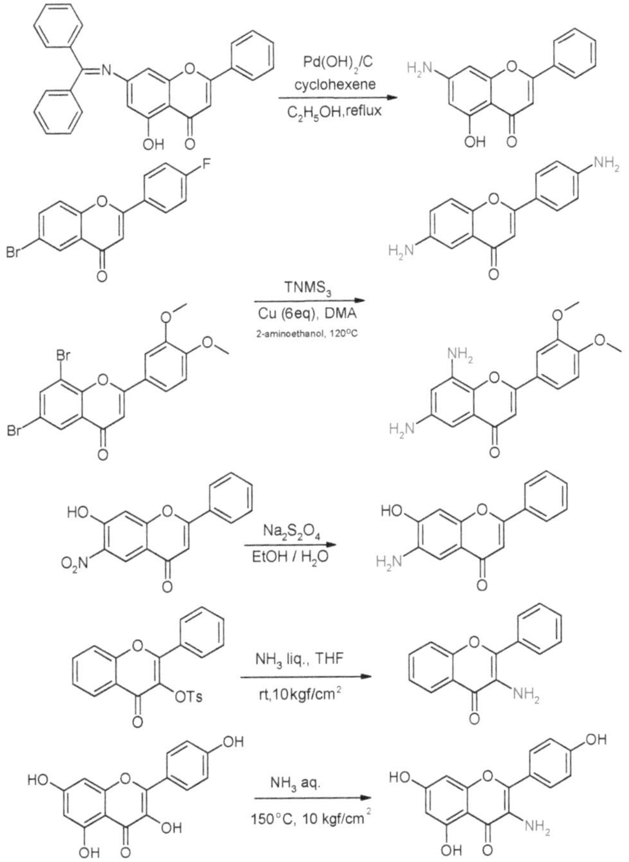

Deng et al. 1999

Shih et al. 2014

Tang et al. 2024

Takechi et al. 2006

Scheme 3. Selected chemical syntheses of aminoflavones [13,14,20–22,24].

Researchers have made attempts to synthesize aminoflavonoids in order to make them available in high amounts or to add some functional groups that would enhance their biological or pharmacological effects [25]. Also, the biotransformation of these compounds is also the current area of interest, aiming to obtain novel biologically and pharmacologically active compounds based on aminoflavones [26], which contain a nitrogen atom in their structure.

3. Biological Activities of Aminoflavones

According to the available literature, the compounds containing the aminoflavone moiety have good potential as anticancer agents. It is suggested that their cytotoxicity is due to the presence of amino and hydroxyl groups located at the ortho position (Figure 1) [14].

Also, aminochalcones, which are the precursors of aminoflavones, with a hydrophobic substituent in ring B proved to be potent antiproliferative agents [27].

One of the important objectives of medicinal chemistry is to design, synthesize, and produce molecules that have potential as human therapeutic agents. The synthesis of flavonoid derivatives containing a nitrogen atom is also an area of significant interest and many synthetic routes have been used to add various moieties to the natural flavonoid nucleus.

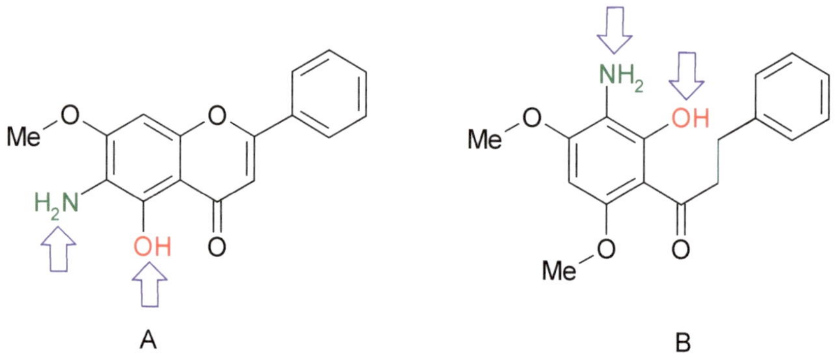

Figure 1. Chemical structure of (**A**) 6-amino-5-hydroxy-7-methoxyflavone and (**B**) 3′-amino-2′-hydroxy-4′6′-dimethoxy-α,β-dihydrochalcone.

In the search for new agents to combat cancer, many compounds with the aminoflavone moiety have been synthesized and tested. Luzzani et al. [28] described the aminoflavone AFP 464, a new anticancer agent, which is currently undergoing phase II clinical trials as an antiproliferative compound active against human renal cancer cells at the molecular level (Section 3).

This compound also modulated the immune response in a mouse model study [29]. Brinkman et al. [30] showed that MDA-MB-468 and Cal51, both Erα-negative human breast cancer cell lines, are sensitive to growth inhibition mediated by another derivative of aminoflavone, which contains two amine groups in the A and B rings (e.g., 5-amino-2-(4-amino-3-fluorophenyl)-6,8-difluoro-7-methylchromen-4-one).

There is also some evidence that combining aminoflavones with other agents can enhance the efficacy of cancer treatment. Reinicke et al. [31] observed the synergistic activity of combinations of fluorinated diaminoflavone (5-amino-2,3-fluorophenyl)-6,8-difluoro-7-methyl-4H-1-benzopyran-4-one) and paclitaxel and some derivatives of topoisomerase I inhibitors (CAMP and SN38—metabolites of irinotecan) against MCF-7 human breast cancer cells.

Dauzonne et al. [32] tested a series of aminoflavones for their cytotoxic activity. 3-Aminoflavones showed in vitro antiproliferative activity against murine L1210 leukemia cells with IC_{50} in the range of 10–163 μM. Among the tested 3-amino derivatives, the most potent were those bearing substituents in the 3′ and/or 4′-positions (Figure 2). The addition of the 3′-NH_2 along with 4′-OCH_3 groups doubled the activity of 3,3′-diamino-4′-methoxyflavone (IC_{50} = 10 μM) compared to the compounds with only the 4′-OCH_3 substituent (3-amino-4′-methoxtflavone: IC_{50} = 22 μM). The presence of more than one methoxy substituent in the B phenyl ring resulted in a decrease in the activity. Substitution with the amino group at the 6-position in ring A (3,6-diaminoflavone) did not improve the anticancer potency.

The compounds where the 3-amino group was substituted with a dialkylamino group demonstrated similar activity to the corresponding 3-amino compounds.

Biological screening of amino-substituted flavone derivatives by Akama et al. [33] showed that 5,4′-diaminoflavone (Table 1) and some of its derivatives exhibited antiproliferative effects against the human breast cancer cell line MCF-7, irrespective of the presence or absence of estradiol (IC_{50} 0.0072 and 0.0098 μM, respectively).

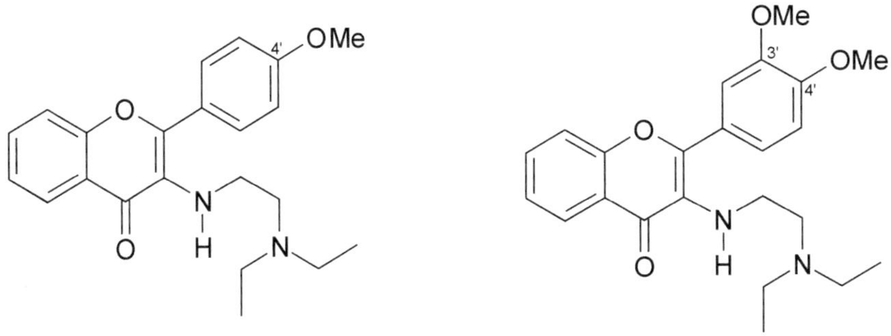

Figure 2. Examples of aminoflavones tested for their anticancer activity [33].

Table 1. Antiproliferative activity of 5-aminoflavone's derivatives against MCF-7 cells.

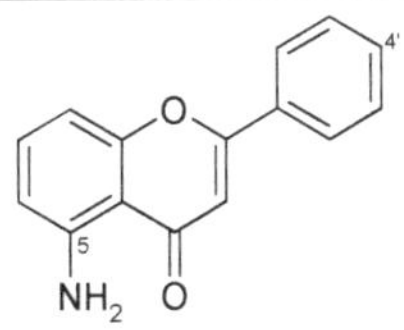

5-aminoflavone (5-AF)

	IC_{50} [μM]	
	Estradiol (10^{-4} μM)	
	—	+
5-AF	>100	13
4′-OH	>100	1.5
4′-OMe	>100	12
4′-Br	9.5	1.0
4′-CN	94	>100
4′-COOH	>10	>10
4′-NH_2	0.0098	0.0072
4′-NHAc	18	13
4′-NHMe	0.0054	0.0023
4′-NHEt	0.0017	0.0013
4′-$NHCH_2Ph$	4.3	4.1
4′-NMe_2	0.0050	0.0040
4′-NEt_2	0.010	0.0020

Chemical modifications of 5-aminoflavone at the C-4′ position of ring B with more electron-withdrawing groups, such as cyano (-CN) and carboxy (-COOH) ones, resulted in a decrease in the activity. Similarly, acetylation of the 4′-amino group or relocation of the 4′-amino group to the 3′-position resulted in a decrease in the activity. Mono- and dialkylation of the 4′-amino group enhanced the antitumor activity; however, alkylation of this group is not essential for exhibiting its antiproliferative potential. Furthermore, relocation of the 5-amino group in the dimethylamino derivative (5′-NH_2 4′-NMe_2) resulted in a decrease in the activity. The tested 5-aminoflavones (5′-NH_2 and 4′-NH_2; 5′-NH_2 and 4′-NHMe; 5′-NH_2 and 4′-NHEt) demonstrated antitumor activity that was highly selective for the ER-positive breast cancer cell line, showing no effects against ER-negative human cancer cell lines, like HeLa S_3, WiD, and MDA-MB-553.

Jin et al. [34] obtained a series of 2′-chloro-4′-aminoflavones and tested them for their antiproliferative activity against hepatocarcinoma cells (HepG2), breast adenocarcinoma cells (MCF-7), and human chronic myelogenous leukemia cells (K562). The introduction of alkyl groups into the A-ring, such as methyl and ethyl ones as weak electron-donating groups, provided better antitumor activities. The best results were noted for 6-ethyl-2′-chloro-4′-aminoflavone (Figure 3) (IC_{50} = 1.8 μM), which was parallel to colchicine, which served as the standard. Furthermore, it was noted that this compound had very low toxicity against normal human liver cells QSG7701 and HL7702. At a concentration of 100 μM, the inhibition rates were 96.4% and 99.2%, respectively. Further study demonstrated that these compounds induced apoptosis in HepG2 cells. Cell-cycle arrest at the G2/M phase was observed in the range from 16 to 31% when increasing the concentration from 5 to 50 μM. All of the compounds had no effect on breast adenocarcinoma cells MCF-7 (with IC_{50} > 50 mM).

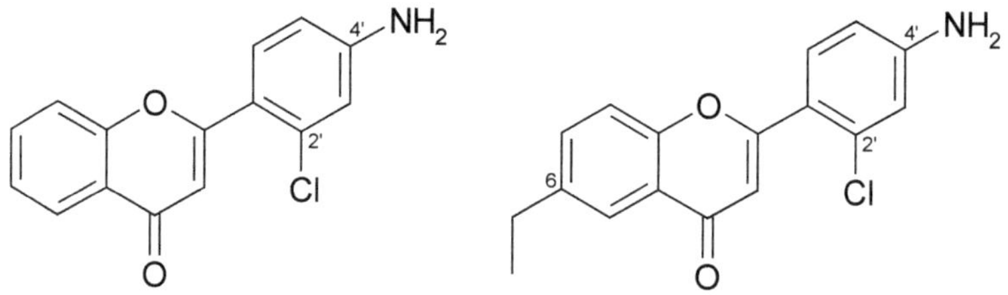

2′-chloro-4′-aminoflavone 6-ethyl-2′-chloro-4′-aminoflavone

Figure 3. Chemical structure examples of 2′-chloro-4′-aminoflavones.

Some of the aminoflavones mentioned below have recently undergone evaluation in clinical trials. Brantley et al. [35] provided a rationale for the continued development of AFP464 (Figure 4), which is a precursor of aminoflavone, as an agent to enhance the therapeutic management of breast cancer via dysregulation of the aryl hydrocarbon receptor (AhR) signaling pathway.

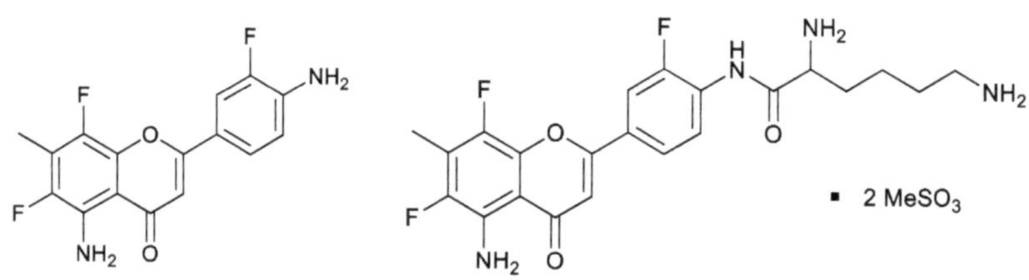

AF **AFP464**

Figure 4. Chemical structure of AF and AFP464.

AFP464 displays cytotoxicity and promotes AhR translocation from the cytosol to the nucleus in LM05 cells. It was found that AFP464 not only reduces bulk tumor cells, similarly to other P450 prodrugs, but also appears to target cells with stem-cell-like properties, at least in part by abolishing α6-integrin expression, which promotes metastasis.

Aminoflavone (AF), the active component of a novel anticancer prodrug (AFP464) in phase I clinical trials, is a ligand of the aryl hydrocarbon receptor (AhR). AhR dimerizes with HIF-1β/AhR, which is shared with HIF-1α, a transcription factor critical for the response of cells to oxygen deprivation. The response of mammalian cells to hypoxia is mediated, at least in part, by a family of transcription factors known as hypoxia-inducible factors (HIF). According to Terzuoli et al. [36], the inhibition of HIF-1α by the aminoflavone AFP464 was independent of a functional AhR pathway. Moreover, this compound was also inactive in MDA-MB-231 cells, yet inhibited HIF-1α in MDA-MB-231 cells transfected with the SULT1A1 gene.

Flavone itself and some aminoflavones have also been reported as inhibitors of protein tyrosine kinase in vitro. Król et al. [37] tested flavone (2-phenylbenzopyran-4-one) and amino-substituted flavones in vitro for the production of nitrite, a chemical product of nitric oxide, in model murine activated peritoneal macrophages. Among the tested compounds, 3′-amino-4′-hydroxyflavone (Figure 5) was the most potent inhibitor of nitrite production. 3′-Amino-4′-methoxyflavone demonstrated the most potent inhibition of the EGF receptor tyrosine kinase's activity derived from A431 cells (42% inhibition at 50 μM).

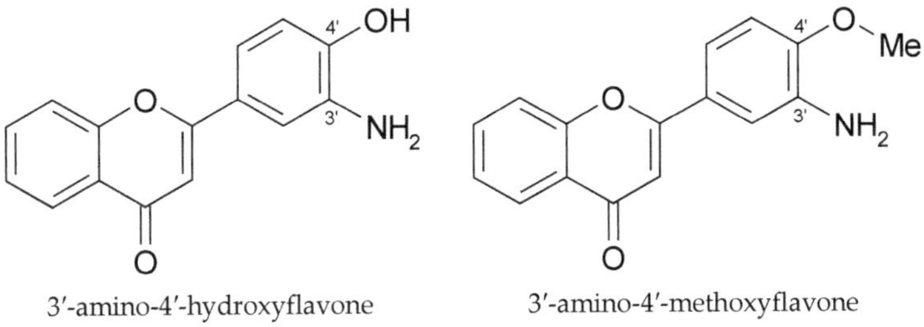

Figure 5. Structure of the 3′-aminoflavone derivatives.

Moreover, Cushman et al. [38] reported a series of 4′-aminoflavones which inhibited the activity of the protein tyrosine kinase. The most potent compound was 4′-amino-6-hydroxyflavone (Figure 6), which inhibited this activity with an IC_{50} = 1.2 μM.

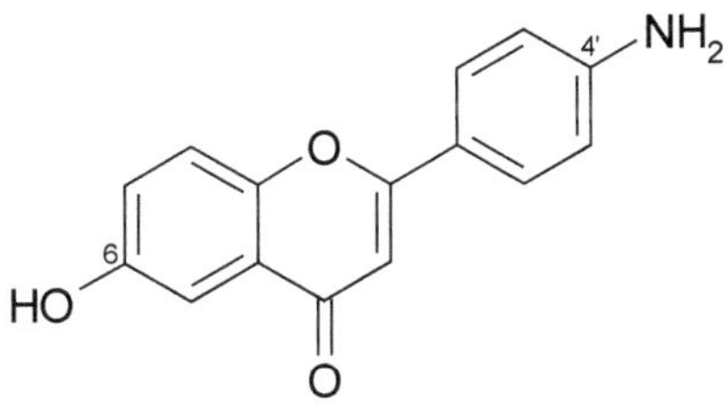

Figure 6. Structure of the 4′-amino-6-hydroxyflavone.

These and further results [18] suggest that aminoflavones may modulate inflammatory and immune responses by controlling the production of nitric oxide. The amino group therefore evidently plays a critical role in allowing recognition, because none of the tested nitroflavones inhibited the activity of the enzymes $p56^{lck}$, EGF_r, and $p60^{v\text{-}src}$ (Figure 7).

The results emphasize the importance of the 5,7-diamino-6-hydroxy substitution pattern, because the most active compounds in the tested series against $p56^{lck}$ (5,7, 4′-triaminoflavone, 6-hydroxyflavone and 5,7,3′-triamino-6-hydroxyflavone) showed an IC_{50} equal to 18 μM and 48 μM, respectively. When tested against EGFr, the compounds 6,4′-diaminoflavone and 5,7,3′-triamino-6-hydroxyflavone showed an IC_{50} of 8.7 and 7.8 μM, respectively. These compounds were also the most active against $p60^{v\text{-}src}$ (IC_{50} between 28.8 and 38.4 μM).

So as to enhance the antitumor activity, heterocyclic analogs have also been synthesized by combining the amino flavonoid ring and the ring of thiazolidine compounds. To study the aminoflavone derivatives, several compounds differing in the number of methoxy groups, such as 6-aminoflavone, 6-amino-3-methoxyflavone, and 6-amino-3,3′, 4′-trimethoxyflavone, were used (Figure 8) [39].

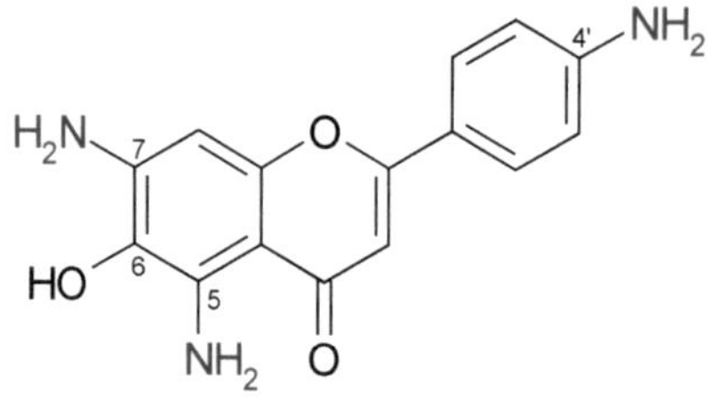

5,7, 4′-triaminoflavone

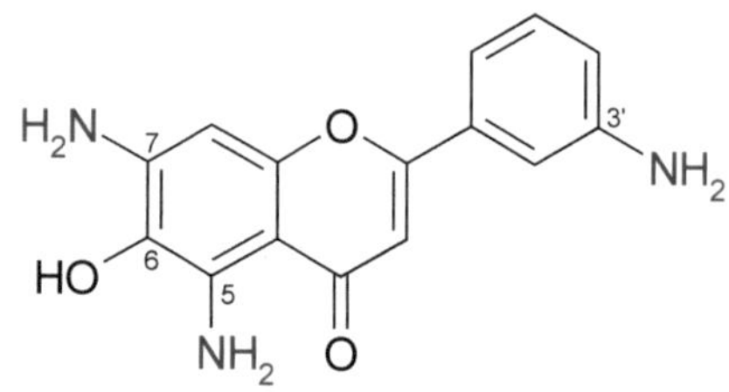

5,7,3′-triamino-6-hydroxyflavone

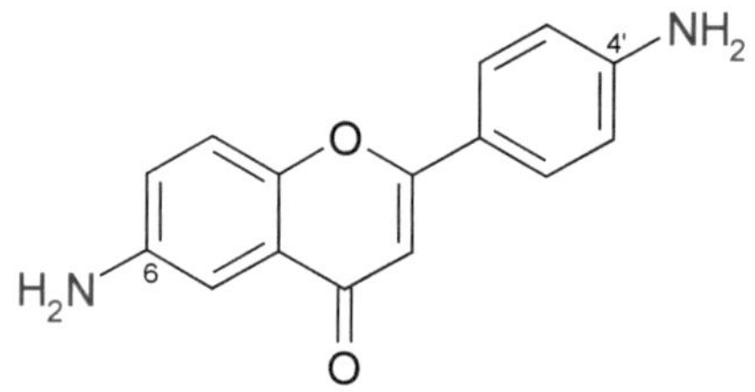

6,4′-diaminoflavone

Figure 7. Chemical structures of aminoflavones with anti-inflammatory action.

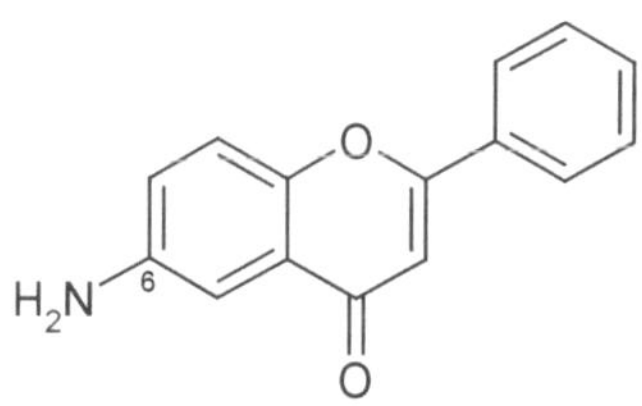

6-aminoflavone

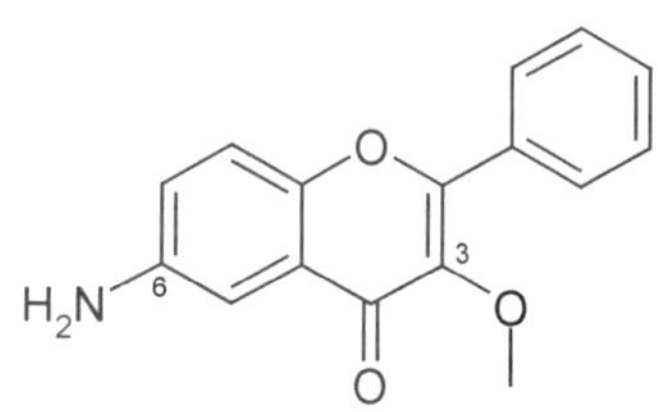

6-amino-3-methoxyflavone

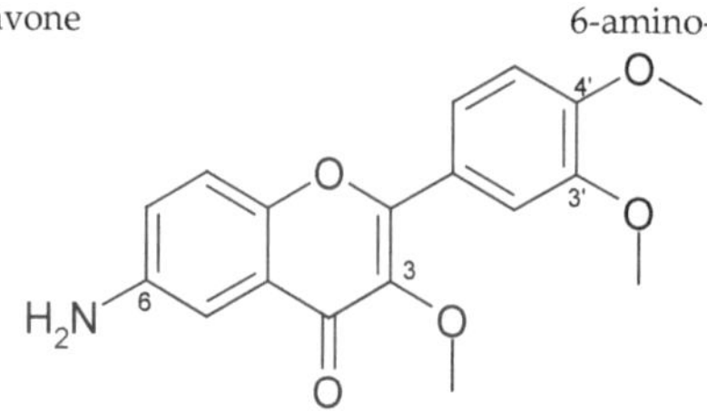

6-amino-3,3′,4′-trimethoxyflavone

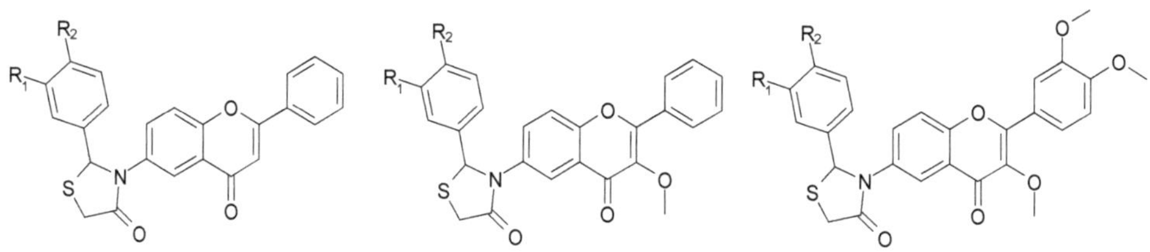

	a	**b**	**c**	**d**	**e**
R₁:	H	H	Cl	OCH_3	H
R₂:	H	OCH_3	Cl	OCH_3	OH

Figure 8. 6-Aminoflavonoids and their thiazolidinone analogs with antitumor effects.

The compounds were tested for their antitumor activity in vitro and in vivo. The research confirmed the cytotoxic potential of new analogs to the human breast adenocarcinoma MDA-MB-435 and HeLa cells. Interestingly, the presence of methoxy groups increased the cytotoxic potential against the cancerous cells without being toxic to healthy ones. As expected, the synthesized compounds increased the lifespan of mice with cancer.

According to recent studies [40], chrysin modified at the 7-position via regioselective alkylation to 7-aminochrysin derivatives showed better anticancer effects than the substrate. Menawhile, the transformation of *N*-phenylacetamides to the corresponding diphenylamine-type compounds decreased the anticancer activity. The promising results were obtained for the product of the reaction of chrysin with 2-chloro-*N*-(3,5-dimethoxyphenyl) acetamide, which showed the highest cytotoxic activity against the MCF-7 cell line of breast cancer (GI_{50} = 30 nM) and the HCT-15 cell line of colon cancer (GI_{50} = 60 nM).

Thorat et al. [41] synthesized a series of novel *N*-benzyl derivatives of 6-aminoflavone (Figure 9) and evaluated them for their anticancer activity (against the MCF-7 and A-549 cell lines) and for their topoisomerase II enzyme inhibition activity. Compounds with 4-chloro and 4-bromo substitution on the beznylamino group 6-(4-chlorobenzylamino)-2-phenyl-4H-chromen-4-one and 6-(4-bromobenzylamino)-2-phenyl-4H-chromen-4-one) were found to be the most potent anticancer agents against MCF-7 cells with an IC_{50} = 9.35 μM and 9.58 μM, respectively. These compounds were also found to exhibit robust inhibition of enzyme topoisomerase II with an IC_{50} of 12.11 and 12.79 μM, respectively. Meanwhile, the compound with a 3,4-dichloro substituent on the benzylamino group (6-(3,4-dichlorobenzylamino)-2-phenyl-4H-chromen-4-one) showed good inhibition (46.48%) of the human lung cancer cell line A-549.

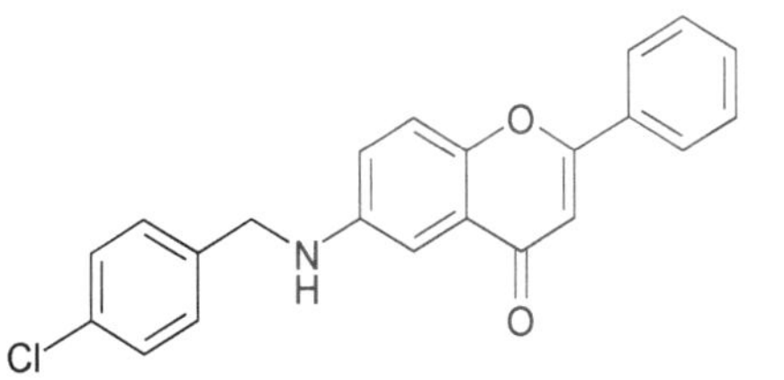

6-(4-chlorobenzylamino)-2 phenyl-4H-chromen-4-one

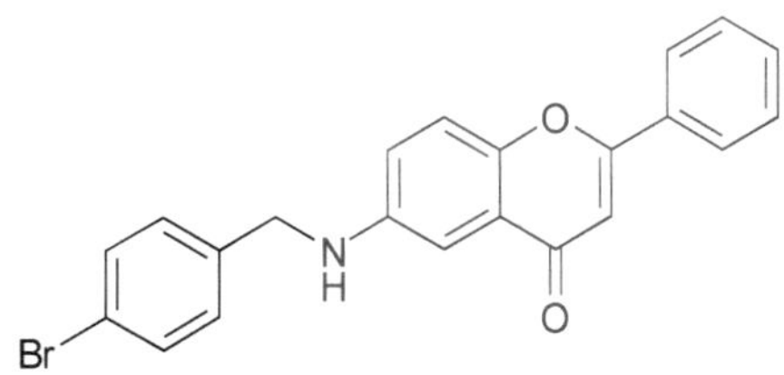

6-(4-bromochlorobenzylamino)-2 phenyl-4H-chromen-4-one

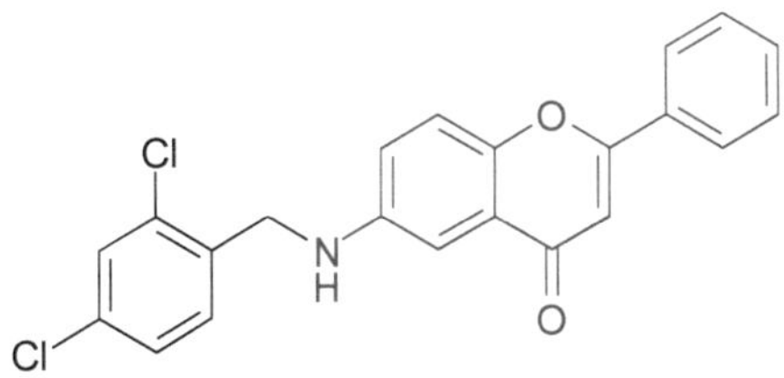

6-(3,4-dichlorobenzylamino)-2-phenyl-4H-chromen-4-one

Figure 9. 6-Benzylaminoflavones.

Another research team came to a similar conclusion. Shelke et al. [42] synthesized a series of new sulphonamide analogs of 6- and 7-aminoflavones (Figure 10) and tested them against cancer cell lines (HepG-2, A-549, and Caco-2). Some of the obtained compounds demonstrated excellent antiproliferative activity (IC_{50} = 0.98 μM), which was comparable to the standard drug adriamycin (IC_{50} = 0.94 μM).

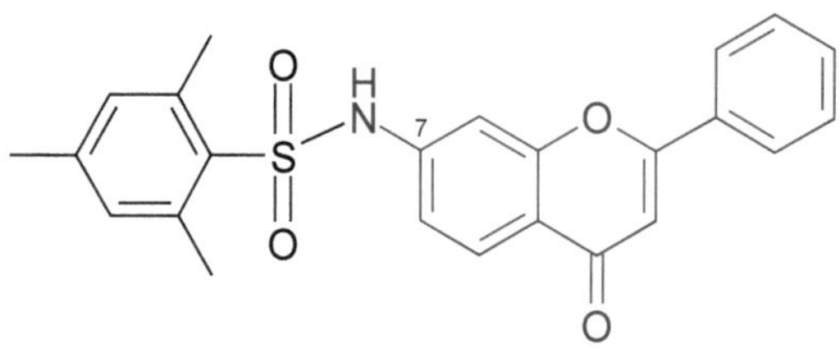

Figure 10. 7-Aminoflavone's sulphonamide structure.

The authors documented that these compounds exhibited anticancer activity via the inhibition of topoisomerase II, thus inducing DNA damage in cancer cells (Table 2).

Table 2. Anticancer activity of aminoflavonoids.

Aminoflavone Derivatives with Anticancer Effects			
Chemical Structure	**Cell Line**	**IC_{50} (μM)**	**Reference**
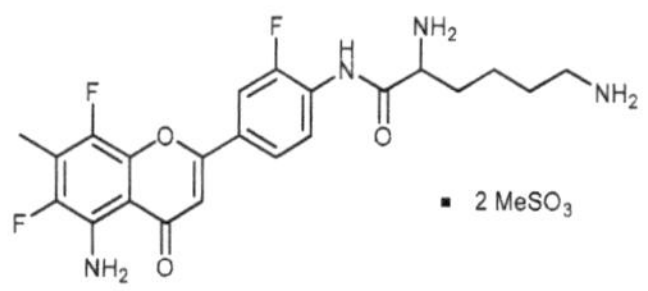	Human renal cancer cells		Luzzani et al. [29] Brantley et al. [36]
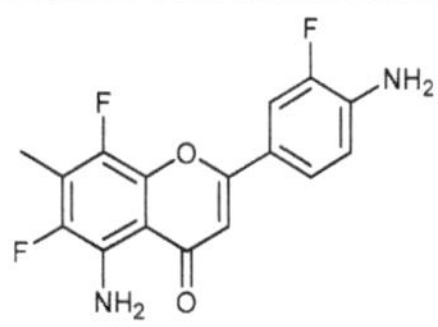	Human breast cancer cell lines		Brinkman et al. [31]
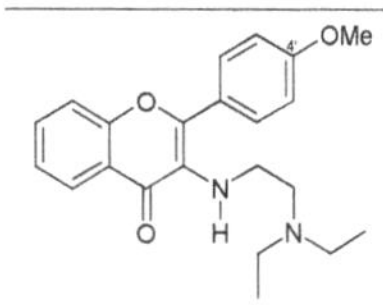	Murine L1210 leukemia cells	22	Dauzonne et al. [33]
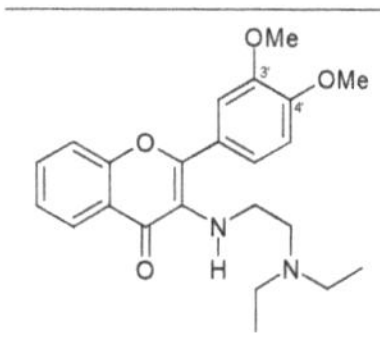	Murine L1210 leukemia cells	10	Dauzonne et al. [33]
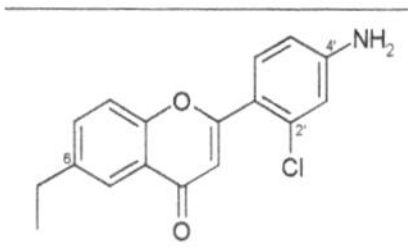	Hepatocarcinoma cells (HepG2), breast adenocarcinoma cells (MCF-7), and human chronic myelogenous leukemia cells (K562)	1.8	Jin et al. [35]
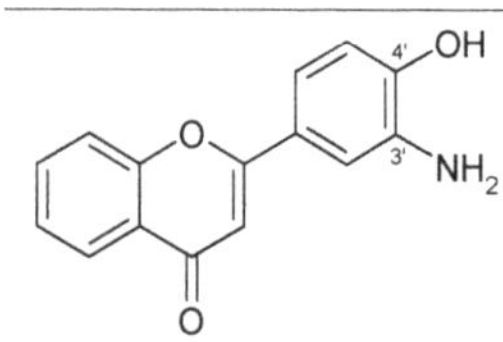	Inhibitors of protein tyrosine kinase		Król et al. [38]

Table 2. *Cont.*

Aminoflavone Derivatives with Anticancer Effects			
Chemical Structure	**Cell Line**	**IC_{50} (μM)**	**Reference**
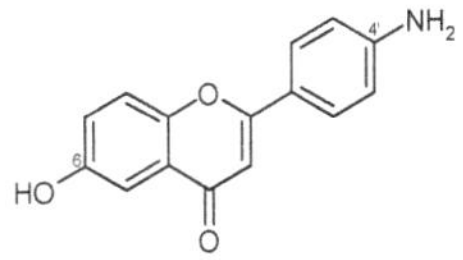	Inhibitors of protein tyrosine kinase	1.2	Cushman et al. [39]
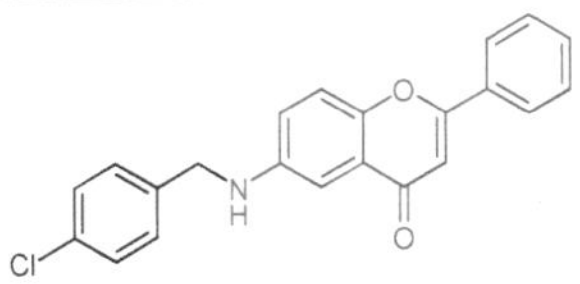	Anticancer activity to MCF-7 cells	9.35	Thorat et al. [41]
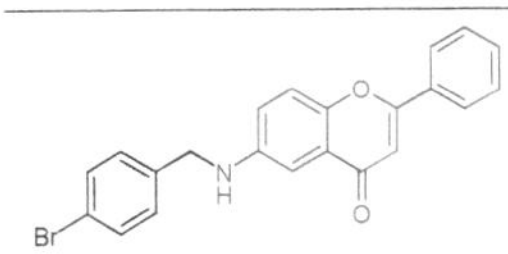	Anticancer activity to MCF-7 cells	9.58	Thorat et al. [41]
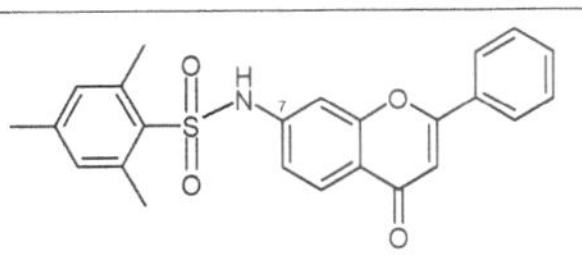	Anticancer to HepG-2, A-549, and Caco-2cell lines	0.98	Shelke et al. [42]

More recently, 6-aminoflavone was proposed by Flores-Flores et al. [43] as a potential agent for the treatment of asthma and related respiratory diseases. It exhibits a significant relaxant activity via calcium channel blockade, and a molecular docking study revealed that it binds to the L-type calcium channel. The same compound was examined by Ahmad et al. [44] as a potent drug that reduces D-galactose-induced oxidative stress, which reduces neuroinflammation. These authors declared that 6-amimoflavone can be viable therapeutic agent for addressing memory loss.

4. Aminoflavone Complexes

The medicinal properties of naturally occurring flavonoids have been well known for many years. However, the discovery that their complexes with metal ions are more effective than flavonoids themselves has changed the course of drug research. Flavonoids act as chelating agents due to the presence of carbonyl oxygen and the substituents with free electron pairs, e.g., hydroxyl or amine groups [45,46]. These results have shown that these complexes can be successfully used in the therapy of a number of diseases, such as diabetes, bacterial infections, and cancer. Their role in the treatment of neurodegenerative diseases, such like Huntington's disease, is also mentioned. Such complexes can also affect the iron balance in living organisms, which is important in the treatment of such diseases as Friedreich's or Thalassa's ataxia [47]. There are also known complexes of aminoflavones with cisplatin (Figure 11), a commonly used anticancer drug. In these complexes, 3-aminoflavone is used as the ligand instead of ammonia. The obtained derivatives are characterized by a high antitumor activity, e.g., in relation to ovarian tumor cells (CAOV 3, OVCAR 3) and lung cancer (A549).

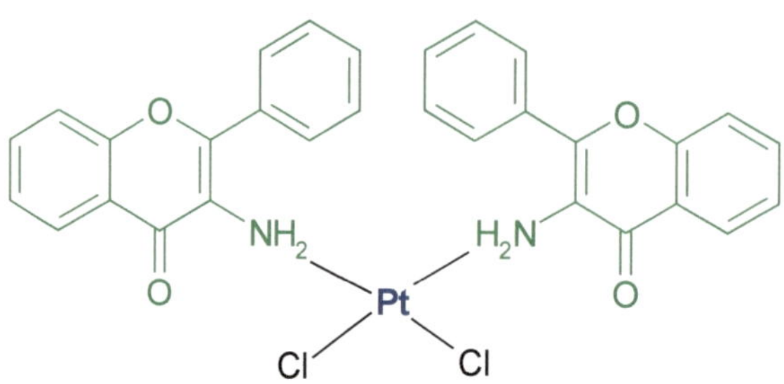

Figure 11. Complexes of 3-aminoflavones with cisplatin.

Antitumor activity was also reported for two new flavanone ruthenium (II) complexes, formed in the reaction of ruthenium (II) chloride with 3-aminoflavone dissolved in aliphatic alcohol. 3-Aminoflavone was first oxidized and then solvolyzed to give 3-imino-2-alkoxyflavanone, which served as the ligand for the ruthenium ion (II) (Scheme 4) [48].

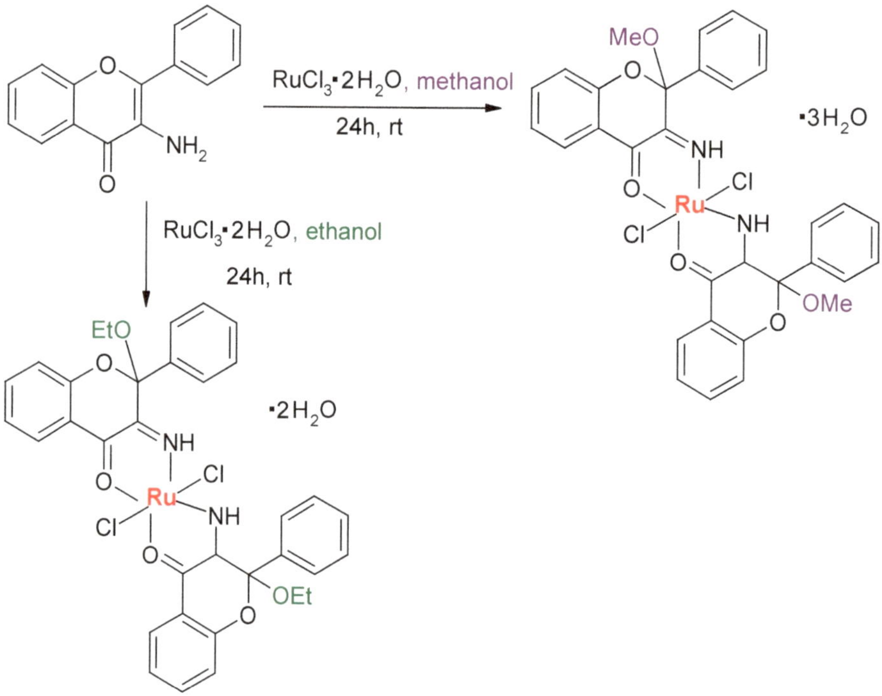

Scheme 4. Metal complexes of aminoflavonoids.

The pharmacological activities of the resulting complexes were evaluated against human bladder cancer cells. This study showed that the obtained compounds are highly cytotoxic against cancer cells. At the same time, the in vitro study proved the low toxicity of the synthesized complexes to healthy human lymphocytes.

Orzechowska et al. [49] analyzed the effect of the *trans*-platinum(II) complex of 3-aminoflavone on the viability and mortality of the OVCAR 3 and CAOV 3 ovarian cancer cells and on the expression of the selected genes involved in the process of apoptosis. It was noted that an increased concentration of the new metal–flavonoid complex resulted in a decrease in viability and an increase in mortality of the ovarian cancer cells. Increased Bax gene expression and decreased expression of Bcl2 and BIRC5 genes in the ovarian cancer cells were observed after the treatment with trans-Pt(3-af)$_2$Cl$_2$.

5. Aminochalcones

Chalcones, intermediates in flavonoid biosynthesis, play an important role as natural pigments found in many plants. Having a simple typical skeleton bearing two phenyls (rings A and B) spaced by a *trans*-enone bridge (1,3-diaryl-2-propen-1-one skeleton) (Figure 12), these compounds are important scaffolds in medicinal chemistry [50].

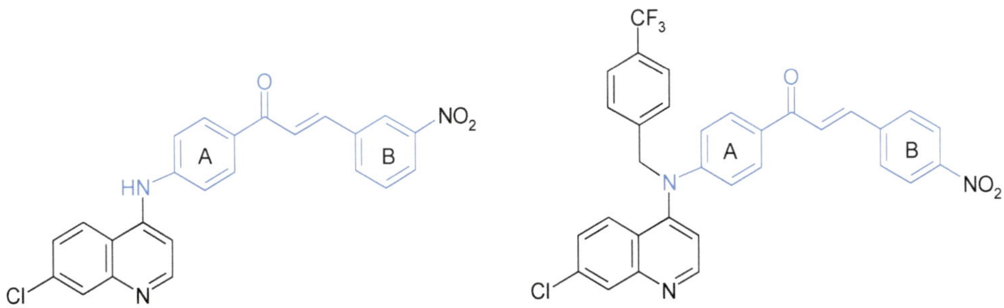

Figure 12. Hybrid derivatives incorporating arylnitro and aminochalcone moieties with antiprotozoal effects.

The presence of an α,β-unsaturated keto functionality (2-propen-2-one chain) seems to be responsible for the medicinal properties of the compounds. Depending on the substitution pattern in the phenolic rings, they differ in their biological properties, which mainly comprise anti-inflammatory, neurodegenerative, antifungal, and antibacterial activity [51–54]. Aminochalcones do not occur naturally. They may be synthesized in a one-pot synthesis involving Claisen–Schmidt condensation (Scheme 5), followed by the chemoselective hydrogenation of the nitro group of the nitrochalcone [55].

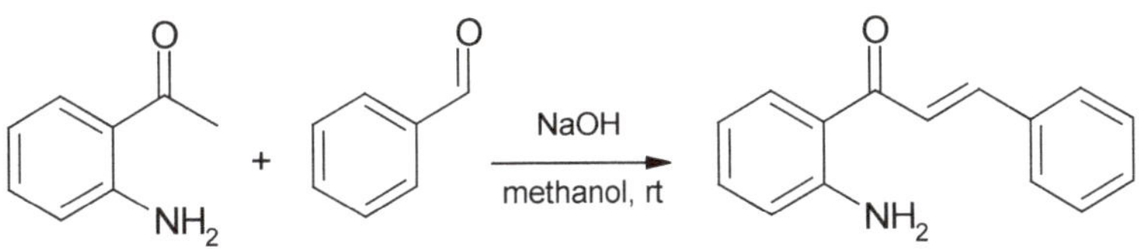

o-aminoaethophenone   benzaldehyde

Scheme 5. One-pot synthesis of aminochalcone involving Claisen–Schmidt condensation.

Potency against cancer cells has been reported for chalcones substituted with amino groups.

A range of chalcones with various electron-withdrawing and electron-donating substituents have been evaluated for their anticancer activity. As a result, a potent and selective cytotoxic effect of 2′-aminochalcone bearing an unsubstituted B ring was observed. Several recent studies have focused on the evaluation of the biological activity of chalcones as potential cytotoxic agents against canine cancer cells. These revealed a potent activity of unsubstituted chalcone and 4′-methoxychalcone against the canine macrophage tumor cell line, which exhibited time-dependent and concentration-dependent cytotoxicity.

Tristão et al. [56] reported a series of chalcone derivatives with a nitrogen atom to be antimicrobial agents. Their study showed that 4-aminochalcones exhibited antifungal effects. The aminochalcones were more toxic than the acetamidochalcones, while the nitrochalcones did not have any toxic effect. The results suggested that the antimicrobial activity to a large extent was governed by the groups attached to the aromatic B ring (furanyl, thiophenyl, chloro and methoxy ones). 4-Amino-4′-chlorochalcone (Figure 13) proved to be an excellent growth inhibitor for the cultured *Trichophyton mentagrophytes* ATCC 9973 and *Trichophyton rubrum* C137 (MIC = 20 μg/mL and 60 μg/mL, respectively).

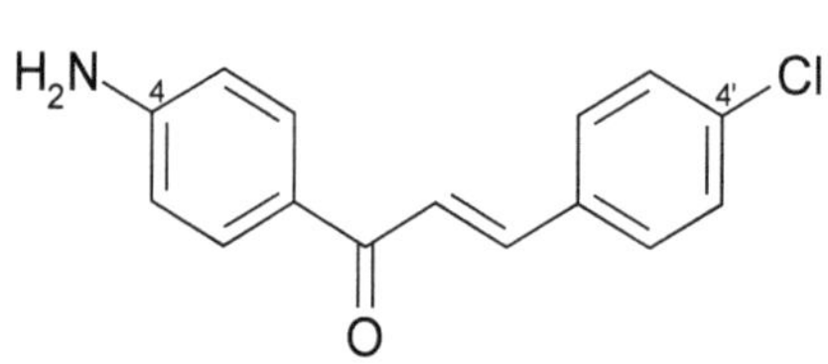

4-amino-4′-chlorochalcone

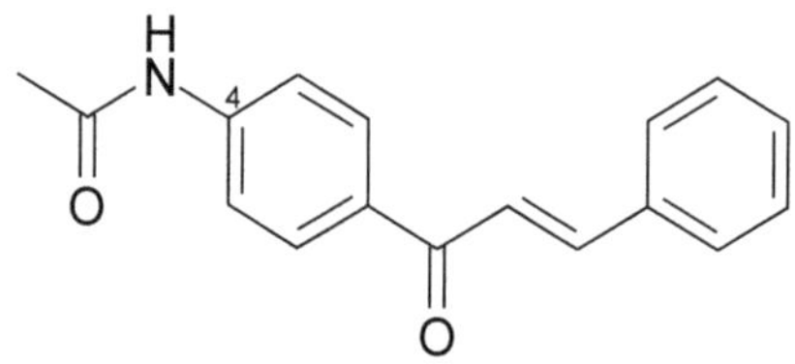

4-acetamidochalcone

Figure 13. Selected chalcones tested against *Trichophyton* species.

Meanwhile, 4-aminochalcones with either furanyl or thiophenyl group at the C-4 position in ring B were found to inhibit the growth of the pathogenic fungus *Aspergillus niger* ATCC 9092. 4-Acetamidochalcone showed measurable levels of minimum inhibitory concentration of 20 μg/mL against *Microsporum canis* C112, *Trichophyton mentagrophytes* ATCC 9972, and *Trichophyton rubrum* C137.

Sulpizio et al. [57] tested 2′-amino-4-hydroxychalcone and its analog 2′-amino-3,4-dihydroxychalcone (Figure 14) in order to assess their antioxidant activity using a DPPH radical assay.

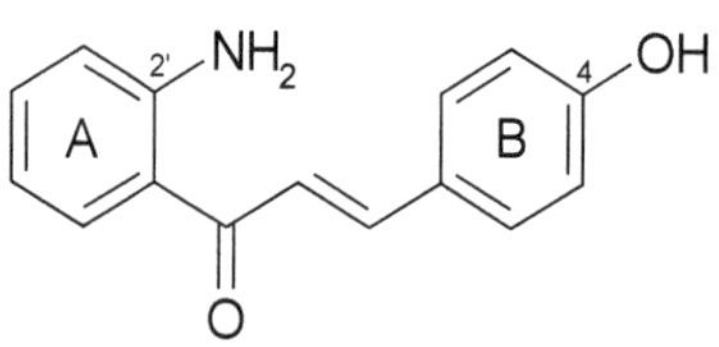

2′-amino-4-hydroxychalcone

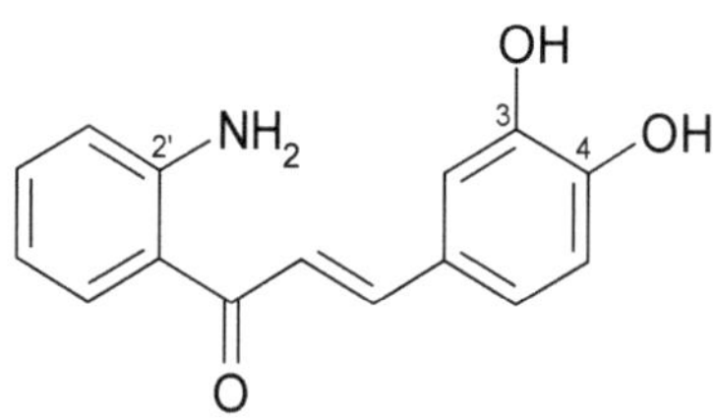

2′-amino-3,4-dihydroxychalcone

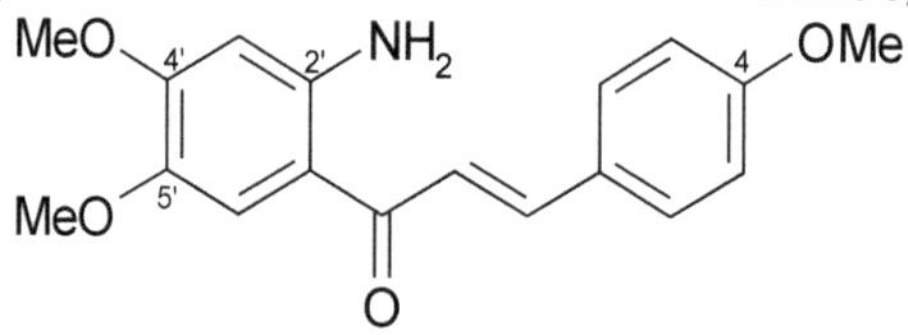

Figure 14. Selected 2′-aminochalcones with antioxidant effects.

The results indicated that the aminochalcone carrying two hydroxyl functionalities in the adjacent meta and para positions in ring B exhibited stronger antioxidant activity than other derivatives. The aminochalcone with two hydroxy groups in ring B showed an IC_{50} value (the compound concentration necessary to decrease the initial amount of DPPH by 50%) of 4.9 ± 1 μM. The results indicated that the presence of the two hydroxyl groups in ring B is important for the antioxidant activity of the aminochalcones. Additionally, it was evidenced that a significant inhibition of DPPH occurs only in the case of the two hydroxyl groups in meta and para positions in ring B, which suggests that the amino group is not involved in the antioxidant activity and that the hydrogen atom transfer comes from the phenolic group rather than from the ketoenolic moiety. The analogs having only one hydroxyl group or the methoxyl substituent in ring B proved to be less effective. The differences in biological activity of 2′-aminochalcones may arise from the formation of a side product via cyclization between the amino group in ring A and the aromatic B ring.

Prasad et al. [58] tested the antioxidant potential of para-substituted aminochalcones. Based on the results of this study, it can be inferred that substitution of the aromatic ring B with electron-donating groups at the ortho and/or para positions will increase the antioxidant activity of 4′-aminochalcones. The study results unequivocally show that 4-amino-3′,4′5′-trimethoxychalcone and 4-amino-4′-dimethylaminochalcone (Figure 15) have significant antioxidant activities, proven both in reactive oxygen species assays and in inhibiting the lipid peroxidation. These results suggest that the electron-releasing pharmacophores, such as the methoxy or dimethylamino groups, may be essential for the antioxidant activity.

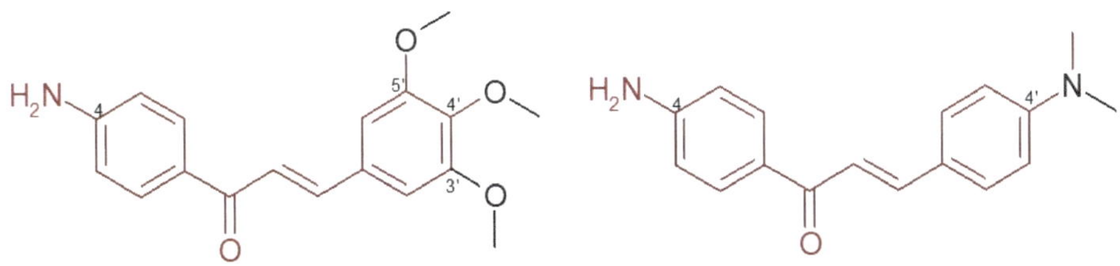

Figure 15. Chemical structures of 4-aminochalcones.

Moreover, Apiraksattayakul et al. [53] recently determined that methoxy derivatives of 4-aminochalcone exhibit neuroprotective effects against H_2O_2-induced oxidative stress in human neuroblastoma SH-SY5Y cells and they have potential as neuroprotective agents.

The study on structure–activity relationships carried out by Liu et al. [59] revealed that the chalcone skeleton is the most preferable for the anticancer activity. The in vitro cytotoxic activity measured by the MTT assay indicated that chalcone derivatives had higher activity than the corresponding flavones, with all IC_{50} values being lower than 10 μg/mL against all tested human tumor cell lines (ESA-109, A-549, HL-60, and PC-3).

Chalcone with a piperidinyl substituent at the C-3′ position (Figure 16A) was the most promising compound due to its high potency against the examined cancer cell lines (the IC_{50} values for ECA-109, A549, HL-60, and PC-3 cells were 1.3, 1.6, 2.6, and 2.5 μM, respectively). The introduction of open-chain aliphatic amino groups at C-3′, affording diethylamino, ethylmethylamino, and dimethylamino derivatives, led to the conclusion that it was a bulkier substituent that was favorable for the antiproliferative activity. The dietylamino derivative (Figure 16B) exhibited excellent antitumor activity against all the tested cell lines, with the IC_{50} ranging from 0.96 μM for HL-60 to 3.9 μM for PC-3 cells. Meanwhile, cyclization of the chalcones to the corresponding flavones and formation of the B ring led to a reduction in the cytotoxic activity.

Zeraik et al. [60] studied 4-aminochalcones substituted with fluorine, methyl, and hydroxyl groups in ring B and showed that three of these compounds were very interesting inhibitors of the chlorinating activity of myeloperoxidase. The compounds 4-aminochalcone, 4-amino-4′-fluorochalcone and 4-amino-4′-methylchalcone presented high oxidation potential, low scavenger capacity, and low inhibition of the respiratory burst of neutrophils. The activity of the tested chalcones was correlated with the lipophilicity of the substituents in ring B. The absence of a C-4′ amino group or its replacement with a hydroxyl one blocked the inhibitory effect. Biological activity screening of 2- and 4-aminochalcones showed the inhibitory activity of the compounds against canine malignant histiocytic cells (DH82). Aminochalcones with a hydrophobic substituent in ring B proved to be potent antiproliferative agents. A comparison of the unsubstituted *trans*-chalcone with 4-aminochalcone led to the conclusion that the amino group improved cytotoxicity against DH82 cells (inhibition of the cell growth by 27 and 59%, respectively, was noted). The presence of halogens and the methyl group at the para-position of the B ring also improved the activity. Replacement of the substituent at C-4′ with the strong electron-donating methoxy group resulted in a decrease in the cytotoxicity compared to 4-aminochalcone [27].

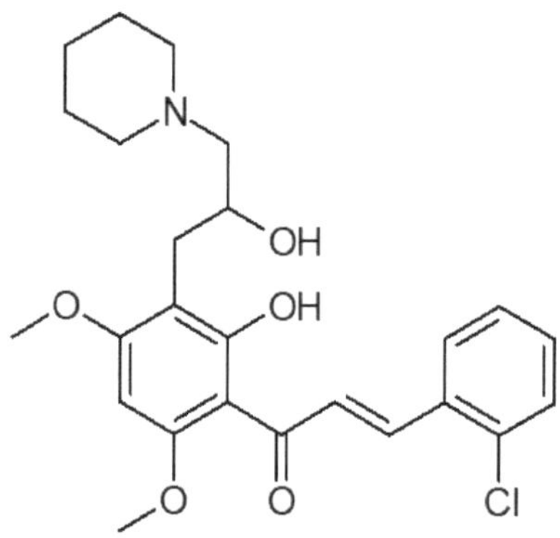

(A)
3-(2-Chlorophenyl)-1-{2-hydroxy-3-[2-hydroxy-3-(piperidin-1-yl)propyl]-4,6-dimethoxyphenyl} prop2-en-1-one

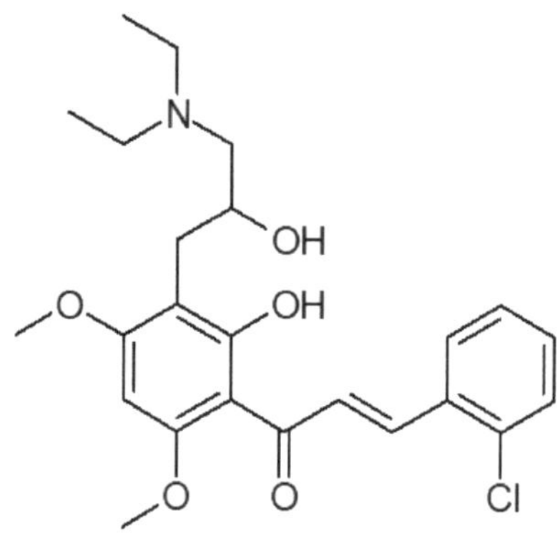

(B)
3-(2-Chlorophenyl)-1-[3-(3-diethylamino-2-hydroxypropyl)-2-hydroxy-4,6-dimethoxyphenyl]prop-2-en-1-one

Figure 16. Chemical structures of selected chalcone derivatives.

The anti-inflammatory and antioxidant properties of 2-aminochalcone, 3-aminochalcone, and 4-aminochalcone and their derivatives were examined by Iqbal et al. [61]. The ranking of activity for regio isomeric aminochalcones was determined as follows: orto > meta > para. It was observed that only the trifluoro-methyl analog showed better antioxidant activity than the unsubstituted chalcone. Good antioxidant potency was also observed for two aminochalcones (with 3-NH_2 and 4-NH_2 group) and for 4-cyanochalcone.

Aminochalcones obtained by Trein et al. [62] were evaluated for their activity against the parasitic protozoan *Trichomonas vaginalis*, which is the most common non-viral sexually transmitted disease worldwide. Among them, 3′-aminochalcone exerted the most potent effect and showed a high cytotoxicity against human vaginal cells. On the other hand, 3′-aminochalcone did not exhibit toxicity against *Galleria mellonella* larvae at a dose of up to 50 mg/kg over 120 h. Similarly, it had no hemolytic effect on human erythrocytes. When the erythrocytes were incubated with 3′-aminochalcone at a concentration of 100 μM, the result was not significantly different from the control. Trophozoites of *T. vaginalis* treated with this compound did not present significant reactive oxygen species (ROS) accumulation, but induced a significantly higher ROS accumulation in human neutrophils after co-incubation.

Xiao et al. [63] prepared a series of 4′-aminochalcones–revastimine hybrids and evaluated them as potential agents for the treatment of Alzheimer's disease. They were tested for their cholinesterase inhibitory activity, antioxidant and metal-chelating effects, inhibitory effects on Aβ aggregation, and their ability to cross the blood–brain barrier (BBB). The results indicated that a compound containing one head group of dimethylamine at the 4′-position and a cyclic amine group at the C-4 position had the best inhibitory potency on acetylcholinesterase (IC_{50} = 4.91 μM) and significant antioxidant activity with a value 2.83-fold greater than that of Trolox. This molecule could also cross the BBB in vitro, possessed the ability to act as a metal chelator, and inhibited self-induced A$\beta_{1\text{–}42}$ aggregation and Cu^{+2}-induced A$\beta_{1\text{–}42}$ aggregation.

The results obtained by Seba et al. [64] indicated that 4′-aminochalcones with the naphthyl moieties 4-amino-4′-naphthylchalcone and 4-amino-4′-methylnaphtylchalcone (Figure 17) have potential anti-metastatic activity mediated by p53, which may be exploited for the treatment of osteosarcoma—the most recurrent malignant bone tumor, with high metastatic potential. It was confirmed that these compounds suppressed the migration of U2OS (p53 wt) cells and that of a SAOS-2 (p53 null) cell line expressing p53.

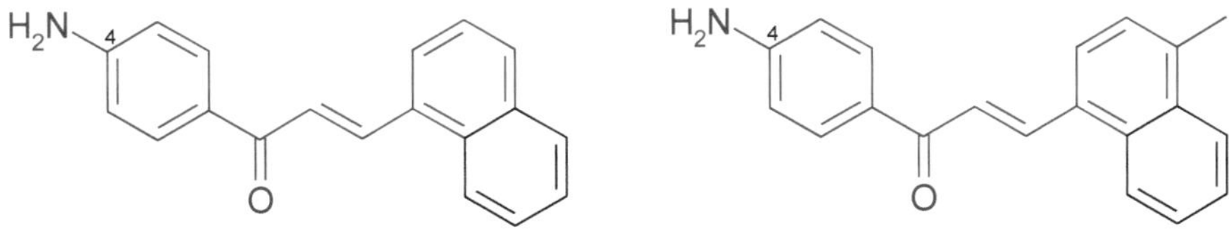

Figure 17. 4-Aminochalcones with naphthyl moieties.

Cavalcante et al. [65] obtained 4-aminochalcone and its derivative with an elongated aliphatic unsaturated chain (Figure 18) that presented trypanocidal effects, causing membrane damage and oxidative stress. These authors stated that the enlargement of the aliphatic unsaturated chain enhances the stability and lipophilicity of p-aminochalcones without affecting the trypanocidal effect. Both molecules induced membrane damage, oxidative stress, and mitochondrial dysfunction, leading to parasitic cell death.

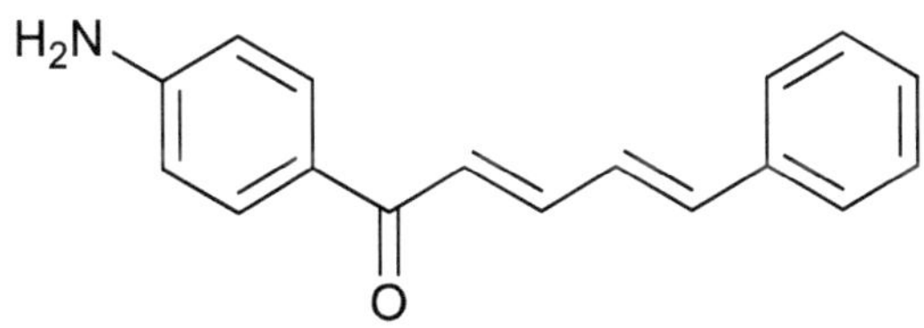

Figure 18. 4-aminochalcone with an elongated aliphatic unsaturated chain.

Prasad et al. [66] synthesized a series of 4-aminochalcones via the Claisen–Schmidt condensation of 4-aminoacetophenone with various substituted aromatic aldehydes and tested them for their antimicrobial and anti-inflammatory activity. The highest antibacterial activity was noted for the compounds with chlorine at C-4′, bromine at C-3′, methoxy groups at C-3′ and C-4′, and also methoxy groups at C-3′, C-4′, and C-5′ in the aromatic ring B (the zone of inhibition was between 17 and 25 mm). All of the synthesized compounds exhibited moderate to considerable antifungal activity when compared to fluconazole, which was used as the reference standard. Some of them were found to also possess significant anti-inflammatory activity, comparable to standard drugs, such as aceclofenac.

However, some studies have revealed that we need to be careful when suggesting these compounds for therapeutic use. According to the toxicological study performed by Mariño et al. [67], all tested 4′-aminochalcones caused lower leucocyte growth compared to the negative control, as well as higher micronucleus generation in the mutagenic test (at 80 μM) than H_2O_2. In this study, the mutagenic and genotoxic effects were determined, which proved to be concentration-dependent.

Aminochalcones can also serve as the basis for obtaining other derivatives with a nitrogen atom, characterized by interesting biological properties. Patel and Dholakiya [68] synthesized new 4′-aminochalcone-based dibromomaleimides with antituberculosis and anti-candida activity.

Although there is some information in the literature on the biological activities of aminochalcone derivatives, additional studies on their safety are essential. Mariño et al. [67]

evaluated the toxicity of 4′-aminochalcones in vitro using the cellular proliferation and viability assays, the micronucleus test, and the comet assay measuring DNA damage. Their results showed that the tested compounds have mutagenic and genotoxic effects, which are concentration-dependent.

6. Conclusions

The goal of the most recent research studies is to design new compounds that may be used in medicine as new pro-health pharmaceuticals, with reduced side effects. The increasing level of drug resistance among patients highlights the urgent necessity of new therapeutic options. Flavonoids containing a nitrogen atom, which constitute a large group of synthetic compounds, may be a promising source of new drugs and health-promoting active substances in the future. The aminoflavone prodrug AFP464 has been evaluated in clinical trials for the treatment of solid tumors, and its ability to suppress tumor-initiating cell (TIC) growth has also been suggested [69]. It not only reduces bulk tumor but also appears to target cells with stem-cell-like properties. On the basis of a recent critical review [70] regarding flavonoid drugs, it is known that nineteen flavonoid-based drugs have been approved and are available as medical prescriptions, and 36 candidates are undergoing or suspended in different clinical phases. Among them, three are undergoing different phases of pharmacological studies or are clinical candidates suspended in different phases of pharmacological tests.

Our summary showed that the synthetic or semi-synthetic modification of compounds with either a flavone or chalcone moiety can change their biological activities. Additionally, good cytotoxic and antioxidant properties of aminoflavonoids arise from their ability to form complexes via the chelation of metal ions. We hope that this review will facilitate the development of more potent active agents based on the scaffold of aminoflavonoids in the future. It is also important to continue the investigation of the mechanisms of their actions.

Author Contributions: Conceptualization and methodology, M.S.-G.; writing—original draft preparation, M.S.-G.; writing—review and editing, A.B.-B., N.P. and I.Z.; visualization, M.S.-G. and A.B.-B.; project administration, M.S.-G.; funding acquisition, M.S.-G. All authors have read and agreed to the published version of the manuscript.

Funding: Monika Stompor-Gorący acknowledges support from the National Science Centre NCN Poland, Grant SONATA 16 number 2020/39/D/NZ9/02023). The funder had no role in study design, data collection and analysis, decision to publish, or preparation of the manuscript.

Institutional Review Board Statement: Not applicable.

Informed Consent Statement: Not applicable.

Data Availability Statement: Data are contained within this article.

Conflicts of Interest: The authors declare no conflicts of interest.

References

1. Tohge, T.; de Souza, L.P.; Fernie, A.R. Current understanding of the pathway of flavonoid biosynthesis in model and crop plants. *J. Exp. Bot.* **2017**, *68*, 4013–4028. [CrossRef] [PubMed]
2. Hollman, P.C.; Bijsman, M.N.; van Gameren, Y.; Cnossen, E.P.; de Vries, J.H.; Katan, M.B. The sugar moiety is a major determinant of the absorption of dietary flavonoid glycosides in man. *Free Radic. Res.* **1999**, *31*, 569–573. [CrossRef] [PubMed]
3. Stompor, M.; Świtalska, M.; Wietrzyk, J. Synthesis and biological evaluation of acyl derivatives of hydroxyflavones as potent antiproliferative agents against resistance cell lines. *Z. Naturforsch C J. Biosci.* **2018**, *73*, 87–93. [CrossRef]
4. Stompor-Gorący, M.; Bajek-Bil, A.; Machaczka, M. Chrysin: Perspectives on contemporary status and future possibilities as pro-health agent. *Nutrients* **2021**, *13*, 2038. [CrossRef]

5. Stompor-Gorący, M.; Włoch, A.; Sengupta, P.; Nasulewicz-Goldeman, A.; Wietrzyk, J. Synergistic proliferation effects of xanthohumol and niflumic acid on Merkel and glioblastoma cancer cells: Role of cell membrane interactions. *Int. J. Mol. Sci.* **2024**, *25*, 11015. [CrossRef] [PubMed]
6. Panche, A.N.; Daiwan, A.D.; Chandra, S.R. Flavonoids: An overview. *J. Nutr. Sci.* **2016**, *5*, e47. [CrossRef]
7. Rammohan, A.; Reddy, J.S.; Sravya, G.; Rao, C.N.; Zyrynov, G.V. Chalcone synthesis, properties and medicinal applications: A review. *Environ. Chem. Lett.* **2020**, *18*, 433–458. [CrossRef]
8. Stompor, M.; Broda, D.; Bajek-Bil, A. Dihydrochalcones: Method of acquisition and pharmacological properties—A first systematic review. *Molecules* **2019**, *24*, 4468. [CrossRef]
9. Stompor, M.; Świtalska, M.; Wietrzyk, J. The influence of a single and double biotinylation of xanthohumol on its anticancer activity. *Acta Biochim. Pol.* **2019**, *66*, 2876. [CrossRef] [PubMed]
10. Stompor, M.; Żarowska, B. Antimicrobial activity of xanthohumol and its selected structural analogues. *Molecules* **2016**, *21*, 608. [CrossRef] [PubMed]
11. Ecker, G.; Huber, D.; Schmid, D.; Chiba, P. The importance of a nitrogen atom in modulators of multidrug resistance. *Mol. Pharmacol.* **1999**, *56*, 791–796. [CrossRef]
12. Casano, G.; Robin, M.; Barbier, P.; Peyrot, V.; Faure, R. Synthesis and complete assignment of the ^{1}H and ^{13}C NMR signals of new acetamido and aminoflavonoid derivatives. *Magn. Reson. Chem.* **2010**, *48*, 738–744. [CrossRef] [PubMed]
13. Barros, A.I.R.N.A.; Dias, A.F.R.; Silva, A.M.S. Reductive coupling reactions of 2-nitrochalcones and their β-hydroxy- analogues: New synthesis of 2-arylquinoline and 2-aryl-4-hydroxyquinoline derivatives. *Monatshefte Fur Chem.* **2007**, *138*, 585–594. [CrossRef]
14. Wanich, S.; Yenjai, C. Amino and nitro derivatives of 5,7-dimethoxyflavone from *Kaempferia parviflora* and cytotoxicity against KB cell line. *Arch. Pharm. Res.* **2009**, *32*, 1185–1189. [CrossRef] [PubMed]
15. Stompor, M.; Dancewicz, K.; Gabryś, B.; Anioł, M. Insect antifeedant potential of xanthohumol and isoxanthohumol and their derivatives. *J. Agric. Food Chem.* **2015**, *63*, 6749–6756. [CrossRef]
16. Dymarska, M.; Janeczko, T.; Kostrzewa-Susłow, E. Biotransformations of flavones and an isoflavone (daidzein) in cultures of entomopathogenic filamentus fungi. *Molecules* **2018**, *23*, 1356. [CrossRef]
17. Stompor, M.; Świtalska, M.; Bajek, A.; Wietrzyk, J. Influence of amide versus ester linkages on the anticancer properties of the new flavone-biotin conjugates. *Z. Naturforsch C* **2019**, *74*, 193–200. [CrossRef]
18. Cushman, M.; Zhu, H.; Geahlen, R.L.; Kraker, A.J. Synthesis and biochemical evaluation of a series of aminoflavones as potential inhibitors of protein-tyrosine kinases $p56^{lck}$, EGFr, and $p60^{v\text{-}src}$. *J. Med. Chem.* **1994**, *37*, 3353–3362. [CrossRef]
19. Deka, N.; Hadjeri, M.; Lawson, M.; Beney, C.; Boumendjel, A. Acetylated dimethoxyaniline as a key intermediate for the synthesis of aminoflavones and quinolones. *Heterocycles* **2002**, *57*, 123–128.
20. Deng, B.; Lepoivre, A.; Lemière, G. Synthesis of 7-vinylflavone and 7-aminoflavone by palladium-catalyzed coupling reactions. *Eur. J. Org. Chem.* **1999**, *10*, 2683–2688. [CrossRef]
21. Tang, L.; Zhang, S.; Yang, J.; Gao, W.; Cui, J.; Zhuang, T. A novel approach to the synthesis of 6-amino-7-hydroxyflavone. *Molecules* **2004**, *9*, 842–848. [CrossRef] [PubMed]
22. Takechi, A.; Takikawa, H.; Miyake, H.; Sasaki, M. Synthesis of 3-aminoflavones from 3-hydroxyflavones via 3-tosyloxy or 3-mesyloxyflavones. *Chem. Lett.* **2006**, *35*, 128–129. [CrossRef]
23. Babu, T.H.; Rao, V.R.S.; Tiwari, A.K.; Babu, K.S.; Srinivas, P.V.; Ali, A.Z.; Rao, J.M. Synthesis and biological evaluation of novel 8-aminomethylated oroxylin A analogues as α-glucosidase inhibitors. *Bioorg Med. Chem. Lett.* **2008**, *18*, 1659–1662. [CrossRef]
24. Shih, T.L.; Chou, C.E.; Liao, W.Y.; Hsiao, C.A. Cooper-mediated trimethylsilyl azide in amination of bromoflavonoids to synthesize unique aminoflavonoids. *Tetrahedron* **2014**, *70*, 3657–3664. [CrossRef]
25. Moorkoth, S.; Srinivasan, K.K.; Kutty, N.G.; Joseph, A.; Naseer, M. Synthesis and evaluation of a series of novel imidazolidinone analogues of 6-aminoflavone as anticancer and anti-inflammatory agents. *Med. Chem. Res.* **2013**, *22*, 5066–5075. [CrossRef]
26. Stompor, M. 6-Acetamidoflavone obtained by microbiological and chemical methods and its antioxidant activity. *J. Biotechnol.* **2016**, *237*, 25–34. [CrossRef]
27. Santos, M.B.; Pinhanelli, V.C.; Garcia, M.A.R.; Silva, G.; Baek, S.J.; França, S.C.; Fachin, A.L.; Marins, M.; Regasini, L.O. Antiproliferative and pro-apoptotic activities of 2′- and 4′-aminochalcones against tumor canine cells. *Eur. J. Med. Chem.* **2017**, *138*, 884–889. [CrossRef] [PubMed]
28. Luzzani, G.; Callero, M.A.; Kuruppu, A.I.; Trapani, V.; Flumian, C.; Todaro, L.; Bradshaw, T.D.; Perez, A.I.L. In vitro antitumor effect of AHR ligands aminoflavone (AFP 464) and benzothiazole (5F 203) in human renal carcinoma cells. *J. Cell Biochem.* **2017**, *118*, 4526–4535. [CrossRef]
29. Callero, M.A.; Rodriguez, C.E.; Sólimo, A.; Bal de Kier Joffé, E.; Loaiza Perez, A.l. The immune system as a new possible cell target for AFP 464 in a spontaneous mammary cancer mouse model. *J. Cell Biochem.* **2017**, *118*, 2841–2849. [CrossRef] [PubMed]
30. Brinkman, A.M.; Wu, J.; Ersland, K.; Xu, W. Estrogen receptor α and aryl hydrocarbon receptor independent growth inhibitory effects of aminoflavone in breast cancer cells. *BMC Cancer* **2014**, *14*, 344. [CrossRef] [PubMed]

31. Reinicke, K.E.; Kuffel, M.J.; Goetz, M.P.; Ames, M.M. Synergistic interactions between aminoflavone, paclitaxel and camptothecin in human breast cancer cells. *Cancer Chemother. Pharmacol.* **2010**, *66*, 575–578. [CrossRef]
32. Dauzonne, D.; Folléas, B.; Martinez, L.; Chabot, G.G. Synthesis and in vitro cytotoxicity of a series of 3-aminoflavones. *Eur. J. Med. Chem.* **1997**, *32*, 71–82. [CrossRef]
33. Akama, T.; Shida, Y.; Sugaya, T.; Ishida, H.; Gomi, K.; Kasai, M. Novel 5-aminoflavone derivatives as specific antitumor agents in breast cancer. *J. Med. Chem.* **1996**, *39*, 3461–3469. [CrossRef] [PubMed]
34. Jin, F.; Zhang, N.; Tan, C.; Gao, D.; Zhang, C.; Liu, F.; Chen, Z.; Gao, C.; Liu, H.; Li, S.; et al. 2′-Chloro-4′-aminoflavone derivatives selectively targeting hepatocarcinoma cells: Convenient synthetic process G_2/M cell cycle arrest and apoptosis triggers. *Arch. Pharm. Chem. Life Sci.* **2012**, *345*, 525–534. [CrossRef]
35. Brantley, E.; Callero, M.A.; Berardi, D.E.; Campbell, P.; Rowland, L.; Zylstra, D.; Amis, L.; Yee, M.; Simian, M.; Todaro, L.; et al. AhR ligand aminoflavone inhibits α6-integrin expression and breast cancer sphere-initiating capacity. *Cancer Lett.* **2016**, *376*, 53–61. [CrossRef] [PubMed]
36. Terzuoli, E.; Puppo, M.; Rapisarda, A.; Uranchimeg, B.; Cao, L.; Burger, A.M.; Ziche, M.; Melillo, G. Aminoflavone, a ligand of the aryl hydrocarbon receptor, inhibits HIF-1α expression in an AhR-independent fashion. *Cancer Res.* **2010**, *70*, 6837–6848. [CrossRef]
37. Król, W.; Czuba, Z.P.; Threadgill, M.D.; Cunningham, B.D.M.; Pietsz, G. Inhibition of nitric oxide (NO) production in murine macrophages by flavones. *Biochem. Pharmacol.* **1995**, *50*, 1031–1035. [CrossRef]
38. Cushman, M.; Nagaratham, D.; Burg, D.L.; Geahlen, R.L. Synthesis and protein-tyrosine kinase inhibitory activities of flavonoid analogues. *J. Med. Chem.* **1991**, *34*, 798–806. [CrossRef] [PubMed]
39. Moorkoth, S. Synthesis and anticancer activity of novel thiazolidinone analogs of 6-aminoflavone. *Chem. Pharm. Bull.* **2015**, *63*, 974–985. [CrossRef]
40. Mayer, S.; Keglevich, P.; Ábrànyi-Balogh, P.; Szigetvàri, Á.; Dèkàny, M.; Szàntay, C.; Hazai, L. Synthesis and in vitro anticancer evaluation of novel chrysin and 7-aminochrysin derivatives. *Molecules* **2020**, *25*, 888. [CrossRef] [PubMed]
41. Thorat, N.M.; Sarkate, A.P.; Lokwani, D.K.; Tiwari, S.V.; Azad, R.; Thopate, S.R. N-benzylation of 6-aminoflavone by reductive amination and efficient access to some novel anticancer agents via topoisomerase II inhibition. *Mol. Diver* **2021**, *25*, 937–948. [CrossRef]
42. Shelke, R.N.; Pansare, D.N.; Sarkate, A.P.; Narula, I.K.; Lokwani, D.K.; Tiwari, S.V.; Azad, R.; Thopate, S.R. Synthesis and evaluation of novel sulfonamide analogues of 6/7-aminoflavones as anticancer agents via topoisomerase II inhibition. *Bioorg. Med. Chem. Lett.* **2020**, *30*, 127246. [CrossRef]
43. Flores-Flores, A.; Estrada-Soto, S.; Millán-Pacheco, C.; Bazán-Perkins, B.; Hernández-Pando, R.; Ibarra-Barajas, M.; Villalobos-Molina, R. Ex vivo and in silico approaches of tracheal relaxation through calcium channel blockade of 6-aminoflavone and its toxicological studies in murine models. *Biomedicines* **2023**, *11*, 1870–1881. [CrossRef] [PubMed]
44. Ahmad, S.; Shah, S.A.; Nishan, U.; Khan, N.; Almutairi, M.H.; Fozia, F.; Jamila, N.; Almutairi, B.B.; Ullah, Z. 6-Aminoflavone activates Nrf2 to inhibit the phosphor-JNK/TNF-α signaling pathway to reduce amyloid burden in an aging mouse model. *ACS Omega* **2023**, *8*, 26955–26964. [CrossRef] [PubMed]
45. Fabijańska, M.; Orzechowska, M.; Rybarczyk-Pirek, A.J.; Dominikowska, J.; Bieńkowska, A.; Małecki, M.; Ochocki, J. Simple trans-platinum complex bearing 3-aminoflavone ligand could be a useful drug: Structure—Activity relationship of platinum complex in comparison with cisplatin. *Int. J. Mol. Sci.* **2020**, *21*, 2116. [CrossRef]
46. Mucha, P.; Hikisz, P.; Gwożdziński, K.; Krajewska, U.; Leniart, A.; Budzisz, E. Cytotoxic effect, generation of reactive oxygen/nitrogen species and electrochemical properties of Cu (II) complex in comparison to half-sandwich complexes of Ru(II) with aminochromone derivatives. *RSC Adv.* **2019**, *9*, 31943–31952. [CrossRef] [PubMed]
47. Grazul, M.; Budzisz, E. Biological activity of metal ions complexes of chromones, coumarins and flavones. *Coord. Chem. Rev.* **2009**, *253*, 2588–2598. [CrossRef]
48. Kasprzak, M.M.; Szmigiero, L.; Zyner, E.; Ochocki, J. Proapoptotic activity in vitro of two novel ruthenium(II) complexes with flavanone-based ligands that overcome cisplatin resistance in human bladder carcinoma cells. *J. Inorg. Biochem.* **2011**, *105*, 518–524. [CrossRef]
49. Orzechowska, M.; Fabijanska, M.; Ochocki, J.; Małecki, M. Anticancer activity of trans-platinum (II) complex of 3-aminoflavone to ovarian cancer cells. *Ginekol. Pol.* **2017**, *88*, 68–74. [CrossRef] [PubMed]
50. Hartman, C.B.; Dube, P.S.; Legoabe, L.J.; Van Pelt, N.; Matheeussen, A.; Caljon, G.; Beteck, R.M. Novel quinoline derivatives with broad-spectrum antiprotozoal activities. *Arch. Pharm.* **2024**, *357*, 2300319. [CrossRef]
51. Santos, M.B.D.; Carvalho Marques, B.; Miranda Ayusso, G.; Rocha Garcia, M.A.; Chiquetto Paracatu, L.; Pauli, I.; Silva Bolzani, V.; Defini Andricopulo, A.; Farias Ximenes, V.; Zeraik, M.L.; et al. Chalcones and their B-aryl analogues as myeloperoxidase inhibitors: In silico, in vitro and ex vivo investigations. *Bioorg. Chem.* **2021**, *110*, 104773. [CrossRef]

52. Sakata, R.P.; Antoniolli, G.; Lancellotti, M.; Kawano, D.F.; Guimarães Barbosa, E.; Almeida, W.P. Synthesis and biological evaluation of 2′-aminochalcone: A multi-target approach to find drug candidates to treat Alzheimer's disease. *Bioorg. Chem.* **2020**, *103*, 104201. [CrossRef]
53. Apiraksattayakul, S.; Pingaew, R.; Leechaisit, R.; Prachayasittikul, V.; Ruankham, W.; Songtawee, N.; Tantimongcolwat, T.; Ruchirawat, S.; Prachayasittikul, V.; Prachayasittikul, S.; et al. Aminochalcones attenuate neuronal cell death under oxidative damage via sirtuin 1 activity. *ACS Omega* **2024**, *8*, 33367–33379. [CrossRef] [PubMed]
54. de Sousa, J.N.; de Andtade Ferreira Barreto, J.; de Alcântara Oliveira, F.A.; Haydée Lima Ferreira, J.; Dias Rufino Arcanjo, D.; Emidio Sampaio Nogueira, C.; Machado Marinho, M.; dos Santos, H.S.; Maria Lins Rolim, H.; de Siqueira-Júnior, J.P.; et al. ADMET study and inhibition of *Staphylococcus aureus* efflux pumps by a synthetic p-aminochalcone. *Results Chem.* **2024**, *7*, 101449. [CrossRef]
55. Climent, M.J.; Corma, A.; Iborra, S.; Marti, L. Process intensification with bifunctional heterogeneous catalysts: Selective one-pot synthesis of 2′-aminochalcones. *ACS Catal.* **2015**, *5*, 157–166. [CrossRef]
56. Tristão, T.C.; Campos-Buzzi, F.; Corrêa, R.; Cruz, R.C.B.; CechinelFelho, V.; Bella Cruz, A. Antimicrobial and cytotoxicity potential of acetamido, amino and nitrochalcones. *Arzneim. Forsch. Drug Res.* **2012**, *62*, 590–594. [CrossRef]
57. Sulpizio, C.; Roller, A.; Giester, G.; Rompel, A. Synthesis, structure, and antioxidant activity of methoxy- and hydroxyl-substituted 2′-aminochalcones. *Monatsh Chem.* **2016**, *147*, 1747–1757. [CrossRef]
58. Prasad, Y.R.; Rani, V.J.; Rao, A.S. In vitro antioxidant activity and scavenging effects of some synthesized 4′-aminochalcones. *Asian J. Chem.* **2013**, *25*, 52–58. [CrossRef]
59. Liu, G.; Ge, Z.; Zhao, M.; Zhou, Y. Design, synthesis and cytotoxic activities of novel aliphatic amino-substituted flavonoids. *Molecules* **2013**, *18*, 14070–14084. [CrossRef] [PubMed]
60. Zeraik, M.L.; Ximenes, V.F.; Regasini, L.O.; Dutra, L.A.; Silva, D.H.S.; Fonseca, L.M.; Coelho, D.; Machado, S.A.S.; Bolzani, V.S. 4′-Aminochalcones as a novel inhibitors of the chlorinating activity of myeloperoxidase. *Curr. Med. Chem.* **2012**, *19*, 5405–5413. [CrossRef] [PubMed]
61. Iqbal, H.; Prabhakar, V.; Sangith, A.; Chandrika, B.; Balasubramanian, R. Synthesis, anti-inflammatory and antioxidant activity of ring-A-monosubstituted chalcone derivatives. *Med. Chem. Res.* **2014**, *23*, 4383–4394. [CrossRef]
62. Trein, M.R.; Rodrigues e Oliveira, L.; Rigo, G.V.; Garcia, M.A.R.; Petro-Silveira, B.; da Silva Trentin, D.; Macedo, A.J.; Regasini, L.O.; Tasca, T. Anti-Trichomonas vaginalis activity of chalcone and amino-analogues. *Parasitol. Res.* **2019**, *118*, 607–615. [CrossRef] [PubMed]
63. Xiao, G.Y.; Li, Y.; Qiang, X.M.; Xu, R.; Zheng, Y.X.Z.; Cao, Z.C.; Luo, L.; Yang, X.; Sang, Z.P.; Su, F.; et al. Design, synthesis and biological evaluation of 4′-aminochalcone-rivastigmine hybrids as multifunctional agents for the treatment of Alzheimer's disease. *Bioorg. Med. Chem.* **2017**, *25*, 1030–1041. [CrossRef]
64. Seba, V.; Silva, G.; Dos Santos, M.B.; Baek, S.J.; França, S.C.; Fachin, A.L.; Regasini, L.O.; Marins, M. Chalcone derivatives 4′-amino-1-naphthyl-chalcone (D14) and 4′-amino-4-methyl-1-naphthyl-chalcone (D15) suppress migration and invasion of osteosarcoma cells mediated by p53 regulating EMT-related genes. *Int. J. Mol. Sci.* **2018**, *19*, 2838. [CrossRef] [PubMed]
65. Cavalcante, J.W.; Maia, M.M.G.; da Silva, F.R.C.; Ali, A.; Marinho, M.M.; Marinho, E.S.; dos Santos, H.S.; Martins, A.M.C.; de Menezes, R.R.P.P.B. Trypanocidal potential of synthetic p-aminochalcones: In silico and in vitro evaluation. *Bioorg. Chem.* **2023**, *141*, 106931–106943.
66. Prasad, Y.R.; Rao, A.S.; Rambabu, R. Synthesis of some 4′-amino chalcones and their anti-inflammatory and antimicrobial activity. *Asian J. Chem.* **2009**, *21*, 907–914.
67. Mariño, P.A.; Pereira, D.B.; Santi, G.; de Souza, R.O.; Faoro, D.; de Oliveira, L.F.S.; Machado, M.M.; Paula, F.R. In vitro and in silico toxicity evaluation of bioactive 4′-aminochalcone derivatives. *Drug Chem. Toxicol.* **2016**, *39*, 147–152. [CrossRef]
68. Patel, J.R.; Dholakiya, B.Z. Synthesis of 1-(4-((E)-3-arylacryloyl) phenyl)-3,4-dibromo-1H-pyrrole-2,5-diones and screening for anti-candida and antituberculosis activity. *Med. Chem. Res.* **2012**, *21*, 1977–1983. [CrossRef]
69. Goetz, M.P.; Reid, J.M.; Qi, Y.; Chen, A.; McGovern, R.M.; Kuffel, M.J.; Scanlon, P.D.; Erlichman, C.; Armes, M.M. A phase I study of once-weekly aminoflavone prodrug (AFP464) in solid tumorpatients. *J. Clin. Oncol.* **2011**, *29*, 2546. [CrossRef]
70. Xu, K.; Ren, X.; Wang, J.; Zhang, Q.; Fu, X.; Zhang, P.-C. Clinical development and informatics analysis of natural and semi-synthetic flavonoid drugs: A critical review. *J. Adv. Res.* **2024**, *63*, 269–284. [CrossRef] [PubMed]

International Journal of
Molecular Sciences

MDPI

Review

COVID-19 in Children and Vitamin D

Teodoro Durá-Travé [1,2,*] and Fidel Gallinas-Victoriano [2,3]

1 Department of Pediatrics, School of Medicine, University of Navarra, 31008 Pamplona, Spain
2 Navarrabiomed (Biomedical Research Center), 31008 Pamplona, Spain; fivictoriano@hotmail.com
3 Department of Pediatrics, Navarra Hospital Universitary, 31008 Pamplona, Spain
* Correspondence: tdura@unav.es

Abstract: In December 2019, the so-called "coronavirus disease 2019" (COVID-19) began. This disease is characterized by heterogeneous clinical manifestations, ranging from an asymptomatic process to life-threatening conditions associated with a "cytokine storm". This article (narrative review) summarizes the epidemiologic characteristics and clinical manifestations of COVID-19 and multi-system inflammatory syndrome in children (MIS-C). The effect of the pandemic confinement on vitamin D status and the hypotheses proposed to explain the age-related difference in the severity of COVID-19 are discussed. The role of vitamin D as a critical regulator of both innate and adaptive immune responses and the COVID-19 cytokine storm is analyzed. Vitamin D and its links to both COVID-19 (low levels of vitamin D appear to worsen COVID-19 outcomes) and the cytokine storm (anti-inflammatory activity) are detailed. Finally, the efficacy of vitamin D supplementation in COVID-19 is evaluated, but the evidence supporting vitamin D supplementation as an adjuvant treatment for COVID-19 remains uncertain.

Keywords: COVID-19; MIS-C; SARS-CoV-2 infection; vitamin D; vitamin D deficiency; vitamin D insufficiency; vitamin D supplementation

1. Introduction

In December 2019, there was an outbreak of a new infectious disease in Wuhan, the capital of the Hubei province (China) caused by a new type of coronavirus: severe acute respiratory syndrome coronavirus 2 (SARS-CoV-2), and later named "coronavirus disease-2019" (COVID-19) [1]. In February 2020, the Chinese Center for Disease Control and Prevention (China CDC) provided important information to the international community about the SARS-CoV-2 infection: "The COVID-19 epidemic had spread very rapidly and it took only 30 days (from 31 December 2019 to 11 February 2020) to spread from Hubei Province to the rest of mainland China" [2]. On 11 March 2020, the World Health Organization (WHO) declared a public health emergency of international concern, but the virus spread rapidly causing a global pandemic. COVID-19 exhibits highly heterogeneous clinical manifestations, ranging from an asymptomatic or paucisymptomatic process to a life-threatening condition characterized by features of interstitial pneumonia, acute respiratory distress syndrome (ARDS), and severe multi-organ failure [3]. It is transmitted from human-to-human in a variety of ways (direct contact or indirect contact (fomites), small airborne droplets or aerosols) but primarily by large, inhaled droplets from coughing or sneezing [4,5]. Although the first data available on COVID-19 suggests possible animal-to-human transmission, probably from bats, epidemiological data have increasingly shown that the virus is transmitted from human-to-human through droplets or direct contact [6,7].

In addition, the introduction of vaccines against SARS-CoV-2 changed the course of the pandemic. In fact, the recent development of vaccines was considered an effective measure to save lives and minimize the impact on health, social systems and the global economy [8]. SARS-CoV-2 genome mutations are known to influence the efficacy of the immune response induced by vaccination. Since the beginning of the COVID-19 pandemic, numerous SARS-CoV-2 mutations have been identified, and regular viral genome sequencing helps to detect new genetic variants circulating in communities. An updated version of the SARS-CoV-2 phylogenetic tree is shared on the GISAID (Global Initiative on Sharing Avian Influenza Data) platform. A variant is recognized by the WHO as a variant of concern (VOC) or variant of interest (VOI) [9].

The China CDC reported the first description of the demographic characteristics of the COVID-19 outbreak of the novel coronavirus SARS-CoV-2 in February 2020. Of the total 44,672 confirmed cases of COVID-19 in China, only 2% were patients aged 0–19 years [2]. Italian data, published on March 2020, reported that only 1.2% of 22,512 Italian cases of COVID-19 were children [10]. Of the total 44,672 COVID-19 cases reported in the United States as of March 2020, only 5% of cases occurred in patients aged 0–19 years [11]. According to a multicenter retrospective observational cohort study of 8886 pediatric patients (0 to 18 years) diagnosed with COVID-19, the most common age group was over 10 years (57.0%). The proportion of disease was 8.4% in children less than 1 year of age, 20.7% in the 1–6 age group, and 13.9% in the 6–10 age group [12]. However, some children may experience a post-infectious inflammatory process temporally related to COVID-19 named "multisystem inflammatory syndrome" (MIS-C) [13–15].

From the early stages of the pandemic, it was observed that there was a significant difference in the severity of COVID-19 in different age groups: elderly patients infected with SARS-CoV-2 are at high risk for ARDS, complications, and death, while children with COVID-19 showed milder cases and a better prognosis than adults, and death was relatively rare [16–22]. The results found in a recent meta-analysis, after analyzing a large population group of patients (611,583 subjects) with COVID-19 from several national and regional registries (China, Italy, Spain, the United Kingdom and New York State), confirm that age has a determining effect on the mortality of COVID-19 patients. According to age, mortality in patients younger than 50 years was 0.3–1.1% and increased exponentially with age: 3.0% (50 to 59 years), 9.5% (60 to 69 years), 22.8% (70 to 79 years) and 29.6% (over 80 years) [23]. In some patients, COVID-19 infection is often accompanied by an aggressive inflammatory response with the release of large amounts of pro-inflammatory cytokines, a so-called "cytokine storm". In addition, several studies suggested that a cytokine storm correlates with lung injury, multi-organ failure, and poor prognosis in severe COVID-19 [24–26]. In addition, recent clinical data [27] have shown that patients with COVID-19 have significantly elevated levels of matrix metalloproteinase (MMPs), which are known to play a role in tissue remodeling and immune responses. In other words, MMPs, together with the cytokine storm, may play an important role in the immunopathogenesis of COVID-19.

At present, vitamin D deficiency is a global health problem that affects all ages and races. The important discovery that vitamin D can act on various cell types through the vitamin D receptor (VDR) opened a wide field for investigation of its role in human health. As a result, VDRs (nuclear transcription factors) have been detected in almost all immune cells, and their activation would regulate the expression of many genes involved in both the innate and adaptive immune systems. Therefore, vitamin D status in COVID-19 could play an important immunomodulatory role by regulating the pathophysiological manifestations of the cytokine storm (anti-inflammatory agent). However, it remains to be seen whether vitamin D supplementation is associated with a potentially favorable outcome of COVID-19 [28–30].

The objective of this review is to provide a comprehensive literature review (narrative review) of (a) recent data on vitamin D and its association with COVID-19 and a cytokine storm and (b) clinical evidence between COVID-19 and/or MIS-C and vitamin D, and finally, (c) evaluate the efficacy of vitamin D supplementation in pediatric patients with COVID-19. This review is based on an electronic search of the literature performed by two independent researchers in the PubMed database of the U.S. National Library of Medicine published between January 2020 and August 2024. The following specific keywords

(Medical Subject Headings) were used alone or in combination for the search: children, COVID-19, MIS-C, SARS-CoV-2 infection, vitamin D or vitamin D deficiency/insufficiency and vitamin D supplementation.

2. The COVID-19 Cytokine Storm

The normal response to a viral infection requires activation of the inflammatory pathways of the immune system. The inflammatory response to the invading virus results in the activation of transcription factors that induce the expression of genes encoding various pro-inflammatory cytokines. Cytokines are produced by different immune cells from both the innate immune system (macrophages, dendritic cells, natural killer cells) and the adaptive immune system (T and B-lymphocytes). The clinical findings are mainly attributed to the action of the pro-inflammatory: interleukin (IL)-1, IL-6, IL-8, tumor necrosis factor-alfa (TNF-α) and interferon-gamma (IFN-γ). This increase in cytokines results in the influx of several immune cells (macrophages, neutrophils, T cells, etc.) from the circulation to the site of infection, with destructive effects on human tissues (damage of vascular barrier and hemodynamic instability, diffuse alveolar damage, multi-organ failure, etc.) and, if left untreated, death [31].

In other words, COVID-19 infection is usually accompanied by an exaggerated inflammatory response with the release of a large number of pro-inflammatory cytokines in an event known as "cytokine storm", which has been directly correlated with lung injury, multi-organ failure, and adverse prognosis [24,25,32]. Few studies provide data on cytokines in children with severe/critical COVID-19 cases, but the most frequently reported elevated cytokines were IL-6 and IFN-γ [32].

Several studies have reported that IL-6 is the cytokine that is most often significantly increased in severe/critical cases compared to mild/moderate cases of COVID-19 in both adults and children [5,24,25,33]. In fact, the optimal results of clinical trials with tocilizumab (an IL-6 inhibitor) in patients with severe and critical COVID-19 led to its approval by the U.S. Food and Drug Administration (FDA) and its inclusion in the current Guidance for the Management of Acute COVID-19 in Children [34].

3. Immune Modulatory Activity of Vitamin D

The widespread expression of VDR in immune cells, including antigen-presenting cells (such as dendritic cells and macrophages), as well as T cells and B cells, suggests that vitamin D plays a critical role in the regulation of both the innate and adaptive immune systems. Therefore, vitamin D deficiency may compromise the integrity of the immune system and lead to inappropriate immune responses [35–38].

Vitamin D is an important regulator of innate immune responses. For example, by modulating the gene expression of potent antimicrobial peptides such as cathelicidin and β-defensin, vitamin D enhances the antimicrobial properties of immune cells (monocytes, neutrophils and natural killer cells) and barrier epithelia, which could reduce viral replication rates. Furthermore, vitamin D has important effects on immune cells, increasing their production of anti-inflammatory cytokines: IL-4 and IL-10, and inhibiting their production of inflammatory cytokines: IL-1, IL-2, IL-6, IL-8, IL-12 and IFN-γ. In this way, it helps prevent an excessive innate response and the resulting tissue damage (systemic inflammation and/or septic shock).

Vitamin D also plays a role in the regulation of adaptive immunity. Vitamin D modulates the activation and differentiation of naïve CD4+ lymphocytes in the lymph nodes, resulting in a shift from a T-helper 1 (Th1) to a Th2 phenotype. This change implies the inhibition of pro-inflammatory cytokine production: IL-2, IFN-γ, and TNF-α, and an increased production of anti-inflammatory cytokines (IL-4, IL-5, and IL-10). Similarly, vitamin D affects differentiation to the Th17 phenotype, leading to a decrease in the production of inflammatory cytokines (IL-17 and IL-21), and facilitates the induction of T regulatory cells (Tregs) with increased production of anti-inflammatory cytokines: IL-10 and transforming growth factor beta (TGF-β) [39–41].

In other words, in COVID-19, a link between vitamin D deficiency and the cytokine storm could be established in view of the anti-inflammatory effect of vitamin D. In fact, several studies have shown that vitamin D can reduce the risk and/or the gravity of certain inflammatory reactions [42]. In severe COVID-19, high levels of pro-inflammatory cytokines have been observed, such as IL-1, IL-6, IL-8, IL-12, IL-17, TNF-α and IFN-γ. It appears that vitamin D, through its immunomodulatory role in both innate and adaptive immunity, could potentially prevent or reduce complications associated with the cytokine storm caused by SARS-CoV-2 infection by increasing the levels of anti-inflammatory cytokines (IL-4, IL-5, IL-10), as well as by reducing the levels of pro-inflammatory cytokines [43,44]. In this context, it would be useful to measure the basal level of vitamin D as soon as a diagnosis of SARS-CoV-2 infection has been made. Sequential measurements should also be made during the course of the disease. Whether vitamin D supplementation is associated with a potentially favorable outcome of COVID-19 remains to be seen.

In short, immunological dysregulation would be the main consequence of hypovitaminosis D. For example, patients with SARS-CoV-2 infection have a tendency to develop a hyperinflammatory state and even an overactive pathological immune response. Uncorrected vitamin D deficiency can lead to a "cytokine storm", increasing the risk of systemic complications (ARDS, multi-organ failure, etc.) and death [45]. In addition, some authors have shown an association between serum vitamin D levels and lymphocyte count in children with COVID-19. This suggests that serum vitamin D concentration may play a role in lymphocyte count, which is very important for the immune system [20].

Vitamin D appears to exert direct and indirect regulation of MMP expression [46] and could therefore act as an inhibitor of MMP proliferation. In fact, vitamin D has been attributed great potential as a therapeutic agent for various human diseases (liver fibrosis, pulmonary emphysema, bone and cartilage diseases, cancer, etc.) by altering the expression and activity of MMPs [47,48].

4. Effect of Pandemic Confinement on Vitamin D Status

Vitamin D is mainly produced by the synthesis of the pro-vitamin dehydrocholesterol in the skin under the influence of ultraviolet radiation (type B). The resulting vitamin D is biologically inert and requires double hepatic and renal hydroxylation to become active, also known as "calcitriol" [28,29]. Inadequate exposure to sunlight is therefore one of the main causes of vitamin D deficiency.

Following the rapid spread of COVID-19 and the recommendations of the WHO, many countries adopted restrictive measures to contain the spread of the virus (total closure of schools, universities, public places and all shops except supermarkets, grocery stores and pharmacies). In other words, a series of social distancing measures were implemented. Travel and movement were restricted, and citizens were asked to stay at home to reduce transmission. In children, for example, confinement meant replacing face-to-face schooling with online education and reducing time spent outdoors. In other words, the COVID-19 pandemic forced confinement in all age groups, leading to a significant reduction in time exposed to sunlight and consequently an increased risk of vitamin D deficiency.

Epidemiological studies conducted in China [49,50], Poland [51], Türkiye [52] and Italy [53] have reported that the COVID-19 pandemic-related confinement changed vitamin D status in pediatric-aged patients.

The first observational study to compare vitamin D status in children before and after pandemic-related confinement was conducted in Guangzhou, China [49]. Their results showed an increase in the proportion of children aged 3–6 years with vitamin D deficiency, which was associated with reduced sunlight exposure. However, another study conducted in Hong Kong found evidence of a progressive decline in serum vitamin D levels in infants and young children aged 2–24 months over the course of social distancing and home confinement during the COVID-19 pandemic [50].

In Warsaw, Poland, mean vitamin D levels in the pediatric population aged 1 month to 18 years were reported to be significantly lower during the COVID-19 pandemic (March

2020 to February 2021) than before the pandemic (January 2019 to February 2020), resulting in an increase in the proportion of children with vitamin D deficiency [51]. In addition, the characteristic pre-pandemic seasonal variability of serum vitamin D levels in relation to changes in sunlight exposure was not observed during the pandemic period. Similar results were also obtained in Izmir (Türkiye). That is, vitamin D levels in schoolchildren (6–12 years) and adolescents (12–18 years) were significantly lower in the pandemic period (April 2019–March 2020) than in the pre-pandemic period (April 2020–April 2021) [52]. Likewise, in Naples, Italy, the incidence of hypovitaminosis D was generally higher during the pandemic period (March 2020 to March 2021) than in the pre-pandemic period (March 2019 to March 2020), with higher incidence in preschoolers (1–5 years) and schoolchildren (6–12 years). In this study, no significant differences in serum vitamin D levels were observed in infants (0–12 months) between the pre- and post-COVID periods [53]. Finally, it should be noted that a recent systematic review and meta-analysis study [54] of a large cohort of children and adolescents found significant decreases in vitamin D levels during the pandemic in all pediatric age groups (<18 years) compared with pre-pandemic levels. However, severe vitamin D deficiency was only observed in preschoolers and schoolchildren, and this condition was not observed in infants (less than 1 year of age), probably because of the recommended vitamin D supplementation in this age group [55].

In contrast, in other studies of late adolescents (aged 18–19 years) living in southern Switzerland [56] or children and adolescents (aged 0–18 years) living on the island of Sardinia, Italy [57], the authors reported that vitamin D concentrations remained in the sufficient range during the pandemic period and did not differ significantly from those observed in the pre-pandemic period. Note that these differences could be due to geographical and/or cultural heterogeneity, national differences in severity/duration of confinement, variability in study periods, pediatrician vitamin D supplementation, intake of vitamin D-fortified foods, etc.

5. Clinical Manifestations of COVID-19 in Children

The incubation period of SARS-CoV-2, which causes COVID-19, is 3 to 7 days. Infectious sources are predominantly respiratory secretions, although SARS-CoV-2 RNA can be isolated from saliva, blood, feces, and urine. A number of systematic reviews and meta-analyses have described the demographic characteristics, clinical symptoms, laboratory findings and radiological features of COVID-19 [7,18,19,21,22,58]. Based on these data, the following criteria for the severity of COVID-19 infection in children were established [55]: asymptomatic and mild, moderate, severe or critical infection.

The prevalence of asymptomatic infection was 14.2% of children (no clinical symptoms and signs, normal chest imaging, but SARS-CoV-2 infection detected by positive nucleic acid test).

Mild infection was present in 36.3% of children (symptoms of acute upper respiratory tract infection such as fever, fatigue, myalgia, cough, sore throat, runny nose and sneezing; some children may have only gastrointestinal symptoms such as diarrhea, vomiting and abdominal pain). There are no auscultatory abnormalities on physical examination.

Moderate infection (features of viral bronchitis and pneumonia such as fever, productive cough, wheezing, or wet crackles without tachypnea and hypoxemia) occurred in 46% of children. In some cases, there may be no clinical symptoms and signs, but a chest computed tomography scan may reveal subclinical pulmonary lesions.

Severe infections occurred in 2.1% of children (the disease usually progresses in about 1 week and dyspnea occurs with central cyanosis and oxygen saturation less than 92%).

Finally, 1.2% of cases had a critical infection (children can rapidly progress to acute respiratory distress syndrome or respiratory failure). Multiple organ dysfunction such as circulatory shock, encephalopathy, heart failure, disseminated coagulopathy and acute renal failure can be life-threatening).

Figure 1 is a summary of the severity of disease in children with COVID-19. This means that about 95% of the children diagnosed with COVID-19 had an asymptomatic, mild or moderate disease and had a good prognosis with recovery within 1 to 2 weeks. However, the first retrospective study on the epidemiological characteristics of COVID-19 in children was reported by the Chinese CDC in March 2020. This study showed that young children, especially infants, are more susceptible to COVID-19 infection. In fact, the proportion of severe and critical disease was 10.6% in children aged <1 year at diagnosis, 1–5 years (7.3%), 6–10 years (4.2%), 11–15 years (4.1%), and 16–17 years (3.0%) [22].

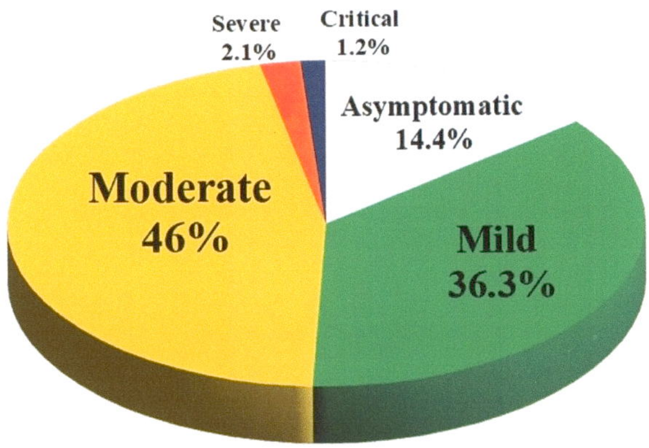

Figure 1. Severity of illness in children with COVID-19.

The most common hematological abnormality in patients with mild COVID-19 was a decreased neutrophil or lymphocyte count, according to laboratory findings reported in several meta-analyses. Biochemical abnormalities were characterized by an increase in the levels of creatine kinase-MB, lactate dehydrogenase (LDH) and liver enzymes: aspartate aminotransferase and alanine aminotransferase. In addition, increased levels of procalcitonin (PCT), D-dimer and C-reactive protein (CRP) were found among laboratory abnormalities related to inflammation/coagulation. Elevated creatine kinase-MB levels may be an indication of cardiac damage, and monitoring of cardiac troponin levels is recommended in hospitalized patients. The higher frequency of elevated PCT levels compared with CRP levels suggests the possibility of viral and bacterial co-infection [18,19,58].

As mentioned above, less than 5% of children with COVID-19 are at risk of developing severe disease; therefore, accumulated data on children with severe COVID-19 are limited. However, leukocyte indices do not appear to be reliable indicators of disease severity in children, according to a pooled analysis and review. In addition, children with severe COVID-19 tended to have elevated levels of LDH, CRP, PCT, D-dimer and serum ferritin, similar to what has been reported in adult patients with COVID-19. Few studies provide data on cytokines in children with severe/critical COVID-19. However, the most commonly reported elevated cytokines were IL-6, IL-8, TNF-α and IFN-γ [58–62].

Chest computed tomography findings of COVID-19 in pediatric patients vary according to the severity of the disease. A recent meta-analysis showed that 41% of cases had normal imaging. The most common lung abnormality was ground-glass opacities (36%) and, to a lesser extent, local (26%) or bilateral (28%) patchy opacities [16,18].

6. Multisystem Inflammatory Syndrome

Initial reports indicated that children had mild or asymptomatic COVID-19 and were hospitalized and died at lower rates than adults did [16,21,22,63]. However, within weeks of the COVID-19 pandemic declaration, clinicians in Western Europe and the United States reported clusters of previously healthy children presenting with cardiovascular shock, fever, and Kawasaki disease-like features in the setting of recent SARS-CoV-2 infection or exposure [13,64,65]. This condition has been referred to as the "multi-system inflammatory syndrome temporally associated with COVID-19" (MIS-C). According to the Centers for

Disease Control and Prevention (CDC) and WHO, the initial definition of MIS-C included persistent fever (lasting more than 4 days), laboratory data showing inflammation and multiple organ dysfunction associated with recent SARS-CoV-2 infection or exposure within the previous 2–6 weeks, and no other likely explanation. The majority of MIS-C patients were school-aged or adolescents (aged 5–13 years), and available data suggest that the incidence of MIS-C was approximately 3 per 10,000 persons aged <21 years infected with SARS-CoV-2. This means that MIS-C is likely to be a post-infectious inflammatory process that is temporally related to COVID-19 [13–15,64,66–68].

According to published descriptive studies and systematic reviews covering the largest cohorts of MIS-C patients to date, it is characterized by persistent fever and multi-organ dysfunction (at least four organ systems are involved), including cardiovascular (hypotension, myocardial dysfunction, pericarditis, coronary artery dilation or aneurysm, myocarditis, etc.), mucocutaneous (skin rash, conjunctival injection), respiratory (pneumonia, acute respiratory distress syndrome, pleural effusion, etc.) and gastrointestinal system (abdominal pain, vomiting or diarrhea). It also includes, to a lesser extent, neurologic (headache, meningism, encephalopathy, etc.), nephrological (renal failure) and hepatological (hepatitis or hepatomegaly) symptoms, and a history of previous asymptomatic or mild COVID-19 or exposure to a person with confirmed COVID-19. Their laboratory tests revealed hematology abnormalities (neutrophilia, lymphopenia, and thrombocytopenia), as well as significant increases in inflammation/coagulation markers (erythrocyte sedimentation rate, CRP, PCT, ferritin, fibrinogen, and D-dimer) and elevated cardiac troponin levels [16,66–69]. Furthermore, these patients had a "cytokine storm" characterized by increased IL-1, IL-6, IL-8, IL-17, TNFα, and IFN-γ [44,70,71]. Although the pathophysiology of MIS-C is still unknown [70], it is thought to be an autoimmune vasculitis in which high levels of anti-SARS-CoV-2 IgG antibodies are associated with a pro-inflammatory cytokine storm. This leads to diffuse endothelial damage by immune complexes and activation of complement pathways [3,71].

More than 80% of MIS-C patients required intensive care for respiratory support and/or hemodynamic instability but recovered with no long-term sequelae and a mortality rate of 1–2% [15,64,67,72,73]. There appears to be a close association between MIS-C and SARS-CoV-2 infection, and MIS-C would therefore be considered a post-infectious manifestation, probably due to an abnormal immune response in most cases.

It is worth noting that previous research has shown elevated baseline levels of some MMPs in children with MIS-C requiring intensive care compared to those with less severe presentations. This means that MMPs may play a role in the pathogenesis of MIS-C and COVID-19 in children and may even serve to differentiate MIS-C from other conditions with overlapping clinical presentations [74,75].

7. Clinical Evidence Between COVID-19 and Vitamin D

Regression models showed that the northernmost countries in the Northern Hemisphere had relatively high COVID-19 mortality (May 2020). That is, there appears to be an association between COVID-19 deaths and countries based on latitude (countries with latitudes below 35° north had low mortality rates). This suggests a possible role for vitamin D in determining COVID-19 outcomes, as people living in regions with latitudes above 35° N would receive less exposure to ultraviolet B radiation (an important source of cutaneous vitamin D synthesis) [76,77]. On the other hand, epidemiologic studies have reported that vitamin D deficiency or insufficiency increases susceptibility to infections, such as acute respiratory infections, and that vitamin D supplementation reduces them [45,78]. There is little evidence that vitamin D can protect against SARS-CoV-2 infection through antiviral activity alone, but it may be very important in preventing the cytokine storm and subsequent acute respiratory distress syndrome or MIS-C, which is often the cause of death [29,40,79].

Based on these premises, several studies were conducted to investigate the clinical significance of vitamin D deficiency in both adults and children with COVID-19. According to a systematic review and meta-analysis conducted in adults, including studies from different European (United Kingdom, Italy and Spain) and Asian (Iran, China, Republic of Korea and Saudi Arabia) countries, vitamin D deficiency was identified as an independent risk factor for COVID-19 (patients with vitamin D deficiency are more susceptible to SARS-CoV-2 infection). In addition, vitamin D deficiency was associated with more severe lung injury, acute respiratory distress and risk of death in elderly COVID-19 patients, probably in relation to an immunological imbalance and hyper-inflammatory state [80].

Few studies, mostly observational, have evaluated the relationship between vitamin D levels and clinical severity and inflammatory markers in pediatric SARS-CoV-2 patients because of the milder clinical course of COVID-19. However, several studies have reported similar results in the pediatric population despite differences in methods and statistical approaches. Most of them indicated that vitamin D status seems to have an impact on the outcome of COVID-19 also in children, as low vitamin D levels were associated with a worse prognosis. In fact, serum vitamin D levels were lower in both MIS-C and COVID-19 patients compared to control groups (healthy children). Furthermore, some authors concluded that vitamin D levels could be valuable in the prediction of severe forms of MIS-C and that correction of its abnormal levels in severe MIS-C could have an influence on its evolution [38,81–86].

These findings were supported by a meta-analysis of pediatric patients with COVID-19, which included studies from Germany, Romania, the United States, Italy, and Türkiye. Indeed, vitamin D-deficient children and adolescents had a higher risk of contracting SARS-CoV-2 and a worse prognosis compared to those with normal vitamin D levels [87]. Additionally, multivariable logistic regression analyses found that vitamin D deficiency and D-dimer and fibrinogen levels are independent predictive biomarkers of moderate-to-severe clinical outcomes [88,89].

8. Why Is COVID-19 Less Severe in Children?

The majority of children infected with COVID-19 are asymptomatic or present with mild or moderate clinical manifestations; namely, symptoms of an acute upper respiratory tract infection or only gastrointestinal symptoms. In the cases with pneumonia, they presented with fever and a dry or productive cough but did not have obvious hypoxemia or respiratory distress [12]. From the early stages of the pandemic, it was observed that children had mild or asymptomatic COVID-19 and had lower rates of hospitalization and death than adults. In fact, since the beginning of the pandemic, age has been identified as an independent predictor of mortality in patients with COVID-19 [16–23,63,90].

Several hypotheses (Table 1) have been proposed to explain the difference in the severity of COVID-19 between children and adults. These hypotheses are based on: (a) factors that would increase the risk in adults, (b) factors that would protect children, or (c) vitamin D status. In reality, these factors are a reflection of the differences in physiological background between children and adults with their clinical consequences [91–93].

(a) Factors increasing risk in adults

Viral entry into cells. Angiotensin-converting enzyme 2 (ACE2) receptors are functional receptors of SARS-CoV-2 that are predominantly expressed in the respiratory epithelium (nasopharynx, oropharynx, pneumocytes, etc.) and facilitate viral entry into cells. The expression of the ACE2 gene in the airway epithelium and the affinity of ACE2 receptors for SARS-CoV-2 increases with age and is lower in children than in adults, which would reduce viral entry in children and, therefore, result in a milder infection.

Endothelial damage and hypercoagulable state. Endothelial damage and an excessive susceptibility to clotting increase with age. SARS-CoV-2 can infect endothelial cells and cause vasculitis, resulting in the activation of coagulation pathways and thrombotic complications such as myocardial infarction and stroke in COVID-19. The endothelium in children is "pre-damaged" to a lesser extent than in adults, and the coagulation system is also different, making children less prone to abnormal clotting.

Comorbidities. Children have a lower prevalence of comorbidities associated with severe COVID-19 in adults (obesity, diabetes mellitus, hypertension, immunosuppression, chronic kidney disease, chronic obstructive pulmonary disease, cardiovascular disease, etc.), probably related to endothelial damage.

Intensity of viral exposure. Adults are at higher risk of exposure to SARS-CoV2 due to their work responsibilities (workplace) and social relationships (shopping, travel, etc.), and viral load influences the severity of COVID-19. Therefore, a lower intensity of viral exposure may be another factor favoring less severe disease in children. In addition, since children are usually infected by an adult, they are infected with a second- or third-generation SARS-CoV-2, and subsequent generations of the virus have reduced pathogenicity in comparison to the first-generation virus.

Immunosenescence. Aging is associated with a gradual decline in the function of the innate and adaptive immune systems, which is likely to contribute to the reduced clearance of SARS-CoV-2 [90]. That is, age-related functional defects in the immune system could result in a failure to control viral replication, which could lead to poor outcomes.

(b) *Factors protecting children*

"Trained" innate immunity. Children infected with SARS-CoV-2 often have co-infections with other viruses (including less pathogenic coronaviruses). Frequent recurrent and concurrent viral infections may interfere with the replication of SARS-CoV-2 by inducing "trained immunity" through epigenetic changes and metabolic reprogramming in innate immune cells, making them more effective in the clearance of SARS-CoV-2.

Off-target effects of live vaccines. Many live vaccines have off-target (non-specific) immunomodulatory effects beyond protection against their target disease. The inclusion of live attenuated vaccines, such as measles, mumps and rubella vaccines and oral polio vaccine, in the childhood immunization schedule would protect against other viral infections by inducing "trained immunity" and/or activating heterologous lymphocytes. This could contribute to differences in the severity of COVID-19.

Age-related differences in immune responses. Children have a stronger innate immune response to cytoplasmic viral pathogens (e.g., SARS-CoV-2) than adults do, with higher natural killer T cell numbers. There is an early type I interferon response (during the incubation period of SARS-CoV-2) and consequently a more rapid activation of host defenses that would lead to more effective containment/elimination of the virus. In terms of adaptive immunity, children also have a higher proportion of T helper 1 lymphocytes (CD4+ T cells), cytotoxic lymphocytes (CD8+ T cells) and activated B lymphocytes than adults, which are essential against viral infections. In addition, it appears that children are less able to generate the pro-inflammatory cytokines storm (lower pro-inflammatory cytokine response); these play an important role in the pathogenesis of severe COVID-19.

(c) *Vitamin D status*

It is noteworthy that vitamin D is an important regulator (genomic pathway) of both the innate and adaptive immune responses and could therefore play a critical role in the host immune response to SARS-CoV-2 infection. As mentioned above, the mechanism by which vitamin D would play a protective role in COVID-19 would be due to its antiviral activity and, in particular, its anti-inflammatory activity, which would moderate the exaggerated inflammatory response (cytokine storm). Clinical evidence suggests that vitamin D status appears to influence the outcome of COVID-19 in both adults and children, as low vitamin D levels are associated with a worse prognosis.

Finally, it should be noted that MMPs might also be a contributor to the pathophysiological differences between adult and pediatric patients with COVID-19. The role of MMPs in COVID-19 stems from the pathogenesis leading to the release of chemokines and pro-inflammatory markers that cause acute lung injury/acute respiratory distress syndrome. Notably, significantly higher serum concentrations of MMP-3, MMP-8 and MMP-9 have been found in adults with COVID-19 than in pediatric patients with the same condition. This means that adult and pediatric patients with COVID-19 appear to have different MMP responses. Although the specific role of each MMP in COVID-19 is not well understood, it could be hypothesized that several MMPs would not be upregulated in children due to a milder course of infection compared to the adult group [75].

Table 1. Differences in the physiological responses of children and adults to SARS-CoV-2.

	Factors That Increase the Risk of COVID-19 in Adults
Receptors and their affinity to SARS-CoV-2	Higher expression of ACE2 in the respiratory tract and higher affinity to SARS-CoV-2 in adults than in children.
Endothelial damage and hypercoagulable state	With age, there is an increase in the damage to the endothelium and an excessive susceptibility to blood clotting. SARS-CoV-2 can cause vasculitis by activating coagulation pathways (thrombotic complications).
Associated comorbidities	Higher prevalence of risk-related comorbidities (obesity, diabetes mellitus, hypertension, immunosuppression, chronic diseases, etc.) in adults.
Immunosenescence	Aging is associated with a gradual decline in immune function, which probably contributes to reduced SARS-CoV-2 clearance.
Intensity of viral exposure	Adults are at higher risk of exposure to SARS-CoV2 given their work responsibilities and social relationships, and viral load influences the severity of COVID-19.
	Factors That Protect Against COVID-19 in Children
"Trained" innate immunity	Frequent recurrent and concurrent viral infections in children could interfere with SARS-CoV-2 replication by inducing "trained immunity".
Off-target effects of live vaccines	Live attenuated vaccines have off-target (non-specific) immunomodulatory effects beyond protection against their target disease by inducing "trained immunity" or heterologous lymphocyte activation.
Age-related differences in immune response	Children have a stronger and faster innate immune response (during the incubation period of SARS-CoV-2) than adults. Children have a higher proportion of lymphocytes (T and B cells) than adults. Children have less capacity to generate the pro-inflammatory cytokine storm than adults.
	Vitamin D Status
Vitamin D, through its immunomodulatory function of the immune system (anti-inflammatory agent), might potentially moderate the cytokine storm, which plays an important role in the pathogenesis of severe COVID-19.	

9. Vitamin D Supplementation in COVID-19

Indeed, low vitamin D is a recognized risk factor for COVID-19 [79,88,94]. Considering that a severe course of COVID-19 is characterized by an overproduction of pro-inflammatory cytokines ("cytokine storm") and that vitamin D has an anti-inflammatory effect and also protects against acute respiratory infections in general [45,78], several clinical trials and observational studies have been conducted to evaluate the efficacy of vitamin D supplementation in improving the clinical conditions of patients with COVID-19.

Randomized controlled trials have been conducted in adult patients and have shown that high-dose oral cholecalciferol (60,000 IU daily for 7 days or a single dose of 600,000 IU) is safe and reduces inflammatory biomarkers, sequential organ failure, and mortality in vitamin D-deficient COVID-19 patients [95,96]. In addition, a multicenter randomized clinical trial conducted in Saudi Arabia showed that oral supplementation with 5000 IU of cholecalciferol daily for 2 weeks reduced cough recovery time and taste sensory loss in patients with suboptimal vitamin D status and mild-to-moderate COVID-19 symptoms, and this dose is recommended as adjuvant therapy for COVID-19 patients [97].

The most recent reviews and meta-analyses of randomized clinical trials suggest that cholecalciferol supplementation may reduce hospital length of stay and intensive care unit admission rates in patients infected with SARS-CoV-2. However, the level of evidence is relatively low [3,98–100]. This is most likely due to the high heterogeneity of the studies analyzed (different study designs, differences in treatment protocols and diagnosis of COVID-19 infection, patients with different severity of disease, vitamin D supplementation in different forms and dosages, etc.). That is, despite some positive findings, their results should be interpreted with caution (the evidence to support the use of vitamin D supplementation as an adjuvant treatment for SARS-CoV-2 infection is still uncertain). Large-scale randomized controlled trials are needed to confirm the potential benefits of vitamin supplementation in both the prevention and treatment of COVID-19.

At present, the available literature on the efficacy and safety of vitamin D supplementation in pediatric patients with COVID-19 is virtually nonexistent. A recent open-label, randomized, controlled, single-blind clinical trial evaluated the efficacy and safety of vitamin D supplementation (intervention: 2000 IU/day in children aged 1 to 17 years for 7–14 days versus control) in children with COVID-19 requiring hospitalization and supplemental oxygen, without excluding patients with comorbidities (obesity, cancer, and chronic diseases). None of the patients receiving vitamin D had adverse effects. In children with vitamin D supplementation, a decrease in baseline CRP was observed (there was no difference in D-dimer and fibrinogen) and seemed to be able to reduce the risk of COVID-19 clinical deterioration and death. The trial was stopped for ethical reasons since none of the patients in either group had normal vitamin D status after receiving the results of the baseline vitamin D status [101]. That is, despite some positive findings, their results should be interpreted with caution because the evidence supporting vitamin D supplementation as an adjuvant treatment for SARS-CoV-2 infection is still uncertain.

10. Conclusions

The clinical manifestations of COVID-19 are highly heterogeneous, ranging from asymptomatic disease to life-threatening conditions (interstitial pneumonia, ARDS or severe multi-organ failure). The normal response to a viral infection requires activation of the inflammatory pathways of the immune system, but COVID-19 infection can generally be associated with an exaggerated release of pro-inflammatory cytokines (cytokine storm), which has been directly correlated with lung injury, multi-organ failure and adverse prognosis. However, some children may experience a post-infectious inflammatory process (MIS-C) associated with COVID-19, probably due to an abnormal immune response. Most of these patients require intensive care for respiratory support and/or hemodynamic instability, but they recover without sequelae and with minimal mortality.

From the early stages of the pandemic, it was observed that most children diagnosed with COVID-19 had asymptomatic, mild or moderate disease and a good prognosis, while older patients infected with SARS-CoV-2 were at high risk of complications and death. Various hypotheses have been proposed for the age-related difference in the severity of COVID-19, including both factors that would increase the risk in adults and factors that would protect children, but there is unlikely to be a single explanation. However, it appears that the immune systems of children are better adapted to clear SARS-CoV-2, and indeed, most children ultimately have mild or asymptomatic disease.

Vitamin D, through its immunomodulatory role in both innate and adaptive immunity, could potentially prevent or reduce complications associated with the cytokine storm caused by SARS-CoV-2 infection by reducing levels of pro-inflammatory cytokines and increasing levels of anti-inflammatory cytokines. Available data would support a role for vitamin D in reducing complications associated with uncontrolled inflammation or the cytokine storm induced by COVID-19. Despite limited clinical data, vitamin D status also appears to affect COVID-19 outcomes in children, as low vitamin D levels have been associated with a worse prognosis.

Although there is no strong evidence to support systematic vitamin D supplementation in patients with SARS-CoV-2 infection, accumulated experience from clinical trials and observational studies suggests that vitamin D supplementation may play a helpful role in the evolution of COVID-19. At present, the available literature on the efficacy and safety of vitamin D supplementation in pediatric COVID-19 patients is practically non-existent. Large-scale randomized controlled trials are needed to confirm the potential benefits of vitamin supplementation in both the prevention and treatment of COVID-19.

Author Contributions: T.D.-T. and F.G.-V. participated in the study design, data collection and analysis. All authors have read and agreed to the published version of the manuscript.

Funding: This research received no external funding.

Institutional Review Board Statement: Not applicable.

Informed Consent Statement: Not applicable.

Data Availability Statement: The datasets generated during and/or analyzed during the current study are available from the corresponding author upon reasonable request.

Conflicts of Interest: The authors declare no conflict of interest.

Abbreviation

ARDS	acute respiratory distress syndrome
China CDC	Chinese Center for Disease Control and Prevention
COVID-19	Coronavirus disease-2019
CRP	C-reactive protein.
IFN-γ	interferon-gamma
IL	interleukin
LDH	lactate dehydrogenase
PCT	procalcitonin
MMPs	matrix metalloproteinases
SARS-CoV-2	Severe acute respiratory syndrome coronavirus 2
TGF-β	growth factor beta
Th1	CD4+ type 1 T helper
Th2	CD4+ type 2 T helper
Th17	CD4+ type 17 T helper
TNF-α	tumor necrosis factor-alfa
Treg	CD4+ T regulatory cells
VDR	vitamin D receptor
WHO	World Health Organization

References

1. Zhu, N.; Zhang, D.; Wang, W.; Li, X.; Yang, B.; Song, J.; Zhao, X.; Huang, B.; Shi, W.; Lu, R.; et al. A Novel Coronavirus from Patients with Pneumonia in China, 2019. *N. Engl. J. Med.* **2020**, *382*, 727–733. [CrossRef] [PubMed]
2. Zhang, Y. The Epidemiological Characteristics of an Outbreak of 2019 Novel Coronavirus Diseases (COVID-19)—China, 2020. *Chin. J. Epidemiol.* **2020**, *2*, 113–122.
3. Rizzi, M.; Avellis, V.; Messina, A.; Germano, C.; Tavella, E.; Dodaro, V.; Vitale, R.; Revelli, A.; Zola, P.; Picone, S.; et al. Vitamin D Supplementation in Neonatal and Infant MIS-C Following COVID-19 Infection. *Int. J. Mol. Sci.* **2024**, *25*, 3712. [CrossRef] [PubMed]
4. Zhao, S.; Zhuang, Z.; Ran, J.; Lin, J.; Yang, G.; Yang, L.; He, D. The association between domestic train transportation and novel coronavirus (2019-nCoV) outbreak in China from 2019 to 2020: A data-driven correlational report. *Travel. Med. Infect. Dis.* **2020**, *33*, 101568. [CrossRef] [PubMed]
5. Chung, Y.S.; Lam, C.Y.; Tan, P.H.; Tsang, H.F.; Wong, S.C.C. Comprehensive Review of COVID-19: Epidemiology, Pathogenesis, Advancement in Diagnostic and Detection Techniques, and Post-Pandemic Treatment Strategies. *Int. J. Mol. Sci.* **2024**, *25*, 8155. [CrossRef]
6. Li, Q.; Guan, X.; Wu, P.; Wang, X.; Zhou, L.; Tong, Y.; Ren, R.; Leung, K.S.; Lau, E.H.; Wong, J.Y.; et al. Early transmission dynamics in Wuhan, China, of novel coronavirus–infected pneumonia. *N. Engl. J. Med.* **2020**, *382*, 1199–1207. [CrossRef]
7. Isoldi, S.; Mallardo, S.; Marcellino, A.; Bloise, S.; Dilillo, A.; Iorfida, D.; Testa, A.; Del Giudice, E.; Martucci, V.; Sanseviero, M.; et al. The comprehensive clinic, laboratory, and instrumental evaluation of children with COVID-19: A 6-months prospective study. *J. Med. Virol.* **2021**, *93*, 3122–3132. [CrossRef]
8. Klimek, L.; Agache, I.; Cooke, E.; Jutel, M.; Akdis, C.A.; O'Hehir, R. COVID-19 vaccines-The way forward. *Allergy* **2022**, *77*, 15–16. [CrossRef]
9. Denaro, M.; Ferro, E.; Barrano, G.; Meli, S.; Busacca, M.; Corallo, D.; Capici, A.; Zisa, A.; Cucuzza, L.; Gradante, S.; et al. Monitoring of SARS-CoV-2 Infection in Ragusa Area: Next Generation Sequencing and Serological Analysis. *Int. J. Mol. Sci.* **2023**, *24*, 4742. [CrossRef]
10. Livingston, E.; Bucher, K. Coronavirus disease 2019 (COVID-19) in Italy. *JAMA* **2020**, *323*, 1335. [CrossRef]

11. Bialek, S.; Boundy, E.; Bowen, V.; Chow, N.; Cohn, A.; Dowling, N.; Ellington, S.; Gierke, R. Severe outcomes among patients with coronavirus disease 2019 (COVID-19)—United States, 12 February–16 March 2020. *MMWR Morb. Mortal. Wkly. Rep.* **2020**, *69*, 343–346.
12. Duman, M.; Şık, N.; Tekşam, Ö.; Akça, H.; Kurt, F.; Çağlar, A.A. COVID-19 Disease in Presenting to the Pediatric Emergency Department: A Multicenter Study of 8886 Cases. *Am. J. Emerg. Med.* **2022**, *59*, 133–140. [CrossRef] [PubMed]
13. Godfred-Cato, S.; Bryant, B.; Leung, J.; Oster, M.E.; Conklin, L.; Abrams, J.; Roguski, K.; Wallace, B.; Prezzato, E.; Koumans, E.H.; et al. COVID-19-Associated Multisystem Inflammatory Syndrome in Children-United States, March–July 2020. *MMWR Morb. Mortal. Wkly. Rep.* **2020**, *69*, 1074–1080. [CrossRef] [PubMed]
14. Payne, A.B.; Gilani, Z.; Godfred-Cato, S.; Belay, E.D.; Feldstein, L.R.; Patel, M.M.; Randolph, A.G.; Newhams, M.; Thomas, D.; Magleby, R.; et al. MIS-C Incidence Authorship Group. Incidence of multisystem inflammatory syndrome in children among US persons infected with SARS-CoV-2. *JAMA Netw. Open* **2021**, *4*, e2116420. [CrossRef]
15. Miller, A.D.; Zambrano, L.D.; Yousaf, A.R.; Abrams, J.Y.; Meng, L.; Wu, M.J.; Melgar, M.; Oster, M.E.; Godfred Cato, S.E.; Belay, E.D.; et al. Multisystem Inflammatory Syndrome in Children—United States, February 2020–July 2021. *Clin. Infect. Dis.* **2022**, *75*, e1165–e1175. [CrossRef]
16. Lu, X.; Zhang, L.; Du, H.; Zhang, J.; Li, Y.Y.; Qu, J.; Zhang, W.; Wang, Y.; Bao, S.; Li, Y.; et al. Coronavirus Study Team.SARS-CoV-2 Infection in Children. *N. Engl. J. Med.* **2020**, *382*, 1663–1665. [CrossRef]
17. Zimmermann, P.; Curtis, N. Coronavirus infections in children including COVID-19: An overview of the epidemiology, clinical features, diagnosis, treatment and prevention options in children. *Pediatr. Infect. Dis. J.* **2020**, *39*, 355–368. [CrossRef]
18. Cui, X.; Zhao, Z.; Zhang, T.; Guo, W.; Guo, W.; Zheng, J.; Zhang, J.; Dong, C.; Na, R.; Zheng, L.; et al. A systematic review and meta-analysis of children with coronavirus disease 2019 (COVID-19). *J. Med. Virol.* **2021**, *93*, 1057–1069. [CrossRef]
19. De Souza, T.H.; Nadal, J.A.; Nogueira, R.J.N.; Pereira, R.M.; Brandão, M.B. Clinical manifestations of children with COVID-19: A systematic review. *Pediatr. Pulmonol.* **2020**, *55*, 1892–1899. [CrossRef]
20. Ozden, A.; Doneray, H.; Hafize-Erdeniz, E.; Altinkaynak, K.; Igan, H. Clinical and Laboratory Findings by Serum Vitamin D Levels in Children with COVID-19. *Eurasian J. Med.* **2022**, *54*, 285–291. [CrossRef]
21. Ludvigsson, J.F. Systematic review of COVID-19 in children shows milder cases and a better prognosis than adults. *Acta Paediatr.* **2020**, *109*, 1088–1095. [CrossRef]
22. Dong, Y.; Mo, X.; Hu, Y.; Qi, X.; Jiang, F.; Jiang, Z.; Tong, S. Epidemiology of COVID-19 among children in China. *Pediatrics* **2020**, *145*, e20200702. [CrossRef]
23. Bonanad, C.; García-Blas, S.; Tarazona-Santabalbina, F.; Sanchis, J.; Bertomeu-González, V.; Fácila, L.; Ariza, A.; Núñez, J.; Cordero, A. The Effect of Age on Mortality in Patients With COVID-19: A Meta-Analysis With 611,583 Subjects. *J. Am. Med. Dir. Assoc.* **2020**, *21*, 915–918. [CrossRef]
24. Chen, G.; Wu, D.; Guo, W.; Cao, Y.; Huang, D.; Wang, H.; Wang, T.; Zhang, X.; Chen, H.; Yu, H.; et al. Clinical and immunologic features in severe and moderate Coronavirus Disease 2019. *J. Clin. Investig.* **2020**, *130*, 2620–2629. [CrossRef]
25. Ragab, D.; Salah Eldin, H.; Taeimah, M.; Khattab, R.; Salem, R. The COVID-19 Cytokine Storm; What We Know So Far. *Front. Immunol.* **2020**, *11*, 1446. [CrossRef]
26. Gao, Y.-D.; Ding, M.; Dong, X.; Zhang, J.-J.; Azkur, A.K.; Azkur, D.; Gan, H.; Sun, Y.-L.; Fu, W.; Li, W.; et al. Risk factors for severe and critically ill COVID-19 patients: A review. *Allergy* **2021**, *76*, 428–455. [CrossRef]
27. da Silva-Neto, P.V.; do Valle, V.B.; Fuzo, C.A.; Fernandes, T.M.; Toro, D.M.; Fraga-Silva, T.F.; Basile, P.A.; de Carvalho, J.C.; Pimentel, V.E.; Perez, M.M.; et al. Matrix Metalloproteinases on Severe COVID-19 Lung Disease Pathogenesis: Cooperative Actions of MMP-8/MMP-2 Axis on Immune Response through HLA-G Shedding and Oxidative Stress. *Biomolecules* **2022**, *12*, 604. [CrossRef]
28. Hossein-Nezhad, A.; Holick, M.F. Vitamin D for health: A global perspective. *Mayo Clin. Proc.* **2013**, *88*, 720–755. [CrossRef]
29. Palermo, N.E.; Holick, M.F. Vitamin D, bone health, and other health benefits in pediatric patients. *J. Pediatr. Rehabil. Med.* **2014**, *7*, 179–192. [CrossRef]
30. Holick, M.F. The vitamin D deficiency pandemic: Approaches for diagnosis, treatment and prevention. *Rev. Endocr. Metab. Disord.* **2017**, *18*, 153–165. [CrossRef]
31. Rhodes, J.M.; Subramanian, S.; Laird, E.; Griffin, G.; Kenny, R.A. Perspective: Vitamin D deficiency and COVID-19 severity—Plausibly linked by latitude, ethnicity, impacts on cytokines, ACE2 and thrombosis. *J. Intern. Med.* **2021**, *289*, 97–115. [CrossRef] [PubMed]
32. Huang, C.; Wang, Y.; Li, X.; Ren, L.; Zhao, J.; Hu, Y.; Zhang, L.; Fan, G.; Xu, J.; Gu, X.; et al. Clinical features of patients infected with 2019 novel coronavirus in Wuhan, China. *Lancet* **2020**, *395*, 497–506. [CrossRef]
33. Sun, D.; Li, H.; Lu, X.-X.; Xiao, H.; Ren, J.; Zhang, F.-R.; Liu, Z.-S. Clinical features of severe pediatric patients with coronavirus disease 2019 in Wuhan: A single center's observational study. *World J. Pediatr.* **2020**, *19*, 1–9. [CrossRef]
34. NIH COVID-19 Treatment Guidelines Panel: Coronavirus Disease 2019 (COVID-19) Treatment Guidelines. 2023. Available online: https://www.ncbi.nlm.nih.gov/books/NBK570371/pdf/Bookshelf_NBK570371.pdf (accessed on 21 March 2024).
35. Cyprian, F.; Lefkou, E.; Varoudi, K.; Girardi, G. Immunomodulatory Effects of Vitamin D in Pregnancy and Beyond. *Front. Immunol.* **2019**, *10*, 2739. [CrossRef]
36. Koivisto, O.; Hanel, A.; Carlberg, C. Key vitamin D target genes with function s in the mmune system. *Nutrients* **2020**, *12*, 1140. [CrossRef]

37. Bilezikian, J.P.; Bikle, D.; Hewison, M.; Lazaretti-Castro, M.; Formenti, A.M.; Gupta, A.; Madhavan, M.V.; Nair, N.; Babalyan, V.; Hutchings, N.; et al. Mechanisms in endocrinology: Vitamn D and COVID-19. *Eur. J. Endocrinol.* **2020**, *183*, R133–R147. [CrossRef]
38. Keleş, Y.E.; Yılmaz, D.; Taşar, S.; Üstündağ, G.; Şahin, A.; Tuz, A.E.; Maden, A.A.; Aksay, A.K.; Çolak, A.; Öncel, E.K. Can Serum 25-Hydroxy Vitamin D Levels Predict the Severity of Multisystem Inflammatory Syndrome in Children and COVID-19? *J. Clin. Res. Pediatr. Endocrinol.* **2023**, *15*, 190–198. [CrossRef]
39. Colotta, F.; Jansson, B.; Bonelli, F. Modulation of inflammatory and immune responses by vitamin D. *J. Autoimmun.* **2017**, *85*, 78–97. [CrossRef]
40. Kalia, V.; Studzinski, G.P.; Sarkar, S. Role of vitamin D in regulating COVID-19 severity—An immunological perspective. *J. Leukoc. Biol.* **2021**, *110*, 809–819. [CrossRef]
41. Sîrbe, C.; Rednic, S.; Grama, A.; Pop, T.L. An Update on the Effects of Vitamin D on the Immune System and Autoimmune Diseases. *Int. J. Mol. Sci.* **2022**, *17*, 9784. [CrossRef]
42. Pamfili, F.M.; Roversi, M.; D'Argenio, P.; Rossi, P.; Cappa, M.; Fintini, D. Possible role of vitamin D in COVID-19 infection in pediatric population. *J. Endocrinol. Investig.* **2021**, *44*, 27–35. [CrossRef] [PubMed]
43. Greiller, C.L.; Martineau, A.R. Modulation of the immune response to respiratory viruses by vitamin D. *Nutrients* **2015**, *7*, 4240–4270. [CrossRef] [PubMed]
44. Carter, M.J.; Fish, M.; Jennings, A.; Doores, K.J.; Wellman, P.; Seow, J.; Acors, S.; Graham, C.; Timms, E.; Kenny, J.; et al. Shankar-Hari: Peripheral immunophenotypes in children with multisystem inflammatory syndrome associated with SARS-CoV-2 infection. *Nat. Med.* **2020**, *26*, 1701–1707. [CrossRef] [PubMed]
45. Wimalawansa, S.J. Infections and Autoimmunity—The Immune System and Vitamin D: A Systematic Review. *Nutrients* **2023**, *15*, 3842. [CrossRef]
46. Vo, H.V.T.; Nguyen, Y.T.; Kim, N.; Lee, H.J. Vitamin A, D, E, and K as Matrix Metalloproteinase-2/9 Regulators That Affect Expression and Enzymatic Activity. *Int. J. Mol. Sci.* **2023**, *24*, 17038. [CrossRef]
47. Li, S.; Niu, G.; Dong, X.N.; Liu, Z.; Song, C.; Leng, H. Vitamin D Inhibits Activities of Metalloproteinase-9/-13 in Articular Cartilage In Vivo and In Vitro. *J. Nutr. Sci. Vitaminol.* **2019**, *65*, 107–112. [CrossRef]
48. Gonga, J.; Gonga, H.; Liub, Y.; Taob, X.; Zhangc, H. Calcipotriol attenuates liver fibrosis through the inhibition of vitamin D receptor-mediated NF-κB signaling pathway. *Bioengineered* **2022**, *13*, 2658–2672. [CrossRef]
49. Yu, L.; Ke, H.J.; Che, D.; Luo, S.L.; Guo, Y.; Wu, J.L. Effect of Pandemic-Related Confinement on Vitamin D Status among Children Aged 0–6 Years in Guangzhou, China: A Cross-Sectional Study. Risk Manag. *Healthc. Policy* **2020**, *13*, 2669–2675. [CrossRef]
50. Wong, R.S.; Tung, K.T.S.; So, H.-K.; Wong, W.H.S.; Wong, S.Y.; Tsang, H.W.; Tung, J.Y.L.; Chua, G.T.; Ho, M.H.K.; Wong, I.C.K.; et al. Impact of COVID-19 Pandemic on Serum Vitamin D Level among Infants and Toddlers: An Interrupted Time Series Analysis and before-and-after Comparison. *Nutrients* **2021**, *13*, 1270. [CrossRef]
51. Rustecka, A.; Maret, J.; Drab, A.; Leszczyńska, M.; Tomaszewska, A.; Lipińska-Opałka, A.; Będzichowska, A.; Kalicki, B.; Kubiak, J.Z. The Impact of COVID-19 Pandemic during 2020-2021 on the Vitamin D Serum Levels in the Paediatric Population in Warsaw, Poland. *Nutrients* **2021**, *13*, 1990. [CrossRef]
52. Beyazgül, G.; Bağ; Yurtseven, I.; Coşkunol, F.; Başer, S.; Çiçek, D.; Kanberoğlu, G..; Çelik, F.; Nalbantoğlu; Özkan, B. How Vitamin D Levels of Children Changed during COVID-19 Pandemic: A Comparison of Pre-Pandemic and Pandemic Periods. *J. Clin. Res. Pediatr. Endocrinol.* **2022**, *14*, 188. [CrossRef] [PubMed]
53. Mosca, C.; Colucci, A.; Savoia, F.; Calì, C.; Del Bene, M.; Ranucci, G.; Maglione, A.; Pepe, A.; Morelli, A.; Vajro, P.; et al. Vitamin D Levels in the Pre- and Post-COVID-19 Pandemic Periods and Related Confinement at Pediatric Age. *Nutrients* **2023**, *15*, 2089. [CrossRef] [PubMed]
54. Cui, X.; Zhai, Y.; Wang, S.; Ding, K.; Yang, Z.; Tian, Y.; Huo, T. Effect of the COVID-19 Pandemic on Serum Vitamin D Levels in People under Age 18 Years: A Systematic Review and Meta-Analysis. *Med. Sci. Monit. Int. Med. J. Exp. Clin. Res.* **2022**, *28*, e935823-1. [CrossRef] [PubMed]
55. Munns, C.F.; Shaw, N.; Kiely, M.; Specker, B.L.; Thacher, T.D.; Ozono, K.; Michigami, T.; Tiosano, D.; Mughal, M.Z.; Mäkitie, O.; et al. Global Consensus Recommendations on Prevention and Management of Nutritional Rickets. *Horm. Res. Paediatr.* **2016**, *85*, 83–106. [CrossRef]
56. Meoli, M.; Muggli, F.; Lava, S.A.; Bianchetti, M.G.; Agostoni, C.; Kocher, C.; Bührer, T.W.; Ciliberti, L.; Simonetti, G.D.; Milani, G.P. Vitamin D Status in Adolescents during COVID-19 Pandemic: A Cross-Sectional Comparative Study. *Nutrients* **2021**, *13*, 1467. [CrossRef]
57. Antonucci, R.; Vacca, N.; Biasia, B.; Locci, C.; Dore, M.P.; Pes, G.M.; Bitti, A. Impact of COVID-19 Pandemic and Associated Restrictions on Vitamin D Status in a Large Cohort of Italian Children and Adolescents. *Medicina* **2024**, *60*, 65. [CrossRef]
58. Henry, B.M.; Benoit, S.W.; de Oliveira, M.H.S.; Hsieh, W.C.; Benoit, J.; Ballout, R.A.; Plebani, M.; Lippi, G. Laboratory abnormalities in children with mild and severe coronavirus disease 2019 (COVID-19): A pooled analysis and review. *Clin. Biochem.* **2020**, *81*, 1–8. [CrossRef]
59. Fang, F.; Chen, Y.; Zhao, D.; Liu, T.; Huang, Y.; Qiu, L.; Hao, Y.; Hu, X.; Yin, W.; Liu, Z.; et al. Recommendations for the diagnosis, prevention and control of coronavirus disease-19 in children-The Chinese perspectives. *Front. Pediatr.* **2020**, *8*, 553394. [CrossRef]
60. Henry, B.M.; Santos de Oliveira, M.H.; Benoit, S.; Plebania, M.; Giuseppe Lippia, G. Hematologic, biochemical and immune biomarker abnormalities associated with severe illness and mortality in coronavirus disease 2019 (COVID-19): A meta-analysis. *Clin. Chem. Lab. Med.* **2020**, *58*, 1021–1028. [CrossRef]

61. Henry, B.M.; Lippi, G.; Plebani, M. Laboratory abnormalities in children with novel coronavirus disease 2019. *Clin. Chem. Lab. Med.* **2020**, *16*, 16. [CrossRef]
62. Moreno-Galarraga, L.; Urretavizcaya-Martinez, M.; Alegria Echauri, J.; Garcia Howard, M.; Ruperez Garcia, E.; Aguilera-Albesa, S.; Alzina de Aguilar, V.; Herranz Aguirre, M. SARS-CoV-2 infection in children requiring hospitalization: The experience of Navarra, Spain. *World J. Pediatr.* **2020**, *16*, 614–622. [CrossRef] [PubMed]
63. Liu, W.; Zhang, Q.; Chen, J.; Xiang, R.; Song, H.; Shu, S.; Chen, L.; Liang, L.; Zhou, J.; You, L.; et al. Detection of COVID-19 in Children in Early January 2020 in Wuhan, China. *N. Engl. J. Med.* **2020**, *382*, 1370–1371. [CrossRef] [PubMed]
64. Dufort, E.M.; Koumans, E.H.; Chow, E.; Rosenthal, E.M.; Muse, A.; Rowlands, J.; Barranco, M.A.; Maxted, A.M.; Rosenberg, E.S.; Easton, D.; et al. Multisystem inflammatory syndrome in children in New York State. *N. Engl. J. Med.* **2020**, *383*, 347–358. [CrossRef] [PubMed]
65. Riphagen, S.; Gomez, X.; Gonzalez-Martinez, C.; Wilkinson, N.; Theocharis, P. Hyperinflammatory shock in children during COVID-19 pandemic. *Lancet* **2020**, *395*, 1607–1608. [CrossRef] [PubMed]
66. Esposito, S.; Principi, N. Multisystem Inflammatory Syndrome in Children Related to SARS-CoV-2. *Pediatric. Drugs* **2021**, *23*, 119–129. [CrossRef]
67. Feldstein, L.R.; Rose, E.B.; Horwitz, S.M.; Collins, J.P.; Newhams, M.M.; Son, M.B.F.; Newburger, J.W.; Kleinman, L.C.; Heidemann, S.M.; Martin, A.A.; et al. Multisystem inflammatory syndrome in U.S. children and adolescents. *N. Engl. J. Med.* **2020**, *383*, 334–346. [CrossRef]
68. Dionne, A.; Son, M.B.F.; Randolph, A.G. An Update on Multisystem Inflammatory Syndrome in Children Related to SARS-CoV-2. Pediatr. *Infect. Dis. J.* **2022**, *41*, e6–e9. [CrossRef]
69. García-Salido, A.; de Carlos Vicente, J.C.; Belda Hofheinz, S.; Balcells Ramírez, J.; Slöcker Barrio, M.; Leóz Gordillo, I.; Hernández Yuste, A.; Guitart Pardellans, C.; Cuervas-Mons Tejedor, M.; Huidobro Labarga, B.; et al. Severe manifestations of SARS-CoV-2 in children and adolescents: From COVID-19 pneumonia to multisystem inflammatory syndrome: A multicentre study in pediatric intensive care units in Spain. *Crit. Care* **2020**, *24*, 666. [CrossRef]
70. Brodsky, N.N.; Ramaswamy, A.; Lucas, C.L. The Mystery of MIS-C Post-SARS-CoV-2 Infection. *Trends Microbiol.* **2020**, *28*, 956–958. [CrossRef]
71. Waltuch, T.; Gill, P.; Zinns, L.E.; Whitney, R.; Tokarski, J.; Tsung, J.W.; Sanders, J.E. Features of COVID-19 post-infectious cytokine release syndrome in children presenting to the emergency department. *Am. J. Emerg. Med.* **2020**, *38*, 2246.e3–2246.e6. [CrossRef]
72. Blat, A.M.; Randolph, A.G. Severe COVID-19 and Multisystem Inflammatory Syndrome in Children in Children and Adolescents. *Crit. Care Clin.* **2022**, *38*, 571–586. [CrossRef] [PubMed]
73. Yilmaz, D.; Keles, Y.E.; Emiroglu, M.; Duramaz, B.B.; Ugur, C.; Kocabas, B.A.; Celik, T.; Ozdemir, H.; Bayturan, S.; Turel, O.; et al. Evaluation of 601 children with multisystem inflammatory syndrome (Turk MISC study). *Eur. J. Pediatr.* **2023**, *182*, 5531–5542. [CrossRef] [PubMed]
74. Kumar, N.P.; Venkataraman, A.; Varadarjan, P.; Nancy, A.; Rajamanickam, A.; Selladurai, E.; Sankaralingam, T.; Thiruvengadam, K.; Selvam, R.; Thimmaiah, A.; et al. Role of matrix metalloproteinases in multi-system inflammatory syndrome and acute COVID-19 in children. *Front. Med.* **2022**, *9*, 1050804.
75. Toczyłowski, K.; Lewandowski, D.; Martonik, D.; Moniuszko-Malinowska, A.; Kruszewska, E.; Parfieniuk-Kowerda, A.; e, al. Differential Inflammatory Responses in Adult and Pediatric COVID-19 Patients: Implications for Long-Term Consequences and Anti-Inflammatory Treatment. *Med. Sci. Monit.* **2024**, *30*, e944052. [CrossRef] [PubMed]
76. Rhodes, J.M.; Subramanian, S.; Laird, E.; Kenny, R.A. Editorial: Low population mortality from COVID-19 in countries south of latitude 35 degrees North supports vitamin D as a factor determining severity. *Aliment. Pharmacol. Ther.* **2020**, *51*, 1434–1437. [CrossRef]
77. Lansiaux, É.; Pébaÿ, P.P.; Picard, J.L.; Forget, J. COVID-19 and vit-D: Disease mortality negatively correlates with sunlight exposure. *Spat. Spatiotemporal Epidemiol.* **2020**, *35*, 100362. [CrossRef]
78. Martineau, A.R.; Jolliffe, D.A.; Hooper, R.L.; Greenberg, L.; Aloia, J.F.; Bergman, P.; Dubnov-Raz, G.; Esposito, S.; Ganmaa, D.; Ginde, A.A.; et al. Vitamin D supplementation to prevent acute respiratory tract infections: Systematic review and meta-analysis of individual participant data. *BMJ* **2017**, *356*, i6583. [CrossRef]
79. Sacco, K.; Castagnoli, R.; Vakkilainen, S.; Liu, C.; Delmonte, O.M.; Oguz, C.; Kaplan, I.M.; Alehashemi, S.; Burbelo, P.D.; Bhuyan, F.; et al. Immunopathological signatures in multisystem inflammatory syndrome in children and pediatric COVID-19. *Nat. Med.* **2022**, *28*, 1050–1062. [CrossRef]
80. Szarpak, L.; Rafique, Z.; Gasecka, A.; Chirico, F.; Gawel, W.; Hernik, J. A systematic review and meta-analysis of effect of vitamin D levels on the incidence of COVID-19. *Cardiol. J.* **2021**, *28*, 647–654. [CrossRef]
81. Feketea, G.; Vlacha, V.; Bocsan, I.C.; Vassilopoulou, E.; Stanciu, L.A.; Zdrenghea, M. Vitamin D in Corona Virus Disease 2019 (COVID-19) Related Multisystem Inflammatory Syndrome in Children (MIS-C). *Front. Immunol.* **2021**, *12*, 648546. [CrossRef]
82. Yılmaz, K.; Şen, V. Is Vitamin D Deficiency a Risk Factor for COVID-19 in Children? *Pediatr. Pulmonol.* **2020**, *55*, 3595–3601. [CrossRef] [PubMed]
83. Miraglia del Giudice, M.; Indolfi, C.; Dinardo, G.; Decimo, F.; Decimo, A.; Klain, A. Vitamin D status can affect COVID-19 outcomes also in pediatric population. *PharmaNutrition* **2022**, *22*, 100319. [CrossRef] [PubMed]

84. Mamishi, S.; Olfat, M.; Pourakbari, B.; Eshaghi, H.; Abdolsalehi, M.R.; Shahbabaie, M.A.; Jalali, F.; Safari, F.; Mahmoudi, S. Multisystem inflammatory syndrome associated with SARS-CoV-2 infection in children: Update and new insights from the second report of an Iranian referral hospital. *Epidemiol. Infect.* **2022**, *150*, 179. [CrossRef] [PubMed]
85. Torpoco-Rivera, D.; Misra, A.; Sanil, Y.; Sabzghabaei, N.; Safa, R.; Garcia, R.U. Vitamin D and morbidity in children with Multisystem inflammatory syndrome related to COVID-19. *Prog. Pediatr. Cardiol.* **2022**, *66*, 101507. [CrossRef] [PubMed]
86. Petrovic, D.; Benzon, B.; Srsen, S.; Polic, B.; Novogradec, A.V.; Milic, P.; Markic, J. The Impact of Vitamin D Levels on Clinical Manifestations of Multisystem Inflammatory Syndrome in Children: A Cross-Sectional Study. *Life* **2023**, *13*, 674. [CrossRef]
87. Shah, K.; Varna, V.P.; Pandya, A.; Saxena, D. Low vitamin D levels and prognosis in a COVID-19 pediatric population: A systematic review. *QJM* **2021**, *114*, 447–453. [CrossRef]
88. Bayramoğlu, E.; Akkoç, G.; Ağbaş, A.; Akgün; Yurdakul, K.; Duru, H.N.S.; Elevli, M. The Association between Vitamin D Levels and the Clinical Severity and Inflammation Markers in Pediatric COVID-19 Patients: Single-Center Experience from a Pandemic Hospital. *Eur. J. Pediatr.* **2021**, *180*, 2699–2705. [CrossRef]
89. Bagiu, I.C.; Scurtu, I.L.; Horhat, D.I.; Mot, I.C.; Horhat, R.M.; Bagiu, R.V.; Capraru, I.D.; Diaconu, M.M.; Adam, O.; Ciornei, B.; et al. COVID-19 Inflammatory Markers and Vitamin D Relationship in Pediatric Patients. *Life* **2023**, *13*, 91. [CrossRef]
90. Zhou, F.; Yu, T.; Du, R.; Fan, G.; Liu, Y.; Liu, Z.; Xiang, J.; Wang, Y.; Song, B.; Gu, X.; et al. Clinical course and risk factors for mortality of adult inpatients with COVID-19 in Wuhan, China: A retrospective cohort study. *Lancet* **2020**, *395*, 1054–1062. [CrossRef]
91. Wong, L.S.Y.; Loo, E.X.L.; Kang, A.Y.H.; Lau, H.X.; Tambyah, P.A.; Tham, E.H. Age-Related Differences in Immunological Responses to SARS-CoV-2. *J. Allergy Clin. Immunol. Pract.* **2020**, *8*, 3251–3258. [CrossRef]
92. Zimmermann, P.; Curtis, N. Why is COVID-19 less severe in children? A review of the proposed mechanisms underlying the age-related difference in severity of SARS-CoV-2 infections. *Arch. Dis. Child.* **2021**, *106*, 429–439. [CrossRef] [PubMed]
93. Kapustova, L.; Petrovicova, O.; Banovcin, P.; Antosova, M.; Bobcakova, A.; Urbancikova, I.; Rennerova, Z.; Jesenak, M. COVID-19 and the differences in physiological background between children and adults and their clinical consequences. *Physiol. Res.* **2021**, *70* (Suppl. S2), S209–S225. [CrossRef] [PubMed]
94. Alpcan, A.; Tursun, S.; Kandur, Y. Vitamin D Levels in Children with COVID-19: A Report from Turkey. *Epidemiol. Infect.* **2021**, *149*, e180. [CrossRef] [PubMed]
95. Rastogi, A.; Bhansali, A.; Khare, N.; Suri, V.; Yaddanapudi, N.; Sachdeva, N.; Puri, G.D.; Malhotra, P. Short term, high-dose vitamin D supplementation for COVID-19 disease: A randomised, placebo-controlled, study (SHADE study). *Postgrad. Med. J.* **2022**, *98*, 87–90. [CrossRef] [PubMed]
96. Singh, A.; Rastogi, A.; Puri, G.D.; Ganesh, V.; Naik, N.B.; Kajal, K.; Kahlon, S.; Soni, S.L.; Kaloria, N.; Saini, K.; et al. Therapeutic high-dose vitamin D for vitamin D-deficient severe COVID-19 disease: Randomized, double-blind, placebo-controlled study (SHADE-S). *J. Public Health* **2024**, *46*, 256–266. [CrossRef]
97. Sabico, S.; Enani, M.A.; Sheshah, E.; Aljohani, N.J.; Aldisi, D.A.; Alotaibi, N.H.; Alshingetti, N.; Alomar, S.Y.; Alnaami, A.M.; Amer, O.E.; et al. Effects of a 2-Week 5000 IU versus 1000 IU Vitamin D3 Supplementation on Recovery of Symptoms in Patients with Mild to Moderate Covid-19: A Randomized Clinical Trial. *Nutrients* **2021**, *13*, 2170. [CrossRef]
98. Sîrbu, A.C.; Sabin, O.; Bocsan, I.C.; Vesa, S.C.; Buzoianu, A.D. The Effect of Vitamin D Supplementation on the Length of Hospitalisation, Intensive Care Unit Admission, and Mortality in COVID-19—A Systematic Review and Meta-Analysis. *Nutrients* **2023**, *15*, 3470. [CrossRef]
99. Sinopoli, A.; Sciurti, A.; Isonne, C.; Santoro, M.M.; Baccolini, V. The Efficacy of Multivitamin, Vitamin A, Vitamin B, Vitamin C, and Vitamin D Supplements in the Prevention and Management of COVID-19 and Long-COVID: An Updated Systematic Review and Meta-Analysis of Randomized Clinical Trials. *Nutrients* **2024**, *16*, 1345. [CrossRef]
100. Sobczak, M.; Pawliczak, R. Effect of Vitamin D3 Supplementation on Severe COVID-19: A Meta-Analysis of Randomized Clinical Trials. *Nutrients* **2024**, *16*, 1402. [CrossRef]
101. Zurita-Cruz, J.; Fonseca-Tenorio, J.; Villasís-Keever, M.; López-Alarcón, M.; Parra-Ortega, I.; López-Martínez, B.; Miranda-Novales, G. Efficacy and safety of vitamin D supplementation in hospitalized COVID-19 pediatric patients: A randomized controlled trial. *Front. Pediatr.* **2022**, *10*, 943529. [CrossRef]

MDPI AG
Grosspeteranlage 5
4052 Basel
Switzerland
Tel.: +41 61 683 77 34

International Journal of Molecular Sciences Editorial Office
E-mail: ijms@mdpi.com
www.mdpi.com/journal/ijms

www.ingramcontent.com/pod-product-compliance
Lightning Source LLC
Chambersburg PA
CBHW041344210926
51850CB00013B/2982
9783725845071